Optical and Fiber Communications Reports

Volume 9

The Optical and Fiber Communications Reports (OFCR) book series provides high-quality monographs, edit volumes and textbooks in the multidisciplinary areas of optics, information science and electrical engineering. Each book is contributed from leading research scientists that gives an up-to-date and broad-spectrum overview of various subjects.
The main topics of this series will include, but not restricted to:

fibers and fiber-based devices
information optics, optical and quantum communication
optoelectronic imaging and multimedia technology
optical metrology and testing
virtual reality and display technologies
broadband lasers
optical switching (MEMS or others)
polarization and chromatic mode dispersion and compensation
long-haul transmission
optical networks (LAN, MAN, WAN)
further topics of contemporary interest.

Including both general information and a highly technical presentation of the results, this series satisfies the needs of scientists and graduates in the academic community as well as professionals and experts from industry. Books in this series establish themselves as comprehensive guides and reference texts following the impressive evolution of this area of science and technology.

More information about this series at http://www.springer.com/series/4810

Zujie Fang · Haiwen Cai
Gaoting Chen · Ronghui Qu

Single Frequency Semiconductor Lasers

上海交通大學出版社
SHANGHAI JIAO TONG UNIVERSITY PRESS

Springer

Zujie Fang
Research Center of Space Laser and Information Technology
Shanghai Institute of Optics and Fine Mechanics, Chinese Academy of Sciences
Shanghai
China

Gaoting Chen
Research Center of Space Laser and Information Technology
Shanghai Institute of Optics and Fine Mechanics, Chinese Academy of Sciences
Shanghai
China

Haiwen Cai
Research Center of Space Laser and Information Technology
Shanghai Institute of Optics and Fine Mechanics, Chinese Academy of Sciences
Shanghai
China

Ronghui Qu
Research Center of Space Laser and Information Technology
Shanghai Institute of Optics and Fine Mechanics, Chinese Academy of Sciences
Shanghai
China

ISSN 1619-1447 ISSN 1619-1455 (electronic)
Optical and Fiber Communications Reports
ISBN 978-981-10-5256-9 ISBN 978-981-10-5257-6 (eBook)
DOI 10.1007/978-981-10-5257-6

Jointly published with Shanghai Jiao Tong University Press, Beijing, China

The print edition is not for sale in China Mainland. Customers from China Mainland please order the print book from: Shanghai Jiao Tong University Press.

Library of Congress Control Number: 2017945219

Printed on acid-free paper

This Springer imprint is published by Springer Nature
The registered company is Springer Nature Singapore Pte Ltd.
The registered company address is: 152 Beach Road, #21-01/04 Gateway East, Singapore 189721, Singapore

Preface

The semiconductor laser is used widely in many important applications, especially optical communications, optical storages, metrology and sensors, etc. New achievements based on laser physics and laser technology emerge continuously, such as the cold atomic clock, the optical clock, the optical coherent communication, and the high-precision interferometers and spectroscopy. These advanced technologies put forward higher and higher technical requirements on semiconductor lasers. The laser must work in a single longitudinal mode, with narrow linewidth, low noises, and high-frequency stability; i.e., it must be a high-quality single-frequency laser.

Numbers of textbooks and monographs on the semiconductor laser have been published, giving a detailed explanation of its theory and characteristics, and description of materials, device structures, and fabrication technologies. The previous publications seemed not to give enough room for recounting and analyzing semiconductor laser's coherence and noise. The research and development of single-frequency laser involves physical mechanisms, designs of different optical systems incorporated with semiconductor lasers, related electronic technologies, and data processing. A book engaged specially in these topics is needed; it is just the aim of this book.

This book is devoted to the explanation of basic concepts and theory of single-frequency laser, description of related technological problems, and introduction to progress in recent years. It can serve as a textbook for students, whose specialty is on researches of the laser and its applications; also as a reference book for the scientists and engineers working in the field. This book puts an emphasis on the principles on optics, explains and discusses the related mechanisms with basic formulas and explicit figures. The book cannot cover all aspects of the laser; detailed references are listed for interested readers.

This book is composed of nine chapters. Chapter 1 is a brief induction to the book. Chapter 2 recounts the fundamentals of semiconductor laser, which is helpful for understanding the contents of other chapters. The focus of Chap. 3 is on the characterization of laser coherence, noise, and stability, giving explanations of the basic concepts and descriptions of the mathematical relations. Chapter 4 is engaged in the monolithically integrated semiconductor lasers, including the distributed feedback semiconductor lasers (DFB), the distributed Bragg reflector semiconductor lasers (DBR), and the vertical cavity surface emitting semiconductor lasers (VCSEL). The external cavity lasers are described in Chap. 5, including those composed of planar grating, Bragg grating, and external cavities. The technologies of frequency stabilization are discussed in Chap. 6, with the details of those based on the saturated absorption spectroscopy. The PDH method, after its inventor's names, is also described. Chapter 7 is devoted to the frequency sweeping of semiconductor lasers, typically with methods of pump current modulation, intra-cavity modulation, and by external modulators. Chapter 8 introduces the frequency transfer and the technology of optical phase-locked loop (OPLL). The optical frequency comb and related applications are also briefly introduced. Chapter 9 gives brief introductions to several attractive applications of single-frequency semiconductor lasers, including the laser cooling and atomic clocks, the coherent optical communications and microwave photonics, the precision metrology and sensors, and the lidar.

The authors of this book, Zujie Fang, Haiwen Cai, Gaoting Chen, and Ronghui Qu, are Professors of Shanghai Institute of Optics and Fine Mechanics, Chinese Academy of Sciences. The authors' team has been working in the field for years. They acknowledge their colleagues, especially Dr. Guofeng Xin, Dr. Dijun Chen, Dr. Fang Wei, and Dr. Fei Yang; without their efforts the publication of this book would not be possible. This book contains a lot of results from journals and monographs, which have been cited in the related text. Authors acknowledge their contributions to the field sincerely.

Shanghai, China

Zujie Fang
Haiwen Cai
Gaoting Chen
Ronghui Qu

The original version of the book was revised: Incorrect additional bibliography and copyright text have been corrected. The erratum to the book is available at https://doi.org/10.1007/978-981-10-5257-6_10

Contents

1 **Introduction** 1
1.1 Historical Review of Semiconductor Laser 1
1.2 Single Frequency Semiconductor Lasers and Their Applications 3
References 7

2 **Fundamentals of Semiconductor Lasers** 9
2.1 Stimulated Emission in Semiconductor 9
2.2 P-N Junction, Heterostructure, and Quantum Well 13
2.2.1 Carrier Injection with P-N Junction 13
2.2.2 Heterostructure and Quantum Well 14
2.3 Cavity Structure and Transverse Modes 16
2.3.1 Basic Structure and Transverse Modes 16
2.3.2 Index Guiding and Gain Guiding 19
2.3.3 Laser Array and Modules 20
2.4 Rate Equation and Output Characteristics 21
2.4.1 Lasing Condition 21
2.4.2 Rate Equations 23
2.4.3 Light–Current Characteristics 24
2.5 Longitudinal Modes and Tunability 27
2.5.1 Longitudinal Mode Characteristics 27
2.5.2 Tunability of Semiconductor Lasers 29
2.6 Transient and Modulation Characteristics 30
2.6.1 Modulation of Output Power 30
2.6.2 Relaxation Oscillation 33
2.6.3 Wavelength Modulation 34

2.7 Thermal Performances 35
2.7.1 Dependence of Threshold and Output Power on Temperature 35
2.7.2 Dependence of Laser Spectrum on Temperature 36
2.7.3 Junction Temperature and Thermal Resistant 37
References 39

3 Noises and Stability of Semiconductor Lasers 41
3.1 Characteristics and Inherent Relations of Laser Noises 41
3.1.1 Mathematical Description of Stochastic Variables 41
3.1.2 Phase Noise and Frequency Noise 43
3.1.3 Intensity Noise 45
3.2 Linewidth and Line Shape of Semiconductor Lasers 46
3.2.1 Linewidth of Semiconductor Lasers 46
3.2.2 Effect of Photon-Carrier Coupling on Line Shape and Noise 49
3.2.3 Measurement of Linewidth 50
3.3 Noises of Semiconductor Laser 54
3.3.1 Characteristics of Frequency Noise in Low Frequency Band 55
3.3.2 Measurement of Laser Noises 58
3.3.3 Intensity Noise and Its Suppression 65
3.4 Frequency Stability and Allan Variance 67
3.4.1 Characterization of Frequency Stability 67
3.4.2 Measurement of Allan Variance 70
3.5 Other Effects and Manifestations of Laser Noises 71
3.5.1 Chaos of Semiconductor Lasers 72
3.5.2 Jitter of Laser Pulses, Mode Partition Noise, and Speckles 75
3.5.3 Noise and Instability Related to Internal Defects 76
References 77

4 Monolithically Integrated Semiconductor Lasers 81
4.1 Distributed Feedback Semiconductor Laser 81
4.1.1 Coupled Mode Theory of DFB Laser 81
4.1.2 Characteristics of DFB Lasers 87
4.1.3 Structures and Fabrication of DFB Laser 92
4.2 Distributed Bragg Reflector Semiconductor Laser 94
4.2.1 Structures and Principle of DBR Laser 94
4.2.2 Reflection Spectrum of Passive Waveguide Grating 95
4.2.3 Characteristics of DBR Lasers 97

4.3 Vertical Cavity Surface Emitting Laser . . . 100
4.3.1 Principles of VCSEL . . . 100
4.3.2 Design and Fabrication of VCSEL . . . 102
4.3.3 Characteristics of VCSEL . . . 107
References . . . 112

5 External Cavity Semiconductor Lasers . . . 117
5.1 General Characteristics and Theory of External Cavity Diode Lasers . . . 117
5.1.1 Basic Model of ECDL . . . 117
5.1.2 Linewidth Reduction of ECDL with Weak Feedback . . . 120
5.1.3 Characteristics of ECDL with Frequency Selective Feedback . . . 122
5.1.4 Injection Locking of Semiconductor Lasers . . . 127
5.2 Planar Grating External Cavity Diode Laser . . . 130
5.2.1 Basic Characteristics of Planar Grating . . . 130
5.2.2 Littrow and Littman ECDL . . . 132
5.2.3 Technical Issues of Grating ECDL . . . 137
5.2.4 Various Cavity Structures and Tuning Schemes . . . 140
5.3 Bragg Grating External Cavity Diode Laser . . . 142
5.3.1 Fiber Bragg Grating ECDL . . . 142
5.3.2 Waveguide Bragg Gating ECDL . . . 149
5.3.3 Volume Bragg Grating ECDL . . . 151
5.4 Diode Laser with Cavity Feedback . . . 156
5.4.1 Feedback from Fabry-Perot Cavity . . . 156
5.4.2 Feedback from Ring Cavity . . . 160
References . . . 163

6 Frequency Stabilization of Semiconductor Lasers . . . 167
6.1 Introduction to Saturated Absorption Spectroscopy . . . 167
6.1.1 Classical Theory of Absorption Spectrum . . . 168
6.1.2 Doppler Broadening and Saturated Absorption Spectroscopy . . . 170
6.2 Frequency Stabilization by Modulation Spectroscopy . . . 175
6.2.1 Spectrum of Frequency Modulated Optical Signals . . . 175
6.2.2 Frequency Stabilization with Modulation . . . 178
6.2.3 Residual Amplitude Modulation and Its Removal . . . 181
6.3 Modulation-Free Frequency Stabilization . . . 184
6.3.1 Frequency Stabilization by Polarization Spectroscopy . . . 184
6.3.2 Frequency Stabilization with Sagnac Interferometer . . . 186
6.3.3 Frequency Stabilization by Magnetic Dichroism . . . 189
6.3.4 Frequency Stabilization by Optical Negative Feedback . . . 191

6.4 Pound–Drever–Hall (PDH) Frequency Stabilization 193
6.4.1 Basic Principle of PDH Method 193
6.4.2 Technical Issues of PDH Frequency Stabilized Laser 197
6.4.3 Different Schemes Based on Resonant Cavities 199
6.5 Noise Reduction by Self-mixing Interference 200
References 201

7 Frequency Sweeping of Semiconductor Lasers 205
7.1 Applications of Frequency-Swept Laser 205
7.2 Frequency Sweeping by LD Current Modulation 208
7.2.1 LD Current Tuning and Its Linearization 208
7.2.2 Linewidth Reduction of Frequency-Swept Lasers 212
7.3 Frequency Sweeping with Intra-cavity Tuning Devices 214
7.3.1 Electro-optic Materials and Devices 214
7.3.2 Frequency Sweeping with Intra-cavity Modulator 219
7.3.3 Frequency Sweeping by Intra-cavity Beam Deflection . . . 220
7.4 Frequency Sweeping with Extra-Cavity Modulators 224
7.4.1 Frequency Shifting and Sweeping with Acousto-Optic Modulators 224
7.4.2 Frequency Sweeping with Electro-optic Modulators 227
References 231

8 Optical Phase Locked Loop and Frequency Transfer 235
8.1 Optical Phase Locked Loop 235
8.1.1 Principles of OPLL 235
8.1.2 Injection Locking of Semiconductor Lasers 238
8.2 Applications of OPLL in Laser Frequency Transfer 239
8.2.1 Coherent Optical Communication 239
8.2.2 Transportation of Time and Frequency Signals 244
8.2.3 Microwave Photonics and Radio Over Fiber 246
8.2.4 Researches on Atomic Physics 249
8.3 Optical Frequency Comb and Its Characteristics 252
8.3.1 Principle of Mode-Locked Laser 252
8.3.2 Application of Four Wave Mixing in Frequency Trimming 254
8.3.3 Control of Carrier-Envelope Phase 255
8.4 Applications of Optical Frequency Comb in Laser Frequency Transfer 258
8.4.1 Optical Frequency Locking and Frequency Synthesizer 258
8.4.2 High Precision Optical Frequency Sweeping 260
References 262

9 Applications of Single-Frequency Semiconductor Lasers 267
9.1 Applications in Laser Cooling and Related Technologies 267
9.1.1 Time Standard and Atomic Clock 267
9.1.2 Laser Cooling of Atoms and Cold Atomic Clock 269
9.1.3 Data for Atomic Physics and Related Spectroscopy 274
9.2 Applications in Optical Communications 278
9.2.1 A Brief Review of Optical Fiber Communications 278
9.2.2 Lasers for Coherent Optical Communications 279
9.2.3 Microwave Photonics and ROF 282
9.2.4 Inter-satellite Optical Communications 283
9.3 Applications in Metrology and Sensors 285
9.3.1 High Precision Interferometry 285
9.3.2 Distributed Optical Fiber Sensors 287
9.3.3 Laser Spectroscopy 293
9.4 Applications in Lidar 296
9.4.1 Basic Concept of Lidar 296
9.4.2 Brief Introduction to Synthetic Aperture Lidar 298
References 299

Erratum to: Single Frequency Semiconductor Lasers E1

Index 303

Chapter 1
Introduction

1.1 Historical Review of Semiconductor Laser

The invention of laser (Light Amplification by Stimulated Emission of Radiation) was an event with epoch making significance. Soon after emergences of the microwave amplification by stimulated emission of radiation (maser) and the ruby laser [1, 2], the stimulated emission from semiconductor was observed [3, 4]. Up-to-now the laser has been greatly developed, being one of the most important achievements of advanced technologies. Among the various kinds of lasers, the semiconductor laser is used most widely in people's daily life, and with the highest degree of industrialization. Its applications cover almost the entire field of opto-electronics. The optical communications with semiconductor lasers as light source become one of the foundation stones in the information infrastructure all over the world. It is the key component for writing and reading signals on/from the optical discs; it is used widely in information acquisition, such as in the peripheral parts of computers. It plays also important roles in areas of military use and fundamental scientific researches.

In the history of semiconductor laser development, the following events have milestone significances.

The mechanism of stimulated emission based on transition between energy bands in semiconductor was explained [5] in 1961. The laser emission was observed from the P-N junction of GaAs in liquid nitrogen temperature in 1962 [3, 4].

The single heterojunction laser of GaAs/AlGaAs emerged in 1969 [6], opening a way to reduce laser's threshold greatly. The double heterostructure laser of AlGaAs/GaAs/AlGaAs was fabricated, and continuous wave (cw) operation in room temperature was realized in 1970 [7, 8]. It became the key device in optical fiber communication systems soon.

The distributed feedback (DFB) laser was fabricated successfully in 1970s [9, 10].

Z. Fang et al., *Single Frequency Semiconductor Lasers*, Optical and Fiber Communications Reports 9, DOI 10.1007/978-981-10-5257-6_1

Research and development of the quantum well laser [11] and the strained layer quantum well laser [12] began in 1980s.

The vertical cavity surface emitting laser (VCSEL) and the quantum cascaded laser (QCL) came into being in 1990s [13, 14].

The wavelength band of semiconductor lasers has been expanded greatly since 1980s. Lasers in the bands of 1300–1500 nm, 980 nm, 650 nm, 2–3 μm, 10–16 μm and in the blue and ultraviolet bands were realized earlier or later based on various semiconductor materials.

The output power of semiconductor lasers was raised tremendously since 1990s; modules with output more than ten thousands watt have been commercialized.

The theory founded by Einstein is no doubt the foundation of semiconductor laser. Before the birth of lasers, John von Neumann presented an assumption for realizing stimulated emission in semiconductor [15]. N.G. Basov, who won the Nobel Prize in Physics 1964 together with C.H. Townes and A.M. Prokhorov, proposed also an idea of semiconductor laser by using P-N junction [16]. The proposal and success of semiconductor heterostructure is one of the most important achievements; Z.I. Alferov and H. Kroemer was awarded the Nobel Prize in Physics 2000 for their contributions on developing semiconductor heterostructures used in high-speed electronics and optoelectronics, including semiconductor lasers [17, 18].

Research and development of materials is the basis of semiconductor lasers. The success of light emitting diode (LED) and laser in blue and ultraviolet bands relies just on the III–V nitrides [19]. For their contribution to the light sources, I. Akasaki, H. Amano, and S. Nakamura won the Nobel Prize in Physics 2014.

The same as other lasers the semiconductor laser features high brightness and high coherence. The light emission in semiconductor occurs by transition between the conduction band and the valence band, different from the transition between two discrete energy levels in solid state and gas lasers. The physical process is related tightly to the movement of electrons and holes in semiconductor. Such a mechanism makes the semiconductor laser unique features different from the other kind lasers.

(1) It works under direct electrical current injection, simply and conveniently; it has higher efficiency of electric to optical energy conversion; whereas gas lasers need electric discharge under high voltage, and solid state lasers work by optical pumping.
(2) It can be modulated by the pumping current with very high speed; whereas modulations of other lasers have to use external modulators. This is of great significance, especially for high speed optical communications and other optoelectronic applications.
(3) It has very small size and very light weight. It is compatible with the optical fiber system and the high density optical storage. It is suitable for mass production, and thus has advantages of low cost, the same as that of microelectronics.
(4) Its working wavelength covers from ultraviolet to middle infrared bands, selectable almost continuously.

On the other hand, its disadvantages, compared with other lasers, are obvious.

(1) It has a widely divergent beam; especially its beam is asymmetric in vertical and horizontal directions, except VCSELs. Although its brightness is similar to that of other lasers, the collimation optics and related techniques are needed in practical uses.
(2) Its spectral linewidth is larger than other lasers; and its lasing wavelength is susceptible to the external disturbances. On the positive hand, it is easily tunable.
(3) Its output power is small. The techniques of combination and packaging are needed.

Because of the advantages its applications are being expanded continuously, especially in the frontier of advanced scientific researches, such as laser cooling, cold atomic clock, coherent optical communications, precision metrology, and apparatus used in the space sciences. These applications put forward further requirements on its characteristics and performances. Generally two trends are notable in recent years. One is for high power output; the other is to enhance and improve its coherence and beam quality greatly. The main symbols of high coherence are the extremely narrow linewidth, and extremely low noises. The purpose of this book is just to expound the principles and technologies of semiconductor lasers with narrow linewidth and low noises.

1.2 Single Frequency Semiconductor Lasers and Their Applications

One of the main features of laser is its high coherence, including spatial coherence and temporal coherence. The former means a high beam quality near the limitation of diffraction, i.e. a high brightness, which is measured as the optical power per unit area and per unit solid angle. The latter means a high mono-chromaticity, i.e. an extremely narrow linewidth. The mono-chromaticity of light is relative. The light resolved by a prism or a grating is monochromatic, compared with light from the sun and incandescent lamps. Radiations generated by transitions between energy levels of atoms or molecules have lines much narrower than that by ordinary monochromator. The laser oscillation makes the line narrowed greatly further. Frequency stabilization of the laser purifies the spectrum extremely. Lasers with frequency linewidth less than 1 Hz have been achieved now.

Laser's mono-chromaticity is measured by its linewidth $\delta\lambda$, or frequency width $\delta\nu$ equivalently. It is also characterized by the coherence time τ_c, proportional to the reciprocal of $\delta\nu$. The coherence length L_c is the distance of light propagation in duration of τ_c, expressed as

$$\tau_c = \frac{1}{2\pi\delta\nu} = \frac{1}{c}\frac{\lambda^2}{\delta\lambda} = \frac{L_c}{c}, \tag{1.1}$$

where c is the light velocity in vacuum. The direct effect of coherence is demonstrated in the performance of optical interferometer. The largest usable optical path difference (OPD) is just the coherence length of source; beyond L_c the interference fringes become more and more indistinct; the precision and the resolution of measurement are thus declined. The precision of interferometer is determined by the clearness of interference fringes; its visibility is then defined as:

$$V = \exp(-\tau/\tau_c) = \exp(-L_{OPD}/L_c), \tag{1.2}$$

where L_{OPD} is the OPD of the interferometer, and $\tau = L_{OPD}/c$ is the delay corresponding to OPD. A high visibility towards unity requires the coherence length of interferometer's source much larger than the OPD length, corresponding to a larger working distance.

The linewidth of a laser is related to its mode structure. Laser oscillation occurs between two cavity mirrors. When the optical path is perpendicular to the mirrors and the phase shift between roundtrip propagations equals integer multiple of 2π, a standing wave optical field is formed, and longitudinal modes result in spectrum. The mode amplitudes are usually not equal, determined by the gain spectrum and mode selectivity mechanism. If a certain mode has amplitude much higher than the others, it is called the main mode; the ratio of its amplitude to that of the second high side mode is called the side mode suppression ratio (SMSR). If SMSR reaches a high level, required by application, e.g. larger than 10, the laser is regarded as working in a single longitudinal mode. The optical wave may propagate in cavity in directions with different angles to the mirror, different field distributions will be formed, determined by cavity structure, which are called the transverse modes. Among them the mode with propagation direction parallel to the cavity length, or with the smallest angle, is called the fundamental mode. The lasing wavelength of high order transverse modes will deviate from that of fundamental mode. The linewidth of longitudinal modes will be broadened by the accompanied transverse modes. Therefore a narrow line signal frequency laser must work in a single longitudinal mode and in the fundamental transverse mode.

The laser linewidth is a reflection of its phase noise. Inherently, spontaneous emission exists inevitably in the laser beam. As the pump to the laser increases, the ratio of spontaneous emission to stimulated emission will go down with the linewidth narrowed. The formula of laser linewidth was given by its inventors, A.L. Schawlow and C.H. Townes, in their original paper [20]:

$$\delta\nu = 2\pi h\nu(\delta\nu_c)^2/P, \tag{1.3}$$

where $\delta\nu_c$ is the half width at half maximum of the passive cavity resonance, which is the reciprocal of photon's lifetime for a roundtrip propagation in cavity, P is the output power. The larger the cavity length and the less the cavity loss, including the transmission from mirrors, the longer the photon lifetime is. The high the power, the lower the fraction of spontaneous content is.

In the growth history of semiconductor lasers, 1960s was the age of its infant; in 1970s the laser's threshold was decreased greatly, benefited from the hetero-junction, and cw operation at room temperature was realized. Laser cavity design and fabrication for the fundamental transverse mode was achieved roughly in 1980s. After many trials of schemes for controlling the longitudinal mode, the DFB laser with performance of dynamically single longitudinal mode became the first choice as the light source for optical fiber communications, and the semiconductor laser entered an era of big scale industrialization at the turn of the century. However, many new applications, especially those in the frontier of advanced science and technology, put forward more and higher requirements. A few of them are listed as follows.

Coherent Optical Communications. The optical communication and the computer push forward the information technology (IT) up to an unprecedented level. However, people's demand on the information increases continuously. The coherent optical communication uses frequency, phase, and polarization state of the optical wave, together with its amplitude, as information carriers; communication capacity and speed are increased greatly. Much higher coherence, lower noises, and higher stability of laser source are then required [21].

Researches of Cold Atomic Physics. Laser cooling of atoms is one of most important achievements in laser technology, which improves the frequency standard and time standard for several orders. For the contributions in development of methods to cool and trap atoms with laser light, Steven Chu, C. Cohen-Tannoudji and W.D. Phillips was awarded the Nobel Prize in Physics 1997 [22]. The development of cold atomic clock pushed emergence of optical clock, which takes the optical frequency comb as the method of frequency counting, another important achievement in laser technology. J.L. Hall and T.W. Hänsch won the Nobel Prize in Physics 2005 for their contributions to the development of laser-based precision spectroscopy, including the optical frequency comb [23]. The two achievements utilize many advanced technology of optics and lasers; among them the high quality single frequency semiconductor laser plays an indispensable role.

High Precision Metrology and Spectroscopy. The optical interferometer played important roles in the history of science. One of the experimental bases of Einstein relativity was the result of measurement of light velocity in "ether" by Michelson interferometer. Since then the interferometry has been used continuously and widely in many areas; e.g. people are using it to measure the inertial wave. Interferometers with semiconductor lasers as the source attract more and more attentions of scientists and industrialists. A typical example is the hydrophone and geophone [24], which is used in different areas from oceanography and geoscience to navigation, fishery, and seismic wave detection.

These applications and more others put forward higher and higher requirements on laser characteristics and performances.

(1) Linewidth. The linewidth of an ordinary semiconductor laser is usually much larger than that of gas lasers and solid state lasers, because the cavity length is much shorter and the reflectivity of cavity mirror is much lower than those of the latter two. The linewidth of DFB laser has been reduced to the order of MHz. In the high precision interferometer and spectroscopy the linewidth is required less than kHz, even in the order of Hz. The first task of single frequency laser is to reduce its linewidth further. The method used mostly is to design and build an external cavity laser with longer cavity length and higher cavity mirror reflectivity.
(2) Frequency Stabilization. In the laser cooling and technologies with cooled atoms, the precision laser spectroscopy and interferometry, and in many other related applications, frequency noises and drifts of the source will degrade the precision and effectiveness of related apparatus definitely. Therefore the laser frequency must be tuned and stabilized exactly at a certain absorption line of the material. The frequency stabilization involves physical mechanisms of linewidth broadening. For the achievement of saturated absorption spectroscopy A.L. Schawlow won the Nobel Prize in Physics 1981 [25]. The frequency width down to Hz and below was realized the frequency stabilization with an ultra-stable Fabry-Perot cavity, i.e. PDH method, invented by R.V. Pound, R.W.P. Drever and J.L. Hall [26].
(3) Tunability and Frequency Sweeping. Practical applications require also the laser frequency swept over a range, fast and linearly or in a certain waveform, with the linewidth kept narrow enough in the sweeping. It is also required in applications that the laser frequency can be transferred to arbitrary frequencies.
(4) Low Noises. The noise level of semiconductor laser is much higher than those of gas and solid state lasers, because of the direct coupling between photons and electrons injected by current, and the susceptibility of semiconductor to the external disturbances. Noise reduction is one of the most important tasks for the single frequency laser, including the intensity noise, the frequency noise, and the frequency stability. For the purpose different schemes were proposed to stabilize laser's frequency, and new technologies are being developed. Obviously, it is necessary to measure and characterize the noise, especially the noise spectrum. Furthermore, the optical phase should be locked under a required error level between lasers, the technology of optical phase locked loop (OPLL) is thus developed.
(5) Polarization and Its Controllability. Polarization variation is one of the noise origins in many laser applications, such as in interferometers. A stable and controllable polarization state is one of the performances of a single frequency laser. The control and improvement of beam quality is required simultaneously. In recent years, research of photon's angular momentum, especially the orbital angular momentum (OAM) of light has attracted people's attention [27].

The semiconductor laser is being developed continuously. Its performances are being improved fast; new experimental phenomena are observed one after the other, and physical mechanisms are being discovered; its applications are expanded rapidly and widely. Different applications may put forward more requirements; many advanced technologies on optics, electronics and mechanics, are necessary for high level devices and apparatuses. This book cannot cover such new developments and all related respects; it will put emphasis on the optical characteristics, their basic mechanisms, and the technologies related mainly to the spectral properties of semiconductor lasers.

References

1. Gordon J, Zeiger H, Townes CH (1955) The maser—new type of microwave amplifier, frequency standard, and spectrometer. Phys Rev 99(4):1264–1274
2. Maiman TH (1960) Stimulated optical radiation in ruby. Nature 187:493–494
3. Nathan MI, Dumke WP, Burns G et al (1962) Stimulated emission of radiation from GaAs P-N junction. Appl Phys Lett 1(3):62–64
4. Hall RN, Fenner GE, Kingsley JD et al (1962) Coherent light emission from GaAs junctions. Phys Rev Lett 9(9):366–368
5. Bernard MGA, Duraffourg G (1961) Laser conditions in semiconductors. Phys Status Solidi 47:699–703
6. Hayashi I, Panish MB, Foy PW (1969) A low-threshold room-temperature injection laser. IEEE J Quantum Electron 5:211
7. Panish MB, Hayashi I, Sumski S (1970) Double-heterostructure injection lasers with room-temperature thresholds as low as 2300 A/cm^2. Appl Phys Lett 16(8):326–327
8. Kressel H, Hawrylo FZ (1970) Fabry-Perot structure $Al_xGa_{1-x}As$ injection lasers with room-temperature threshold current densities of 2530 A/cm^2. Appl Phys Lett 17(4):169–171
9. Kogelnik H, Shank CV (1972) Stimulated emission in a periodic structure. Appl Phys Lett 18:152–154
10. Scifres DR, Burnham RD, Streifer W (1974) Distributed-feedback single heterojunction GaAs diode laser. Appl Phys Lett 25:203–206
11. Holonyak N, Kolbas RM, Dupuis RD et al (1980) Quantum-well heterostructure lasers. IEEE J Quantum Electron 16(2):170–186
12. Adams AR (2011) Strained-layer quantum-well lasers. IEEE J Sel Top Quantum Electron 17 (5):1364–1373
13. Jewell JL, Harbison JP, Scherer A et al (1991) Vertical-cavity surface-emitting lasers: design, growth, fabrication, characterization. IEEE J Quantum Electron 27(6):1332–1346
14. Faist J, Capasso F, Sivco DL et al (1994) Quantum cascade laser. Science 264(5158):553–556
15. Townes CH (1999) How the laser happened: adventures of a scientist. Oxford University Press, Oxford
16. Guo Y, Shen H (2002) The nobel prize in physics (a centenary volume). Shanghai Press of Science Popularization, Shanghai (in Chinese)
17. Kroemer H (1963) A proposed class of hetero-junction injection laser. Proc IEEE 51 (12):1782–1783
18. Alferov ZI (2000) The double heterostructure: concept and its applications in physics, electronics and technology. http://www.nobelprize.org/mediaplayer/index.php?id=970. Accessed 14 Nov 2016
19. Akasaki I, Amano H, Koide Y et al (1989) Effects of AlN buffer layer on crystallographic structure and on electrical and optical properties of GaN and $Ga_{1-x}Al_xN$ ($0 < x <= 0.4$) films grown on sapphire substrate by MOVPE. J Cryst Growth 98(1–2):209–219
20. Schawlow AL, Townes CH (1958) Infrared and optical masers. Phys Rev 112:1940–1949

21. Ip E, Lau APT, Barros DJF et al (2008) Coherent detection in optical fiber systems. Opt Express 16(2):753–791
22. Chu S (1998) The manipulation of neutral particles. Rev Mod Phys 70(3):685–703
23. Hänsch TW (2006) Nobel lecture: passion for precision. Rev Mod Phys 78(4):1297–1309
24. Nash P (1996) Review of interferometric optical fibre hydrophone technology. IEEE Proc—Radar Sonar Navig 143:204–209
25. Schawlow AL (1982) Spectroscopy in a new light. Rev Mod Phys 54(3):697–707
26. Drever RWP, Hall JL, Kowalski FV et al (1983) Laser phase and frequency stabilization using an optical resonator. Appl Phys B 31(2):97–105
27. Padgett M, Courtial J, Allen L (2004) Light's orbital angular momentum. Phys Today 57 (5):35–40

Chapter 2
Fundamentals of Semiconductor Lasers

2.1 Stimulated Emission in Semiconductor

The principles of semiconductor laser have been expounded in textbooks and monographs [1–8]. A brief introduction to its basic physics and characteristics is given in this section.

(1) **Population Inversion in Semiconductor**

Light emission in semiconductors occurs due to transitions between the conduction band and the valence band, i.e., the emission by inter-band recombination of carriers. The active particles in semiconductor are nonequilibrium carriers, including electrons in the conduction band and holes in the valence band. The electron population obeys Fermi-Dirac distribution [1, 2]:

$$f_V(E_1) = \frac{1}{\exp[(E_1 - F_1)/k_B T] + 1}, \tag{2.1a}$$

$$f_C(E_2) = \frac{1}{\exp[(E_2 - F_2)/k_B T] + 1}, \tag{2.1b}$$

where the subscripts V and C stand for the valence band and the conduction band, E_1 and E_2 are the electron energies, and F_1 and F_2 are their Fermi energy levels in the two bands, respectively; k_B is Boltzmann constant, T is Kelvin temperature. The probability of holes' population is $[1 - f_V(E_1)]$.

The electrons in the conduction band and the holes in the valence band can move freely in the volume. The Fermi levels of two bands are equal with each other under equilibrium, $F_1 = F_2$, determined by the electrical neutrality, including the discharged donors and acceptors. At the nonequilibrium state, the neutrality is maintained dynamically; but the Fermi levels are not equal, $F_1 \neq F_2$, and vary with injected carrier density, called quasi-Fermi levels. At high injected carrier densities,

Z. Fang et al., *Single Frequency Semiconductor Lasers*, Optical and Fiber Communications Reports 9, DOI 10.1007/978-981-10-5257-6_2

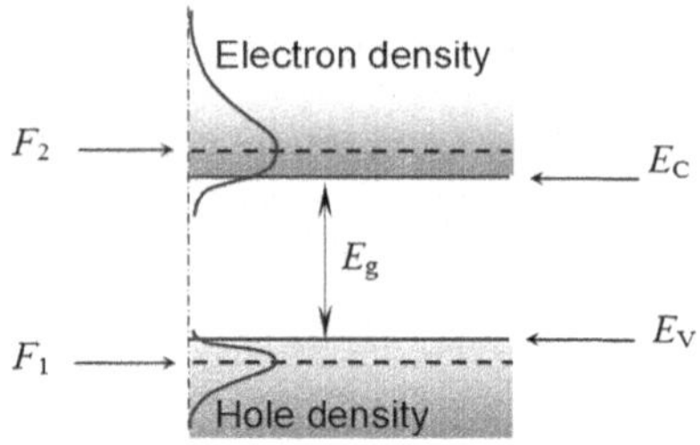

Fig. 2.1 Distributions of electrons and holes in conduction band and valence band

F_1 and F_2 will move towards the top of valence band, E_V, and the bottom of conduction band, E_C, respectively; and even enter the bands, as shown in Fig. 2.1.

The recombination probability of electrons and holes is proportional to their concentrations for the spontaneous emission, and also to the photon density for stimulated emission or absorption. The transition rates are expressed as

$$r_{sp} = A_{21} f_C (1 - f_V), \tag{2.2a}$$

$$r_{12} = B_{12} f_V (1 - f_C) P(E_{21}), \tag{2.2b}$$

$$r_{21} = B_{21} f_C (1 - f_V) P(E_{21}), \tag{2.2c}$$

where r_{sp} , r_{12} and r_{21} are the rates of spontaneous emission, absorption and stimulated emission, respectively; $P(E_{21})$ is the density of photons, whose energy is $E_{21} = E_2 - E_1 = hv$; A_{21}, B_{12} and B_{21} are Einstein coefficients. Under condition of thermal equilibrium, the transition rates satisfy relation of $r_{sp} + r_{21} = r_{12}$, resulting in photon density expressed as [1, 2]:

$$P(E_{21}) = \frac{A_{21} f_C (1 - f_V)}{B_{12} f_V (1 - f_C) - B_{21} f_C (1 - f_V)} = \frac{A_{21}}{B_{12} \exp(E_{21}/k_B T) - B_{21}}. \tag{2.3}$$

According to Plank black body radiation law, the radiation density is expressed as

$$P(v, T) = \frac{8\pi n^3 h v^3}{c^3} \frac{1}{\exp(hv/k_B T) - 1}, \tag{2.4}$$

with the unit of energy per volume per Hz. Einstein coefficients are then obtained as [1, 2, 7]

$$B_{12} = B_{21}, \tag{2.5a}$$

$$A_{21} = \frac{8\pi n^3 h v^3}{c^3} B_{21}. \tag{2.5b}$$

When nonequilibrium carriers are injected into the material the stimulated emission rate will increase. When it exceeds the absorption rate, i.e., $r_{21} > r_{12}$, the medium will amplify the incident light. The condition of stimulated emission is thus derived [1, 2] as

$$f_C(1 - f_V) > f_V(1 - f_C), \tag{2.6a}$$

or equivalently

$$F_2 - F_1 > E_2 - E_1 = h\nu \approx E_C - E_V = E_g. \tag{2.6b}$$

It is just the condition of population inversion of the semiconductor laser, with a form different from that of the conventional solid state and gas lasers. The net stimulated emission rate is

$$r_{\text{st}} = r_{21} - r_{12} = B_{21}(f_C - f_V)P(h\nu); \tag{2.7}$$

and the gain coefficient per unit path is obtained as

$$g = r_{\text{st}}/[v_g P(h\nu)] = B_{21}(f_C - f_V)/v_g, \tag{2.8}$$

where v_g is the group velocity of light. With $\Delta F = F_2 - F_1$, the spontaneous emission rate is expressed as

$$r_{\text{sp}} = A_{21} f_C(1 - f_V) = \frac{8\pi n^2 h\nu^3}{c^2} \frac{g}{1 - e^{(h\nu - \Delta F)/k_B T}}. \tag{2.9}$$

It is indicated that the spontaneous emission spectrum is different from the gain spectrum. In region of $h\nu < F_2 - F_1$ both of the gain and spontaneous emission rate are positive; in region of $h\nu > F_2 - F_1$ the gain is negative while the spontaneous emission remains positive.

The electrons and holes are actually distributed in sub-energy levels of the bands, so that the emission and absorption rates should be calculated by integration over the energy weighted with the state density. The state density function depends on the material structure and the physical situation, such as temperature. The related theory can be found in the literature, e.g., in Ref. [9]. Figure 2.2 shows the spontaneous emission spectrum and gain spectra of different injected carrier densities schematically [1, 2, 10, 11].

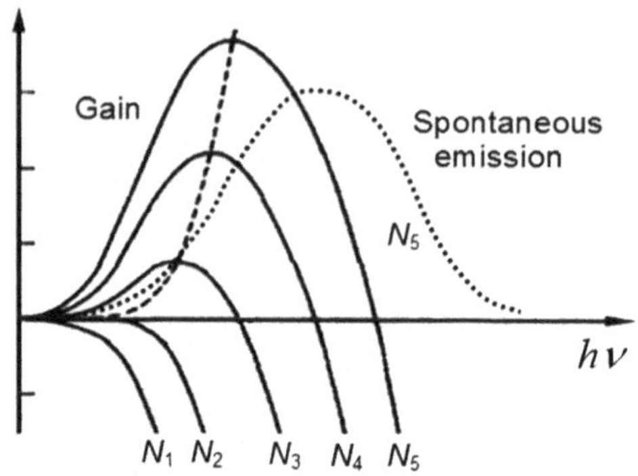

Fig. 2.2 Gain spectra and spontaneous emission spectrum (with $N_1 < N_2 < N_3 < N_4 < N_5$)

(2) **Requirement on Band Structure of Materials**

In inter-band transitions, apart from the energy conservation of $E_2 - E_1 = h\nu$, the momentum conservation law must be satisfied simultaneously, that is

$$\boldsymbol{p}_2 - \boldsymbol{p}_1 = \boldsymbol{p}_{\mathrm{ph}}, \tag{2.10}$$

where $\boldsymbol{p}_1$ and $\boldsymbol{p}_2$ are the carrier momentums before and after transition; $\boldsymbol{p}_{\mathrm{ph}}$ is the photon's momentum. The carrier's momentum in crystals is given by $\boldsymbol{p}_{1,2} = \hbar\boldsymbol{K}_{1,2}$, where $\boldsymbol{K}_{1,2}$ is the wavevector of carriers' wave function; whereas the photon's momentum is $\boldsymbol{p}_{\mathrm{ph}} = \hbar\boldsymbol{k}$ with its wavevector $\boldsymbol{k}$. $\boldsymbol{K}_{1,2}$ is in the order of crystal lattice constant, which is much smaller than photon's wavelength. Therefore, the photon's momentum is much smaller than carrier's, so that the carrier's momentum is required unchanged after transition, $\boldsymbol{p}_1 = \boldsymbol{p}_2$.

The band structures of semiconductor are divided into two categories: the direct bandgap and the indirect bandgap as shown in Fig. 2.3. The transition in direct bandgap materials occurs near the zero point of wavevector space with the momentum conservation law satisfied. The valley of conduction band of the indirect bandgap is far away from the zero point. The transition in indirect bandgap materials must be incorporated with a phonon, which is the quantized particle of crystal lattice vibrations and has a larger momentum and a lower energy than those of photons. That is to say, the optical transition involves three particles, resulting in much lower transition probability. Such a property can be characterized by the

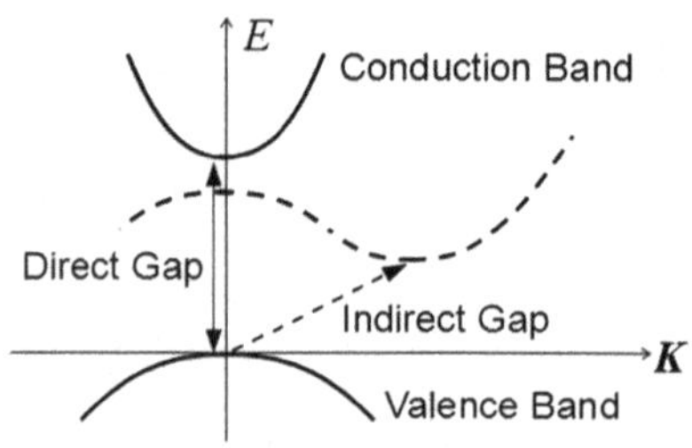

Fig. 2.3 Energy band structures of direct and indirect bandgap semiconductor

carrier's lifetimes: it is around nanosecond for direct bandgap materials, whereas in the order of millisecond for indirect bandgap materials. Therefore, the stimulated amplification can hardly occur in the indirect materials. Most of III–V compounds belong to the direct bandgap material, satisfy the precondition of lasing.

It is seen from the above analysis that the photon energy of semiconductor laser is equal to the difference of quasi-Fermi levels, which corresponds to the bandgap. Semiconductor laser in a very broad spectral band, from ultraviolet through middle infrared, can be obtained by using different semiconductors and their solid solutions with different compositions of compounds. Figure 2.4 shows the lattice constants and bandgaps of semiconductors, used most frequently for lasers, light emitting diodes, and some detectors [12, 13]. Apart from III–V compounds, some IV–VI group compounds have direct bandgaps, such as PbS, which are used to make lead-salt semiconductor lasers working in middle infrared bands.

2.2 P-N Junction, Heterostructure, and Quantum Well

2.2.1 Carrier Injection with P-N Junction

It is necessary for lasing in semiconductors to inject carriers with density high enough to get the quasi-Fermi levels entering the conduction band and the valence band, respectively. One of the methods is the optical pump; the other is electron beam injection in vacuum. The efficiency of these methods is very low, and not convenient in practice. Fortunately, semiconductor P-N junction provides an excellent way to inject high concentration carriers. P-N junction is composed of P-type and N-type regions with lattice constants matched with each other. Carriers can move between the two regions by both of drifting under electric field and diffusion under carrier concentration gradient. Figure 2.5 depicts the band structure of P-N junction schematically.

At the equilibrium without voltage applied, electrons and holes diffuse towards the opposite directions over the interface of P-N junction. Thus two carrier depleted layers on both sides of the junction result with discharged positive donors and

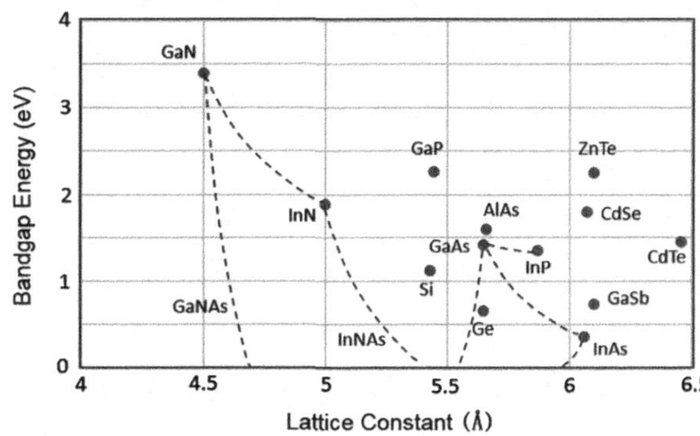

Fig. 2.4 Crystal lattice constants and bandgaps of semiconductors

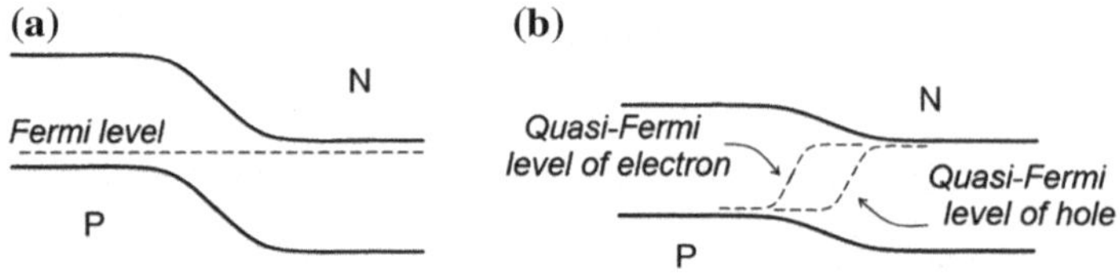

Fig. 2.5 **a** P-N junction at equilibrium; **b** Non-equilibrium under positive voltage bias

negative acceptors remained. The built-in Coolum field equilibrates the carrier's diffusion, leading to the energy bands of the two regions shift with each other, as shown in Fig. 2.5a. When a positive bias is applied on *P*-region, its energy level (with respect to electrons) goes down; the electrons are injected into *P*-region from *N*-region and holes injected in opposite direction, generating the positive injected current. The nonequilibrium carriers are thus concentrated on two regions near the junction; photons emit then due to the recombination of nonequilibrium carriers.

Fermi levels near the junction are now separated into two quasi-Fermi levels corresponding to electrons and holes as shown in Fig. 2.5b. When the applied voltage increases further towards the bandgap, the injected carriers are accumulated higher and higher, and the quasi-Fermi levels will enter the bottom of conduction band and the top of valence band, respectively; the stimulated amplification of emission will then occur. Therefore, the working voltage of semiconductor laser in *Volt* is roughly equal to the bandgap

$$V \geq E_g/e = 1.24/\lambda, \tag{2.11}$$

where E_g is the bandgap in eV, λ is the lasing wavelength in μm. The working voltage of most commonly used near-infrared lasers is around 1 V. The actual voltage measured on the electrodes will be a little higher than bandgap, since the series resistance exists apart from the junction voltage, which includes the Schottkey barriers on metal–semiconductor contact between laser material and electrodes and the resistance of volume materials. Compared with other lasers, the low working voltage is an outstanding advantage.

Obviously, the laser should be made of highly doped low resistivity materials, which benefit not only a low series resistance, but also a higher injected carrier concentration. It is because the equilibrium Fermi levels are close to the bottom of conduction band and the top of valence band respectively; the quasi-Fermi levels are easier to be shifted to satisfy the lasing condition.

2.2.2 Heterostructure and Quantum Well

The carrier's diffusion is a limitation to increase of carrier concentration for homo-junction composed of the same semiconductors. If the injected carrier can be

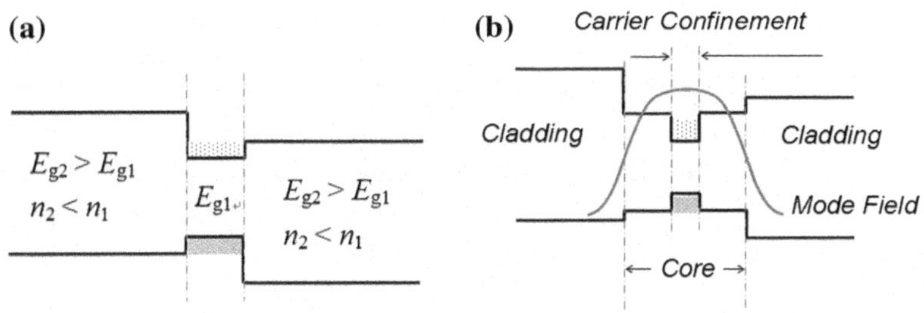

Fig. 2.6 **a** Double heterojunction; **b** Separate confinement heterostructure

confined in a thin layer, its concentration will be enhanced effectively. H. Kroemer and Z. Alferov proposed a new concept of heterojunction [14, 15], that is, to build a barrier close to the junction by two semiconductors with bandgaps a little difference but lattice constants nearly the same. The barrier prevents the injected carriers from diffusion, so that a much higher carrier concentration can be obtained under the same injected current, resulting in a much lower threshold. Figure 2.6a shows the band diagram of double heterostructure schematically; in a small region the electrons and holes injected from larger bandgap semiconductors are accumulated with higher concentration and higher gain.

The refractive index of semiconductor is dependent on its band structure, especially the bend gap; for compound semiconductor, roughly $n \propto E_g^{-1/4}$ [16]. The index of active region with smaller bandgap is larger than two sides, as depicted in Fig. 2.6a. Therefore, the heterostructure forms a planar optical waveguide also. To increase the injected carrier concentration the carrier confinement layer is designed as thin as several nanometers, for example, which may be too thin to be the core of optical waveguide. Thus a separate confinement heterostructure (SCH) was proposed as shown in Fig. 2.6b, which is widely used in many semiconductor laser structures.

When the thickness d of active layer is decreased to nearly equal or less than the electron's de Broglie wavelength, the layer becomes a quantum well (QW), where the energy of carriers is quantized into sub-levels, roughly expressed as: $E = n^2h^2/(8m^*d^2)$, where $n = 1, 2, \ldots$ is integers, m^* is the effective mass of carriers [17], as shown schematically in Fig. 2.7a. Compared with the ordinary

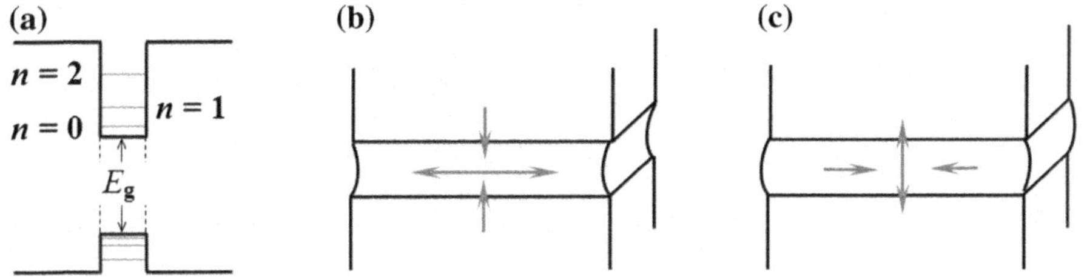

Fig. 2.7 **a** Energy levels of a square potential quantum well; **b** Stretched strained layer QW; **c** Compressed strained layer QW

heterostructure lasers, the quantum well laser enhances further the population inversion, showing excellent characteristics, such as higher gain coefficient, higher modulation response, and lower temperature sensitivity.

The quantum well can be made not only of semiconductors with matched lattice constants, but also of materials with slightly different lattice constants. In the latter case, the quantum well is deformed asymmetrically in directions perpendicular and parallel to the layer, while the junction is maintained as a mono-crystal under certain conditions. The material is called strained layer quantum wells, including that compressed in plane of crystal growth and stretched in the plane, as depicted in Fig. 2.7b, c [18]. The achievement of strained layer quantum well broke through the limitation of lattice matching, and the spectral range of semiconductor laser is expanded greatly.

Due to the strains, the energy band structure is modified and new properties with unique features are exploited. The design and fabrication of strained layer quantum well is termed the energy band engineering [19]. Based on the related theory and technology, a new kind of semiconductor laser, quantum cascaded laser (QCL), is invented, which makes use of transition between sublevels of the quantum well, instead of transitions between the valence band and the conduction band [20]. The wavelength range is thus further expanded, especially in infrared band, whereas it is not easy for the ordinary LD based on inter-band transition to work at room temperature, because the concentration of thermal excited equilibrium carriers is too high to realize the population inversion in such a low bandgap materials.

The fabrication of semiconductor lasers relies on a series of advanced technologies, which can be found in monographs and journals. Nowadays, semiconductor lasers and LEDs from ultraviolet to middle infrared bands have been realized, and most of them are commercially available [21].

2.3 Cavity Structure and Transverse Modes

2.3.1 Basic Structure and Transverse Modes

A typical structure of semiconductor laser is shown in Fig. 2.8. It is a semiconductor P-N diode; that is why the semiconductor laser is called also laser diode (LD), or diode laser. The active region is a thin core layer of optical waveguide, its thickness is usually less than 1 μm; its upper and beneath layers are the P- and N-type claddings. The active region is confined laterally by burying materials with lower index and high resistivity, forming a stripe in width of only 2–5 μm to reduce the pump current and to maintain the fundamental transverse mode. The cavity is composed of the two cleaved facets of semiconductor crystal. Most of semiconductors for lasers are III–V compounds with cubic crystal structure. Their feasible cleaved facet is surface [110], which is perpendicular to the active layer grown on surface [100]. The cavity length is typically in range of 0.3–1.0 mm. The electrodes

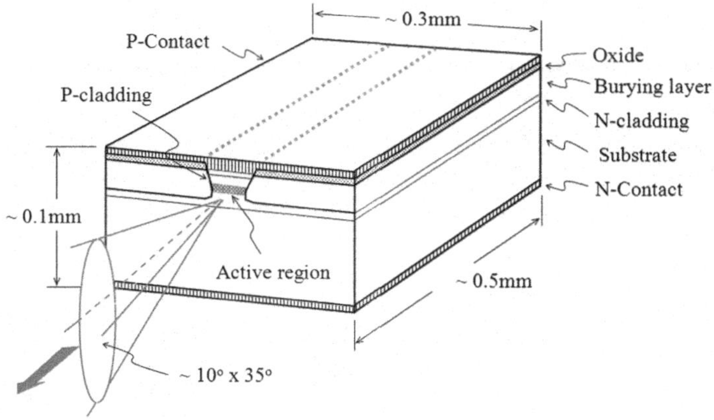

Fig. 2.8 Schematic structure of a typical semiconductor laser

are fabricated on its upper and bottom surfaces. The LD chip is with thickness of about 0.1 mm, and width of sub-millimeter, showing its outstanding advantages of small size and small weight.

The multiple layers are grown on the substrate, usually by liquid phase epitaxy (LPE), or molecular beam epitaxy (MBE), or metal organic vapor phase epitaxy (MOVPE). The processing of material growth, device fabrication and packaging are very similar to that of ordinary microelectronic devices, except for the cavity mirror by cleaving and for the output via windows or fiber coupling. Therefore, the semiconductor laser is suitable for mass production with the advantage of low cost.

As stated above, the heterostructure plays a role of planar optical waveguide; the larger bandgap and lower index materials are the cladding, and the smaller bandgap, higher index layer is the core, in which the P-N junction is contained. The waveguide determines the transverse modes in dimension perpendicular to the junction. For a symmetric three region waveguide as shown in Fig. 2.9 the electric field in the junction plane (TE mode) is expressed as [1, 2].

$$E_y = E_y(0)\begin{cases} \cos(\kappa d/2)e^{\gamma(x+d/2)} & x \leq -d/2 \\ \cos(\kappa x) & -d/2 < x < d/2, \\ \cos(\kappa d/2)e^{\gamma(d/2-x)} & x \geq d/2 \end{cases} \tag{2.12a}$$

$$E_y = E_y(0)\begin{cases} \sin(\kappa d/2)e^{\gamma(x+d/2)} & x \leq -d/2 \\ \sin(\kappa x) & -d/2 < x < d/2, \\ \sin(\kappa d/2)e^{\gamma(d/2-x)} & x \geq d/2 \end{cases} \tag{2.12b}$$

where $\kappa = \sqrt{n_1^2 k^2 - \beta^2}$, $\gamma = \sqrt{\beta^2 - n_2^2 k^2}$; β is the optical wavevector in z-direction. (2.12a) and (2.12b) correspond to even modes and odd modes; their eigen equations are $\tan(\kappa d/2) = \gamma/\kappa$ and $\tan(\kappa d/2) = -\kappa/\gamma$, respectively. For TM

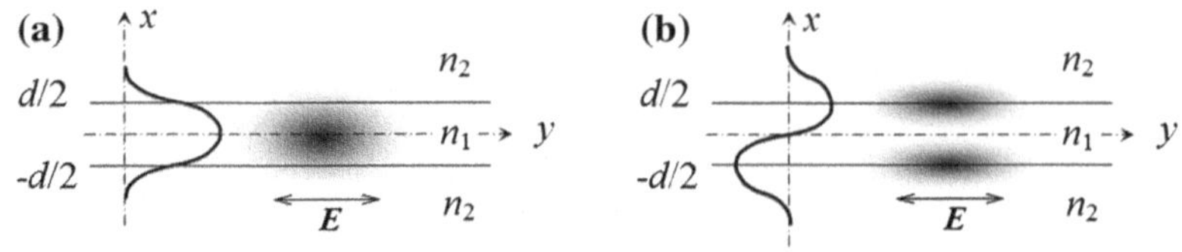

Fig. 2.9 Transverse TE modes. **a** Even mode; **b** odd mode

mode, the magnetic field is in junction plane, with $H_y(x)$ expressed the same as (2.12a) and (2.12b), but the eigen equations should be written as tan $\kappa d/2 = n_2^2\gamma/n_1^2\kappa$ and tan $\kappa d/2 = -n_1^2\kappa/n_2^2\gamma$. Multiple solutions are given from the two transcend equations, with β determined as a function of thicknesses and indexes of core and cladding.

Most of semiconductors work in TE modes, which is with the electric field in the plane of P-N junction. It is attributed to the factor that the facet reflectance for TE modes is a little higher than that of TM modes [1, 2], resulting in lower threshold of TE modes. The degree of polarization increases with the output power increasing. By special designs, the stained layer quantum well can be made with gain coefficient of TM mode higher than that of TE mode; a TM mode laser can then be realized. Since the core thickness is very small, usually less than 1 μm, only the fundamental mode with the largest β works, while all the higher order transverse modes are suppressed. Figure 2.9a, b depict the lowest even and odd TE modes schematically.

In the dimension of junction plane, multiple transverse mode operation is easy to occur, which is not welcome to most applications. In solid-state lasers and gas lasers, the cavity is composed of optical components, the transverse mode can be designed. The LD cavity is made of cleaved facets; its lateral boundary has to be formed by semiconductor processing, at the stage of material growth and/or chip fabrication. Figure 2.10 shows several typical structure of lateral confinement, such as the burred heterostructure, the ridge waveguide, the channeled substrate planar waveguide.

In the burred heterostructure, a stripe mesa is formed and then burred by higher bandgap semiconductor. The ridge waveguide is probably the mostly often used structure, where the thickness of upper cladding out of the stripe is thinned in processing, while the thickness above the stripe remains the original size, resulting in an effective index difference between the two regions. The channeled substrate

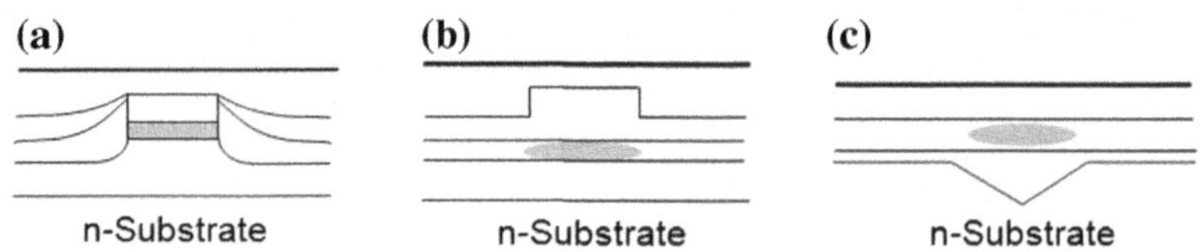

Fig. 2.10 Typical LD structures in lateral dimension: **a** Burred heterostructure; **b** Ridge waveguide; **c** Substrate planar structure

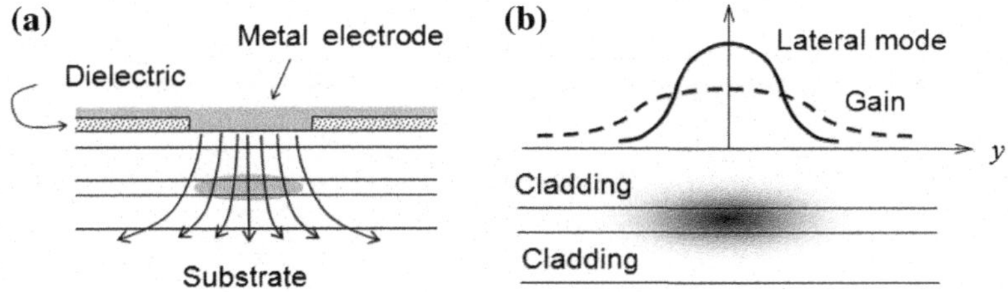

Fig. 2.11 **a** A planar stripe structure; **b** Diagram of gain-guided waveguide

planar waveguide can be regarded as an upper-and-down reversed ridge waveguide. To ensure the laser operated at the fundamental lateral mode, the index difference should be made small, and the stripe width should also be small enough, such as 2–5 μm and less.

2.3.2 Index Guiding and Gain Guiding

The complex expression of electro-magnetic wave may take either form of $\exp[j(\omega t - kz)]$ or $\exp[j(kz - \omega t)]$. The latter is used in this book, except for cases specified otherwise. The index of a medium with gain or loss should be expressed as a complex $n = n_r + jn_i$; and its imaginary part is thus written as $n_i = -g/2k$. The effective index difference can be generated either by the real part of complex index, as in the structures of Fig. 2.10, or by its imaginary part. The former is called index guiding, whereas the latter is called gain guiding. The gain guiding is a unique effect in semiconductor lasers. Figure 2.11a shows a planar stripe structure with current distribution confined by dielectric isolator, Fig. 2.11b gives a picture of gain-guided lateral mode.

The distribution of gain coefficient in y-direction can be described basically by a parabolic function $n_i(y) = \delta n(1 - y^2/y_0^2)$. Substitute it into Maxwell equations, a lateral mode field in Gaussian–Hermit function can be obtained as

$$\begin{aligned} E_y(y) &= H_q(\kappa_y y)\exp[-(q+1/2)\kappa_y^2 y^2] \\ &\propto E_0 \exp[-(1+j\chi)y^2/y_1^2]. \end{aligned} \tag{2.13}$$

where H_q is the qth Hermit polynomial, $\kappa_y^2 = (1-j)\sqrt{n\delta n}k/y_0$. The second expression of (2.13) is the fundamental lateral mode with $H_1 = const.$ χ is introduced to describe the different curvatures of real part and imaginary part of index with the real index distribution taken into account.

It is seen from Exp. (2.13) that the optical wave in the waveguide has a Gaussian amplitude distribution, and the wavefront is no longer a plane but a parabolic surface. Figure 2.12 shows its wavefront and that of index-guided lasers for comparison.

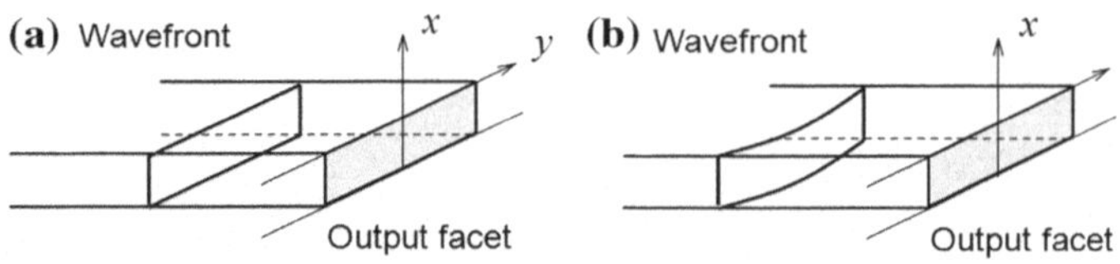

Fig. 2.12 Wavefront of index-guided laser (**a**) and gain-guided laser (**b**)

The different wavefront inside the waveguide will lead to different characteristics of the beams emitted from laser cavity. The beam is approximately described by a Gaussian beam with different divergences in directions of perpendicular and parallel to the junction plane, which are usually called the fast and slow axes, corresponding to the modes with a lower and a higher effective index. In the slow axis, the positions of beam waist for gain guiding and for index guiding are different. For the former, it is located behind the laser facet because of the convex wavefront, whereas it is at the laser facet for the latter. In the fast axis, the wave is guided by pure index distribution, and the beam waist is just at the facet. Therefore, an astigmatism exists in gain guide lasers, whereas it is negligibly small in the index-guided laser. The strength of astigmatism is characterized by K factor, defined as [22, 23].

$$K = \frac{\left[\int |E(y)|^2 \mathrm{d}y\right]^2}{\left|\int E^2(y)\mathrm{d}y\right|^2} \tag{2.14}$$

It is deduced from Exp. (2.13) $K = \sqrt{1+\chi^2}$. For index guiding, $K = 1$; for pure gain guiding with $\chi = 1$, $K = \sqrt{2}$. Since the derivative of real index with respect to carrier density is negative in semiconductors, a nonuniform carrier density distribution in y-direction will induce also a real index guiding. A parabolic carrier density distribution exists in practice, corresponding to a concave index distribution, giving the effect of anti-guiding. In such a case, $K >> 1$ may result under some conditions.

Since the sizes of fundamental mode LD are very small, the divergence angles of output beam are quite large, typically $10 \times 40°$, much larger than that of solid-state laser and gas laser, which are in the order of milliradian. Therefore, an optical collimation system is usually necessary in LD applications; the beam characteristics, especially its astigmatism, must be taken into consideration in the optics design.

2.3.3 Laser Array and Modules

The narrow stripe LD is very useful, because it can work in the fundamental transverse and lateral mode, which is one of the basic conditions of single

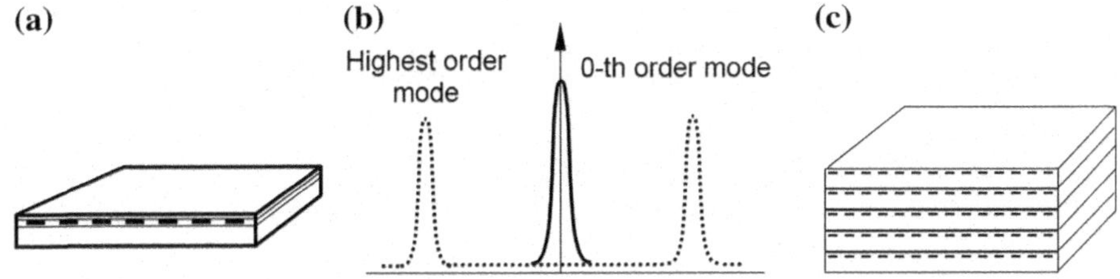

Fig. 2.13 **a** LD array; **b** Far field patterns of a phase locked LD array; **c** LD stack

frequency operation. Another advantage is its capability of coupling with the optical fiber. Besides, there is a big area of applications which need the laser power as high as possible, such as for pumps of all-solid-state lasers and for medical applications. The high power laser is usually composed of multiple broad area stripes to build an array; and even build a stack with several LD array bars. Figure 2.13a depicts such an array. When the spacing between stripes is narrowed to a certain degree, optical coupling between stripes will occur, resulting in optical phase locked. Such an array is called the phased array. Its mode behavior should be analyzed by the super-mode theory [24]. The mode with zero phase difference between adjacent stripes is the 0th order super-mode, whose output beam is a single narrow lobe. The mode with π phase difference between adjacent stripes is the highest order super-mode with output beam of two narrow lobes, as shown in Fig. 2.13b. Generally the highest super-mode has the lowest threshold. The divergence angle is greatly narrowed by the phase locking. However, the thermal performance is more important for high power lasers; a wide spacing between stripes is beneficial for reducing temperature rising at high pump current. In order to obtain higher power, LD stacks are also developed, as shown in Fig. 2.13c, which is suitable for pulsed operations.

2.4 Rate Equation and Output Characteristics

2.4.1 *Lasing Condition*

The LD cavity is composed of two cleaved facets, which are parallel with each other naturally for single crystal materials. Such a cavity is called Fabry-Perot (F-P) cavity. By the same method as that used in analyzing F-P resonator [1, 2], its transmission can be obtained as

$$\frac{E_{\text{out}}}{E_{\text{in}}} = \frac{t_1 t_2 \exp(j\beta L)}{1 - r_1 r_2 \exp(j2\beta L)}, \tag{2.15}$$

where $\beta = (n_r + jn_i)k$ and $n_i = (\alpha_c - g)/2k$ with loss coefficient α_c inside the cavity. $r_{1,2} = \sqrt{R_{1,2}}$ are the field reflectance of the two facets; $t_{1,2} = (1 - r_{1,2}^2)^{1/2}$ in case of no loss at facets. When the device gives a finite output E_{out} under condition of a null input field E_{in}, the laser is lasing. The lasing condition is thus defined as

$$r_1 r_2 \exp(j2\beta L) = 1. \tag{2.16}$$

From its real part, $r_1 r_2 \exp(-2n_i kL) = 1$, the threshold gain is obtained

$$g_{\text{th}} = \alpha_c + \frac{1}{2L}\ln\left(\frac{1}{R_1 R_2}\right). \tag{2.17}$$

The phase condition of lasing is from the imaginary part of (2.16), $2n_r kL = 2m\pi$, i.e.

$$2n_r L = m\lambda_m. \tag{2.18}$$

It is seen that the lasing wavelengths are a series of discrete values, termed the longitudinal modes. The spacing between the adjacent longitudinal modes is deduced as

$$\Delta\lambda = \lambda^2/(2n_g L), \tag{2.19}$$

where $n_g = n - \lambda(\partial n/\partial\lambda)$ is its group index, with the chromatic dispersion of medium index taken into account. The typical phase index of semiconductor is near 3.6, and the group index is around 4.0.

Semiconductor lasers have noticeable features different from solid-state lasers and gas lasers

(1) High gain coefficient. It is because the doping concentration in semiconductor is as high as in the order of 10^{18} cm^{-3}, much higher than densities of active particles in solid and gas; and every doped atoms can contribute one carrier, resulting in higher gain coefficient.
(2) Due to the high gain coefficient, it is not necessary to use cavity mirrors with high reflectance. That is why a plain cleaved facet can serve the mirror. According to Fresnel formulas, the reflectance of optical intensity is about 0.32, much lower than that used in other lasers, which is often near unity.
(3) Due to the high gain coefficient, a short cavity length is enough to get high single trip gain. The cavity length of ordinary LD is only 0.3–0.5 mm; for high power LD, it is around 1 mm. The short cavity length results in large longitudinal mode spacing.
(4) The gain spectrum is much wider than that of gas lasers and solid-state lasers. It is because the transition occurs between energy bands, rather than between narrow levels.

2.4.2 Rate Equations

The rate equations of semiconductor laser have been discussed in detail in monographs and papers [6, 25] and used often for analyzing its characteristics. Compared with gas and solid-state lasers with three or four energy levels involved, the rate questions of semiconductor laser have different features.

(1) The active particles at the lower level are not electrons but holes. The concentrations of non-equilibrium electrons and hole are equal with each other to meet the requirement of electric neutrality. Therefore only electrons in the conduction band needs to be taken into account.
(2) The photon density inside the cavity along the propagation direction is nonuniform, because of the low mirror reflectance and high gain coefficient.
(3) The active carriers are generated by current injection. The pump term in the rate equation is directly written as the injected electron density.

The rate equation of carrier concentration N and photon density p is thus written as

$$\frac{\partial N}{\partial t} = \frac{J}{ed} - R_{\text{sp}} - R_{\text{nr}} - v_g \Gamma \sum_m g_m p_m^{\pm} - D\nabla^2 N, \tag{2.20a}$$

$$\frac{\partial p_m^{\pm}}{\partial t} = v_g (\Gamma g_m - \alpha_c) p_m^{\pm} + \gamma_{\text{sp}} R_{\text{sp}} \pm v_g \frac{\partial p_m^{\pm}}{\partial z}. \tag{2.20b}$$

where J is the injected current density, d is the thickness of carrier confinement layer, J/ed is the injection rate of electrons; v_g is the group velocity of light; $D\nabla^2 N$ stands for diffusion of carriers, which can be neglected in usual cases. $p_m^{\pm}$ are the photon densities of the mth longitudinal mode propagating in $\pm z$ directions. $\Gamma = \int_d |E|^2 \mathrm{d}s / \int_\infty |E|^2 \mathrm{d}s = d/d_M$ is called the confinement factor, reflecting the ratio of gain distribution over intensity distribution, where d_M is the effective thickness of optical mode. Because the active layer d is very thin, the confinement factor is small, usually only about 5%. $R_{\text{sp}} = B[(N_0 + N)(N_{\text{h0}} + N) - N_0 N_{\text{h0}}] = N/\tau_{\text{sp}}$ is the recombination rate of carriers, i.e., the spontaneous emission rate, where B is the coefficient of recombination between bands, N_{h0} and N_0 are the equilibrium hole and electron concentration,τ_{sp} is the carrier lifetime, approximately $\tau_{\text{sp}} \approx [B(N + N_0 + N_{\text{h0}})]^{-1}$. If there exist certain imperfections in the material, the carrier lifetime is rewritten as $\tau = 1/(\tau_{\text{sp}}^{-1} + \tau_{\text{nr}}^{-1})$ with non-radiation recombination rate of $R_{\text{nr}} = N/\tau_{\text{nr}}$. The ratio of $\tau/\tau_{\text{sp}} = \eta_i$ is termed the internal quantum efficiency.

The gain coefficient of semiconductor lasers is a function of injected carrier concentration. The theoretical analysis and experimental studies indicate that the gain can be basically described as $g = g_0 \ln(N/N_{\text{tr}})$ [6]. Within a certain range of carrier concentration variation, it is approximated as $g = g_N(N - N_{\text{tr}})$, where N_{tr} is the transparent carrier concentration, $g_N = \partial g/\partial N$ is the differential gain coefficient. The gain spectrum cannot be expressed by an analytic expression exactly; it is determined by the complicated band structure. A Lorentzian line shape is usually used as an approximation

$$g_m = \frac{g_N(N - N_{\text{tr}})}{1 + (\lambda_m - \lambda_p)^2/(\Delta\lambda)^2}, \tag{2.21}$$

where $\Delta\lambda$ is the spectral width of gain, λ_p is the peak wavelength of gain spectrum.

α_c in (2.20b) is the loss coefficient inside the cavity. γ_{sp} in the rate equations is termed the spontaneous emission factor, whose physical meaning is the proportion of spontaneous emission entering the laser modes. It is proportional to the ratio of laser mode linewidth to spontaneous emission linewidth and the ratio of their divergences: $\gamma_{\text{sp}} \propto (\Omega_{\text{laser}}\Delta\lambda_{\text{laser}})/(\Omega_{\text{sp}}\Delta\lambda_{\text{sp}})$ [26], where Ω is the solid angle of beam divergences. γ_{sp} is much smaller than unity, typically in the order of 10^{-3}–10^{-4} for gain-guided lasers and 10^{-5}–10^{-6} for index-guided lasers. The spontaneous emission factor in solid-state lasers and gas lasers is even several orders smaller. Although it is so small, γ_{sp} plays an important role, affecting some of the laser behaviors greatly, especially the longitudinal mode behavior, as discussed later in Sect. 2.5.

2.4.3 Light–Current Characteristics

Let us discuss the light versus current characteristics (*L-I*) of a signal longitudinal mode at the continuous wave (cw) operation. The rate equations for the stationary case are written as

$$\frac{J}{ed} - \frac{N}{\tau} - v_g \Gamma g_N(N - N_{\text{tr}})(p^+ + p^-) = 0, \tag{2.22a}$$

$$\pm v_g \frac{\partial p^\pm}{\partial z} = v_g[\Gamma g_N(N - N_{\text{tr}}) - \alpha_c]p^\pm + \frac{\gamma_{\text{sp}} N}{\tau}. \tag{2.22b}$$

The solution should meet the boundary conditions of $p^+(0) = R_1 p^-(0)$ at $z = 0$ and $p^-(L) = R_2 p^+(L)$ at $z = L$, where L is the cavity length. Obviously p and N have nonuniform distributions in z-direction. Since the diffusion of carriers is fast, the carrier concentration can be regarded a constant. By such an approximation the rate equations can be solved for a uniform gain coefficient. With $g_{\text{net}} = \Gamma g - \alpha_c$, the photon densities are obtained

$$p^+ = \frac{\gamma_{sp}N}{v_g\tau_p g_{net}} \frac{(1-R_1)+(1-R_2)R_1\exp(g_{net}L)}{1-R_1R_2\exp(2g_{net}L)}\exp(g_{net}z), \tag{2.23a}$$

$$p^- = \frac{\gamma_{sp}N}{v_g\tau_p g_{net}} \frac{(1-R_2)+(1-R_1)R_2\exp(g_{net}L)}{1-R_1R_2\exp(2g_{net}L)}\exp[g_{net}(L-z)]. \tag{2.23b}$$

The threshold condition is thus deduced as $g_{net} = g_{th} = (2L)^{-1}\ln(1/R_1R_2)$ from condition of the denominator towards zero. Here condition of (2.17) is modified with Γ factor. Generally the averaged photon density $\bar{p}$ is used as a parameter for characterization, deduced as

$$\bar{p} = \frac{1}{L}\int_0^L (p^+ + p^-)\mathrm{d}z = \frac{p_{out}}{g_{net}L}, \tag{2.24}$$

where $p_{out} = p^+(L) - p^-(L) + p^-(0) - p^+(0)$ is the total output. The ratio of outputs from two facets is deduced as

$$\frac{p^+_{out}}{p^-_{out}} = \frac{(1-R_2)\sqrt{R_1}}{(1-R_1)\sqrt{R_2}}. \tag{2.25}$$

The photon lifetime in cavity is defined as the ratio of photon number inside the cavity to the loss rate, including the loss inside cavity and the loss of output

$$\tau_p = \frac{L\bar{p}}{v_g(p_{out} + \alpha L\bar{p})} = \frac{1}{v_g(\alpha + g_{th})} \tag{2.26}$$

By using the averaged photon density, the rate equations are simplified as

$$\frac{\eta_{inj}J}{ed} - \frac{N}{\tau_{sp}} - v_g\Gamma g_N(N - N_{tr})p = 0, \tag{2.27a}$$

$$v_g\Gamma g_N(N - N_{tr})p - \frac{p}{\tau_p} + \frac{\gamma_{sp}N}{\tau_{sp}} = 0, \tag{2.27b}$$

where $\bar{p}$ is written as p for simplicity; η_{inj} is the pump efficiency, taking the internal quantum efficiency and lateral spreading of the injection current into account. It is noticed that at the stationary state the gain plus the contribution of spontaneous emission is equal to the loss; therefore the gain coefficient is not equal to the threshold gain, but a little less than the latter.

Assuming τ_{sp} is approximated as a constant in a certain range of carrier concentration, denoting two auxiliary parameters of $N_J = \eta_{inj}J\tau_{sp}/ed$ and $N_g = 1/(v_g\Gamma g_N\tau_p)$, the rate equations can be solved analytically

$$N = \frac{1}{2}[(N_J + N_{\rm tr} + N_g) - \sqrt{(N_J - N_{\rm tr} - N_g)^2 + 4\gamma_{\rm sp}N_J N_g}], \tag{2.28a}$$

$$p = \frac{\tau_p}{2\tau_{\rm sp}}[(N_J - N_{\rm tr} - N_g) + \sqrt{(N_J - N_{\rm tr} - N_g)^2 + 4\gamma_{\rm sp}N_J N_g}]. \tag{2.28a}$$

The *L–I* characteristics below the threshold, under condition of $N_J \ll N_{\rm tr} + N_g$, is obtained as $p \approx (\eta_{\rm inj} J\tau_p/ed)[\gamma_{\rm sp}N_g/(N_{\rm tr} + N_g)]$ and $N \approx \eta_{\rm inj} J\tau_{\rm sp}/ed$. The *L–I* above threshold is $p \approx \eta_{\rm inj} J\tau_p/ed - (\tau_p/\tau_{\rm sp})N_{\rm th}$ under condition of $(N_J - N_{\rm tr} - N_{\rm g})^2 \gg 4\gamma_{\rm sp}N_J N_1$. It is worthy of noticing that the injected carrier concentration is towards a constant $N \approx N_{\rm tr} + N_{\rm g} = N_{\rm th}$, no longer increasing with injected current linearly, almost all injected electrons transfer to photons. Such a case is called carrier concentration pinning. The output power, the threshold current density and the external differential quantum efficiency are then written as

$$P_{\rm out} = \eta_{\rm d} V_J (I - I_{\rm th}), \tag{2.29}$$

$$J_{\rm th} = ed(N_{\rm tr} + N_g)/\eta_i \tau_{\rm sp}, \tag{2.30}$$

$$\eta_d = \eta_i \frac{\alpha_m}{\alpha_c + \alpha_m}, \tag{2.31}$$

where $V_{\rm J} \approx hv/e$ is the P-N junction voltage, and $\alpha_m = g_{\rm th}$ is in meaning of the output loss from the cavity mirrors. It is indicated that a higher mirror reflectance will reduce the threshold, but does not benefit the external efficiency. It is therefore necessary to design the mirror according to the application requirement.

In practice, people pay attention more to the power conversion efficiency

$$\eta_P = \frac{\eta_d V_J (I - I_{\rm th})}{I V_J + I^2 R_s}, \tag{2.32}$$

where R_s is the series resistance. It is seen that η_P varies with the injected current, going up first and then down. The maximum efficiency is obtained at $I = I_{\rm th}(1 + \sqrt{1 + V_J/I_{\rm th}R_s})$ [27]:

$$\eta_{P,max} = \frac{\eta_{\rm d} V_J}{(\sqrt{V_J + I_{\rm th}R_{\rm s}} + \sqrt{I_{\rm th}R_{\rm s}})^2}. \tag{2.33}$$

The *L–I* characteristics around the threshold is affected greatly by the spontaneous emission factor $\gamma_{\rm sp}$. The curvature $c_{\rm LI}$ of *L–I* there is a function of $\gamma_{\rm sp}$:

$$c_{\rm LI} \propto \frac{{\rm d}^2 p}{{\rm d}I^2} \propto \frac{\tau_p/\tau_{\rm sp}}{\sqrt{\gamma_{\rm sp}N_{\rm g}(N_{\rm tr} + N_{\rm g})}}. \tag{2.34}$$

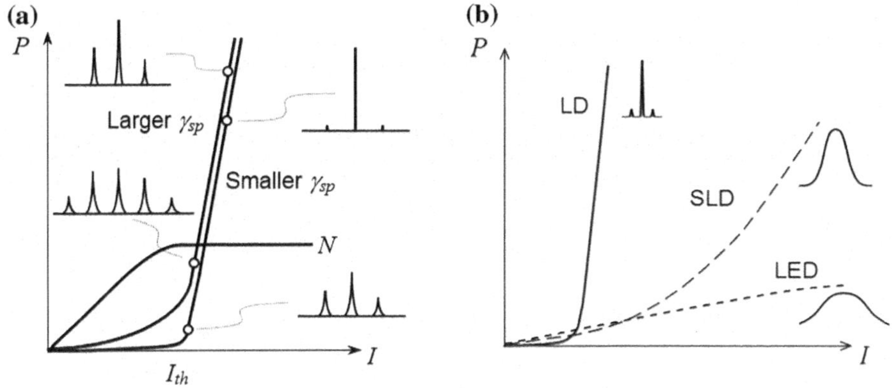

Fig. 2.14 **a** *L–I* and *N–I* curves with different $\gamma_{\rm sp}$. **b** *L–I* curves and line shapes of LD, LED and SLD

As stated above, the gain-guided laser has much higher $\gamma_{\rm sp}$ than the index-guided laser; therefore the *L–I* curve of the latter shows a sharp turn-on at the threshold, whereas the gain-guided laser has a soft turn-on *L–I* curve. Figure 2.14a depicted schematically *L–I* curves and the curve of carrier concentration versus current. The semiconductor laser gives usually a much higher spontaneous emission power below threshold than gas lasers and solid-state lasers; one of the reasons for the difference is just its larger spontaneous emission factor.

Figure 2.14b shows also *L–I* curves of the light emission diode (LED) and super-luminescence diode (SLD) for comparison. The LED and SLD can be made of the same materials as LD, and with the structures similar to LD, but no cavity exists in LED and very low reflective facet is used in SLD. Usually a totally anti-reflective coating or the active stripe tilted to the cleaved facet is used in LED, so that no lasing oscillation occurs. For SLD, the spontaneous emission is amplified, giving output power higher than LED, but still below the threshold due to the weak feedback from the mirror. The spectral linewidth of LD and SLD is much larger than that of LD, which benefits suppression of coherent noises in some applications.

The LD active region can be utilized as an amplifier if both of its two facets are coated by anti-reflective films to eliminate the round-trip oscillations. Such a device is called the semiconductor optical amplifier (SOA), used often in optical communications, in ring-cavity lasers and other applications.

2.5 Longitudinal Modes and Tunability

2.5.1 Longitudinal Mode Characteristics

The longitudinal modes are determined by (2.18), that is, the cavity length is an integer multiple of half wavelength. It is necessary to notice that the mode spacing

$\Delta\lambda = \lambda^2/(2n_gL)$ is not a constant but dependent on wavelength because of the index dispersion. The dispersion obeys the universal Kramers—Kronig relations; for the electric susceptibility it is expressed as [7, 11].

$$\chi_r(\omega) = \frac{1}{\pi} P \int_{-\infty}^{+\infty} \frac{\chi_i(\omega')}{\omega' - \omega} \mathrm{d}\omega', \tag{2.35a}$$

$$\chi_i(\omega) = -\frac{1}{\pi} P \int_{-\infty}^{+\infty} \frac{\chi_r(\omega')}{\omega' - \omega} \mathrm{d}\omega', \tag{2.35b}$$

where P stands for the Cauchy principal of integral; χ_r and χ_i are the real part and imaginary parts of electric susceptibility. By using the relation to index $\varepsilon/\varepsilon_0 = n^2 = n_r^2 - n_i^2 + j2n_rn_i = \chi + 1$ the variation of index with gain's change is deduced as

$$\Delta n(\lambda) = \frac{1}{2\pi^2} P \int_0^{\infty} \frac{\Delta g(\lambda')\mathrm{d}\lambda'}{1 - (\lambda'/\lambda)^2}, \tag{2.36}$$

The mode spacing is thus dependent on the gain spectrum; and the mode wavelength will move with the variation of pump levels. Such a phenomenon is called the frequency pulling. The effect in semiconductor lasers is much stronger than that of gas lasers and solid-state lasers because of its higher gain coefficient and higher dispersion.

It is known that the gain spectrum of a medium is dependent on its linewidth broadening mechanism. If all active atoms make the same contributions to the linewidth broadening, the broadening is called homogeneous; oppositely, if atoms make different contributions, respectively, it is called inhomogeneous. The typical example of the latter is the Doppler broadening induced by different velocity of atom's movement in gas. The related line shape is in Gaussian. Intrinsically, the excited state of atoms will decay and emit spontaneously with a finite probability, leading to a homogeneous broadening with a Lorentzian line shape. It is considered that the broadening in semiconductors is basically homogeneous with a gain spectrum as described by (2.21).

Because of the large linewidth of gain coefficient, the semiconductor laser works likely with multiple longitudinal modes, especially for the F-P cavity LD without measures of mode selection. The intensities of the modes are dependent on the gain spectrum. Roughly, the mode intensity increases inversely with the difference between the mode gain and the threshold gain [3]

$$I_m \propto f\{[g_{\mathrm{th}} - g(\lambda_m)]^{-1}\}, \tag{2.37}$$

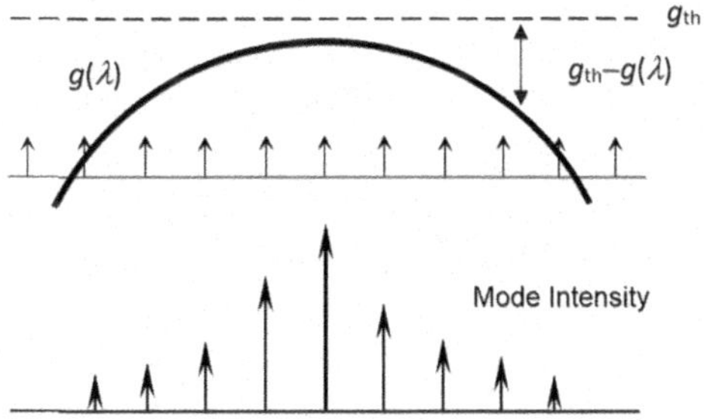

Fig. 2.15 Gain spectrum and longitudinal mode intensities

where *f*(*x*) is a certain ascending function. As pump current increases, this difference of main mode decreases towards zero, the side mode suppression ratio (SMSR) will increase. Figure 2.15 gives a schematic illustration of the mode intensity distribution. It is seen that the number of lasing modes depends on the mode spacing, i.e., the cavity length; and also on the working point. The intensities of side modes decrease with increasing of the main mode intensity for the homogeneous broadening medium; that is so-called mode competition. Therefore the envelope linewidth of LD decreases with the pump current increasing.

The envelope linewidth is also dependent on the spontaneous emission factor. A smaller γ_{sp} means a sharper transition from spontaneous emission to lasing and a short process from multiple modes to a dominant main mode. It is observed experimentally that the index-guided laser tends to operate with fewer modes than the gain-guided laser. It is because the latter has a convex wavefront and thus a larger K factor as described by (2.14), and a larger spontaneous emission factor $\gamma_{sp} \propto \Omega_{laser}/\Omega_{sp} \propto K$. The different longitudinal mode behaviors are depicted schematically in Fig. 2.14a.

Practical applications of semiconductor laser put forward different requirements on its spectral characteristics. For high-speed optical fiber communications, lasers are required with dynamically single longitudinal mode; whereas its linewidth may not be very small. Many special semiconductor laser structures have been developed. The distributed feedback laser (DFB-LD), the distributed Bragg reflector laser (DBR-LD), and the vertical cavity surface emitting laser (VCSEL) are thought the best lasers meeting the requirements, and used most widely. Single longitudinal mode lasers with linewidth narrow enough are demanded in many important and newly emerging fields. More technologies are developed for linewidth reduction and frequency stabilization.

2.5.2 *Tunability of Semiconductor Lasers*

One of the advantages of semiconductor laser is its tunability by changing the injected current and/or the temperature. In the range of linear approximation, the variation of lasing wavelength with temperature T and current I is written as

$$\lambda(T, I) = \lambda_0 + \frac{\partial \lambda}{\partial T}(T - T_0) + \frac{\partial \lambda}{\partial I}(I - I_0). \tag{2.38}$$

The refractive index of semiconductor is a function of temperature, carrier concentration and wavelength: $n = n(T, N, \lambda)$. The temperature tuning is attributed to the temperature coefficient of index and the thermal expansion of cavity length

$$\frac{\partial \lambda}{\partial T} = \frac{\lambda}{n}\frac{\partial n}{\partial T} + \frac{\lambda}{L}\frac{\partial L}{\partial T}. \tag{2.39}$$

The current tuning is attributed to the carrier coefficient of index and thermal effect of current

$$\frac{\partial \lambda}{\partial I} = \frac{\lambda}{n}\frac{\partial n}{\partial N}\frac{\partial N}{\partial I} + \frac{\partial \lambda}{\partial T}\frac{\partial T}{\partial I}. \tag{2.40}$$

The temperature tuning involves the thermal characteristics, as discussed in Sect. 2.7.

2.6 Transient and Modulation Characteristics

2.6.1 Modulation of Output Power

Another important advantage of the semiconductor laser is its capability of direct modulation by the injected current. For understanding its modulation response and transient characteristics it is needed to start from Eqs. (2.20), which are rewritten for a single longitudinal mode as [3–10]:

$$\frac{\mathrm{d}N}{\mathrm{d}t} = \frac{\eta_{\mathrm{inj}} J}{ed} - \frac{N}{\tau_{\mathrm{sp}}} - v_g \Gamma g_N (N - N_{\mathrm{tr}}) p, \tag{2.41a}$$

$$\frac{\mathrm{d}p}{\mathrm{d}t} = v_g \Gamma g_N (N - N_{\mathrm{tr}}) p - \frac{p}{\tau_p} + \frac{\gamma_{\mathrm{sp}} N}{\tau_{\mathrm{sp}}}. \tag{2.41b}$$

Usually the laser is modulated at a certain bias DC current, where the static photon density and carrier concentration has relation of $(v_{\mathrm{g}} \Gamma g_{\mathrm{N}} p_0 + \gamma_{\mathrm{sp}} \tau_{\mathrm{sp}}^{-1}) N_0 = (\tau_p^{-1} + v_{\mathrm{g}} \Gamma g_N N_t) p_0$, where the subscript 0 stands for the DC value. γ_{sp} can be neglected for bias high enough; thus $v_{\mathrm{g}} \Gamma g_{\mathrm{N}} (N_0 - N_{\mathrm{tr}}) \approx \tau_p^{-1}$. The photon density is then expressed as

$$p_0 = \frac{\eta \tau_p J_0}{ed} - \frac{N_{\rm th}\tau_p}{\tau_{\rm sp}} = \frac{\eta \tau_p}{ed}(J_0 - J_{\rm th}). \tag{2.42}$$

Adding a single frequency modulation current on the DC bias, $J = J_0 + J_1 e^{j2\pi ft}$; under the approximation of small signal, $N = N_0 + N_1 e^{j2\pi ft}$ and $p = p_0 + p_1 e^{j2\pi ft}$. Substituting them into the equations, with the products of small quantities neglected, the modulated photon density is obtained

$$p_1 = \frac{\eta J_1 \tau_p}{ed} \frac{f_R^2}{f_R^2 - f^2 + j\gamma_R f}, \tag{2.43a}$$

where $f_R = \sqrt{v_g \Gamma g_N p_0/\tau_p}/2\pi$, $\gamma_R = (v_g \Gamma g_N p_0 + \tau_{\rm sp}^{-1})/2\pi$. It is shown that the modulation response is a complex, with a phase shift existing between the modulated output power and the modulating current. (2.43a) can be rewritten as

$$p_1 = |p_1| e^{j\phi} = \frac{f_R^2}{\sqrt{(f^2 - f_R^2)^2 + \gamma_R^2 f^2}} \frac{\eta J_1 \tau_p}{ed} e^{j\phi}. \tag{2.43b}$$

The phase factor is $\phi = \tan^{-1}[\gamma_R f/(f^2 - f_R^2)]$. The maximum of modulated photon density is obtained at resonance

$$|p_1|_{\max} = \frac{f_R^2}{\gamma_R \sqrt{f_R^2 - \gamma_R^2/4}} \frac{\eta_{inj}\tau_p}{ed} J_1 \approx \frac{f_R \eta_{inj}\tau_p}{\gamma_R ed} J_1, \tag{2.44}$$

with resonance frequency of

$$f_{\rm osc} = \sqrt{f_R^2 - \gamma_R^2/2}. \tag{2.45}$$

The last approximation of (2.44) is for $f_R \gg \gamma_R$ and $f_{\rm osc} \approx f_R$, which holds usually. The phase shift at resonance is near $\pi/2$:

$$\phi_R = \tan^{-1} \frac{\sqrt{f_R^2 - \gamma_R^2/2}}{\gamma_R/2} \approx \tan^{-1} \frac{2f_R}{\gamma_R} \approx \frac{\pi}{2}. \tag{2.46}$$

In the low frequency range,

$$p_1 \approx \frac{f_R^2}{f_R^2 - f^2} \frac{\eta_{inj}\tau_p J_1}{ed} \exp \frac{-j\gamma_R f}{f_R^2}. \tag{2.47}$$

Figure 2.16 shows a calculated modulation depth of $M = |\partial p_1/\partial J|/p_0$.

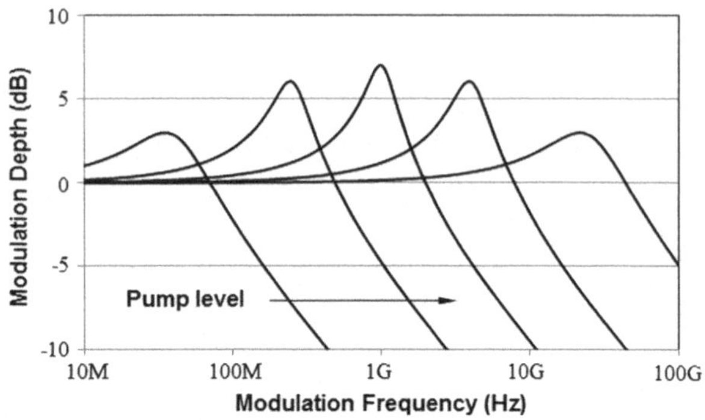

Fig. 2.16 Modulation depth spectra for different bias levels

The 3 dB bandwidth of response, defined as $M(f_{3\mathrm{dB}}) = M(0)/2$, is a characteristic parameter; It is deduced from (2.43) $f_{3\mathrm{dB}} \approx \sqrt{3} f_R$. However, people pay more attention to the bandwidth of detected electrical power $P_E \propto i_{\mathrm{PD}}^2 \propto p_1^2$; the 3 dB bandwidth is reduced to [6].

$$f_{3\mathrm{dB}} \approx \sqrt{1 + \sqrt{2}} f_R \approx 1.55 f_R. \tag{2.48}$$

It is concluded from the expression of resonance frequency that to get a higher modulation frequency it is necessary to shorten the photon lifetime τ_p, to enhance the differential gain coefficient g_N, and to increase the biased level. A short cavity length and a lower reflectance of cavity mirror will benefit a short photon lifetime. However, a higher threshold is the price; a tradeoff should be taken for practical applications. The effect of biased power is understandable, since higher bias means higher carrier recombination rate by the stimulated emission, and thus higher response speed. The response of LED is much lower than LD, because only spontaneous emission occurs in LED and the response frequency is determined solely by the carrier lifetime.

The above analysis is only for an LD chip. It is necessary to consider some other factors for a real packaged LD, especially the electrode wires and the can structure. Their capacity and inductance will lower the response frequency, with the modulation depth multiplied by $(1 + \omega^2 C^2 R^2)^{-1}$, where C and R are the characteristic capacity and resist of the packaging.

Since the resonance frequency can be written as $f_R^2 = \eta_{\mathrm{inj}} v_g \Gamma g_N (J - J_{\mathrm{th}})/(4\pi^2 ed)$ the modulation current efficiency factor (MCEF) is used to judge and compare modulation response performances of different lasers, defined as [27].

$$MCEF = \frac{f_{3\mathrm{dB}}}{\sqrt{I - I_{\mathrm{th}}}} \propto \frac{1.55}{2\pi} \sqrt{\frac{\eta v_g g_N}{e V_M}}. \tag{2.49}$$

The second expression shows the relation with the internal parameters of LD chip with the mode volume of $V_M = LWd_M$, where W is the mode width.

The resonance frequency of LD is usually measured several GHz. Compared with gas laser and solid-state laser, the capability of direct modulation with high speed response is one of the important advantages of semiconductor lasers.

The modulation will also change laser's longitudinal mode behavior. The LD with good index guiding structure may work in a state of high side mode suppression ratio (SMSR) under DC current pumping; but it will be degraded greatly under high frequency modulation. Such an effect can be understood based on the multi-mode rate equations and with the gain spectrum of (2.21). The effect must be overcome for many applications, especially the high-speed optical communications. The dynamically single longitudinal mode lasers will be introduced in Chapt. 4.

2.6.2 Relaxation Oscillation

Apart from the single frequency modulation, characteristics of LD under pulse modulation are of practical significance. It is indicated experimentally and theoretically that if a step current is applied on the LD, a short time delay is needed for it to output stimulated emission, because the carrier concentration cannot reach the threshold immediately. The delay can be obtained from the rate equations [1, 2, 4–6]

$$t_d \approx \tau_{sp} \ln \frac{I - I_b}{I - I_{th}}, \tag{2.50}$$

where I is the amplitude of current step, I_b is the DC bias. When I_b is set equal to, or higher than, I_{th}, the delay will be null, that is the case used mostly in modulated lasers. If a step-like current increment is added to the DC bias with amplitude $I > I_{th}$, the output will oscillate after t_d , called the relaxation oscillation, and damp towards a level corresponding to the case of cw operation. If the carrier concentration change is neglected, the oscillation is described by [6]:

$$\Delta p(t) = \Delta p(\infty)[1 - e^{-\pi\gamma_R t}\cos(2\pi f_{osc}t) - \frac{\gamma_R}{2f_{osc}} e^{-\pi\gamma_R t}\sin(2\pi f_{osc}t)], \tag{2.51}$$

where the frequency of relaxation oscillation is just the resonant frequency given by (2.45). Figure 2.17 shows schematically the waveform of relaxation oscillation.

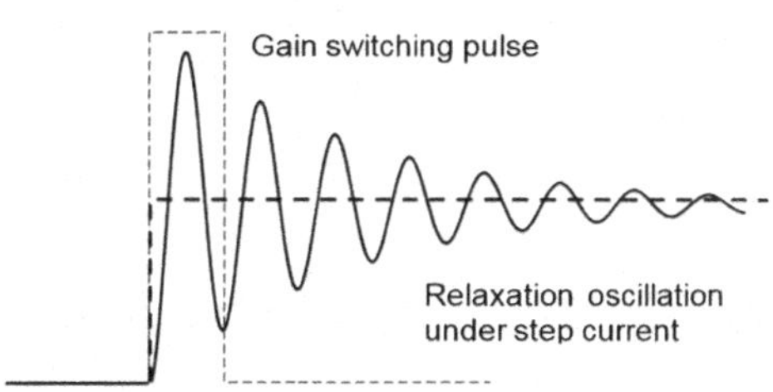

Fig. 2.17 Relaxation oscillation of LD and gain switching

If a pulsed current with pulse width less than the period of relaxation oscillation is applied, the second period and the following will be cut off; and only the first period emission outputs as a short pulsed laser beam, as shown in Fig. 2.17 by a thin dashed line. This method of short optical pulse generation is called the gain switch, which is based on the capability of direct modulation, the unique feature of semiconductor lasers. It is expected that the pulse width of the gain switched pulses can be narrowed by higher injection current due to a higher resonance frequency. Compared with the other method of short pulse operation, e.g., the mode-locking, the gain switch method has advantages of simplicity, flexibility, and low cost, though the pulse width is usually not as short as that by mode-locking. The mechanism of mode-locking will be explained briefly in Chap. 8.

Even under continuous wave operation, the output of semiconductor laser may oscillate in some cases. Such a phenomenon is called the self-sustained pulsation [28]. It is one of the phenomena of laser's instability.

2.6.3 Wavelength Modulation

The lasing wavelength of LD will be modulated at the same time as the power being modulated due to its tunability, described by (2.39). For the single frequency modulation with injected current of $I = I_{\text{bias}} + \Delta I \sin \Omega t$ the modulated wavelength is written as

$$\begin{aligned} \lambda &= \bar{\lambda} + \Delta\lambda \sin(\Omega t + \phi) \\ \Delta\lambda &= \left(\frac{\lambda}{n}\frac{\partial n}{\partial N}\frac{\partial N}{\partial I} + \frac{\partial \lambda}{\partial T}\frac{\partial T}{\partial I}\right)\Delta I. \end{aligned} \tag{2.52}$$

The frequency modulation response depends on the modulation frequency [29]. In the low frequency band, the thermal effect of injected current is the main mechanism. The cavity length expands with temperature; and the index increases also due to the thermo-optic effect. As discussed in Sect. 2.7, the temperature modulation is with a delay; and the modulation amplitude decreases with the increasing of modulation frequency. In the high frequency band, the temperature of the active region can hardly follow the modulation of injected current; the dominant effect is the variation of index with carrier concentration. Because the index variations with temperature and with carrier concentration have opposite effects,$\partial n/\partial T > 0$, whereas $\partial n/\partial N < 0$, the laser's frequency modulation in low frequency band is in-phase, $\mathrm{d}\nu/\mathrm{d}I > 0$; it is out of phase in high frequency band, $\mathrm{d}\nu/\mathrm{d}I < 0$. Their combination will cause the frequency modulation rate towards a minimum at some frequency in middle. Since the modulation of carrier concentration obeys the rate equation, its response reaches the maximum at the resonance frequency. Such a behavior is illustrated schematically in Fig. 2.18, where the ordinate is modulation rate of $|\mathrm{d}\nu/\mathrm{d}I|$.

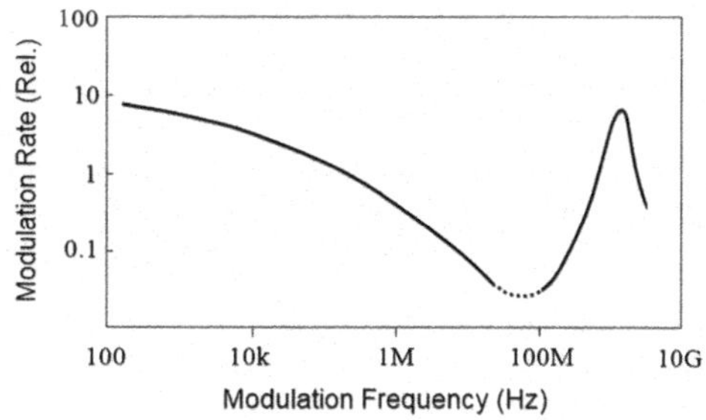

Fig. 2.18 Schematic characteristics of current tuning of LD

It is noticed that the power modulation and the frequency modulation will occur at the same time in the current modulation. In many applications of single-frequency lasers, a pure frequency modulation is often needed, where the power modulation, called the residue amplitude modulation (RAM), should be avoided. Several methods have been proposed for RAM control and compensation, which will be introduced later in the following chapters.

2.7 Thermal Performances

Characteristics of LDs have strong temperature dependences, attributed to the intrinsic temperature sensitivity of semiconductor materials. It is a shortcoming from view of stability. From a different angle, however, it provides a possibility of adjusting and controlling the characteristics by the temperature effect.

2.7.1 Dependence of Threshold and Output Power on Temperature

Semiconductor lasers in the earlier stage could work only at low temperature, such as 77 K of the liquid nitrogen for the threshold increased very fast with temperature. The main reason of the phenomenon is that the thermally excited carrier concentration increases with the temperature, and its distribution in the band tends to spread in a wider range. A higher injected current is thus needed to reach the population inversion, whereas increasing of the current heats the devise stronger, causing a higher temperature. Generally, the dependence can be described by an empirical formula:

$$I_{\mathrm{th}}(T) = I_0 \exp[(T - T_0)/T_0] \propto \exp(T/T_0), \tag{2.53}$$

where T_0 is termed the characteristic temperature of threshold–temperature relation, T is the real temperature of P-N junction. Due to the heating of the device at work, the junction temperature will be higher than the environmental temperature. Only working under very low duty cycle pulsed current, the junction temperature is regarded equal to the environmental temperature. However the junction temperature cannot be measured directly by thermometers, the characteristic temperature has to be measured in operation with narrow pulse and low repetition rate. The LD characteristic temperature depends on the laser materials. The typical values are 100 K for conventional double heterostructure lasers, and more than 200 K for quantum well lasers.

Temperature rising will also cause increase of intra-cavity loss and non-radiation recombination, resulting in decrease of differential quantum efficiency η_{d}. The temperature dependences of the threshold and efficiency cause decreasing of the output power with temperature increase. The *L–I* curve becomes sub-linear instead of an ideal straight line, since the laser's temperature increases with the current. In current modulation operations, the nonlinearity generates higher order harmonic modulation; when a single frequency modulating current $\Delta I \sin \Omega t$ is applied, the output power should be expressed as

$$P = \eta_d(I_B - I_{\mathrm{th}}) + \eta_{\mathrm{d}}\Delta I \sin(\Omega t + \phi_1) + \eta_2 \Delta I \sin(2\Omega t + \phi_2). \quad (2.54)$$

2.7.2 Dependence of Laser Spectrum on Temperature

The dependence of semiconductor laser spectrum on temperature is attributed to two mechanisms, temperature dependences of band gap and of refraction index. Generally, the band gap of semiconductor decreases with temperature increasing, resulting in red shift of gain peak wavelength. For example, it is $\partial E_g/\partial T \approx -5 \times 10^{-4}$eV/K for GaAs [2]. Correspondingly, the temperature coefficient of gain peak wavelength is about 0.3 nm/K. On the other hand, the index increases with temperature; and the cavity length will expand with temperature. From $\lambda_m = 2n\mathrm{L/m}$, the longitudinal mode wavelength is an ascend function of temperature:

$$\frac{\delta\lambda}{\lambda_m} = \frac{\delta L}{L} + \frac{\delta n}{n} = \alpha_L \delta T + \alpha_n \delta T, \quad (2.55)$$

where α_L is the thermal expansion coefficient, and $\alpha_n = n^{-1}\partial n/\partial T$ is the thermal coefficient of index. It is measured experimentally that $\partial\lambda_m/\partial T \approx 0.06 - 0.07\, nm/K$ for GaAs-based laser in 800 nm band, and $\partial\lambda_m/\partial T \approx 0.1\, nm/K$ for In *P*-based lasers in 1550 nm band.

Figure 2.19a is a schematic diagram of lasing wavelength versus temperature, where the dashed line stands for the red shift of gain peak, and the solid lines are for the different longitudinal modes. The lasing wavelength varies also with the injected current. Two effects are involved. One is that the difference between electron's and hole's quasi-Fermi levels increases with the injected carrier concentration, called the model of band filling. The effect causes the gain peak blue-shifting with the current. The carrier concentration increasing causes also the

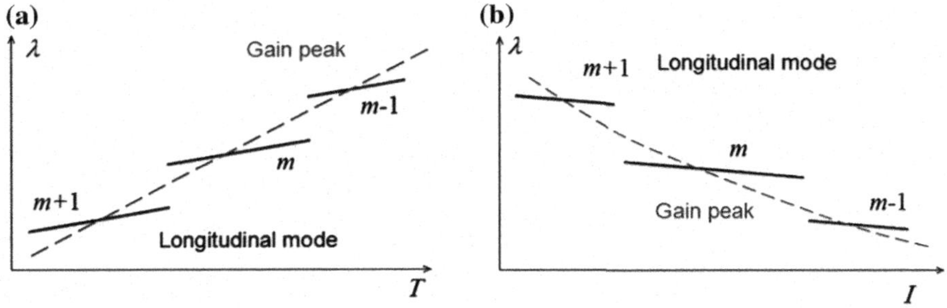

Fig. 2.19 Variations of lasing wavelength with temperature (**a**) and with driving current (**b**)

index change, as mentioned in Sect. 2.5. The second effect is the heating of junction by the injected current, resulting in red shift of both gain peak and longitudinal mode. It is shown experimentally and by simulation that the composite effect is blue-shifting both of the gain peak and the modes in a certain range of current variation, as shown in Fig. 2.19b schematically [30].

2.7.3 *Junction Temperature and Thermal Resistant*

When a semiconductor laser is working, it will be heated by the input electric power. Basically, two factors generate the heating. One is by the series resistance R_s composed of semiconductor material resistance and resistance of Schottkey barriers at the metal-semiconductor contacts of electrodes. The other is the incomplete electro-optic conversion at the junction, for example, by recombination through non-radiation centers and carrier leakage over the junction. The voltage applied on the laser is expressed as $V = V_\mathrm{J} + IR_\mathrm{s}$; the input electric power is IV. The heat generated in unit time is thus written as $Q = IV - P_\mathrm{out} = (1 - \eta_P)IV$, where P_out is the output optical power and η_P is the power conversion efficiency. The relation of temperature rising ΔT_J and generated heat is characterized by the thermal resistance, defined as

$$R_H = \Delta T_J / \Delta Q. \tag{2.56}$$

The temperature rising ΔT_J can be measured from the variation of lasing wavelength, as stated above; then the thermal resistance is calculated.

When a pulsed current is injected in an LD, its temperature cannot change immediately, but with a certain delay, the same as the common thermal phenomenon. The temporal variation of the junction temperature can be described by the following equation [31]:

$$\frac{\mathrm{d}T}{\mathrm{d}t} = \frac{Q(t)}{C} - \frac{T - T_0}{\tau_H}, \tag{2.57}$$

where C is the heat capacity of the laser, determined by the specific heat of materials and the sizes of laser structure; τ_H is called thermal relaxation time. From (2.56) and the stationary solution of (2.57), $\tau_H = R_H C$ is deduced. When the laser is working at a square wave pulsed current with pulse width of u, the solution of (2.57) is obtained as

$$T - T_0 = \begin{cases} R_H Q(1 - e^{-t/\tau_H}) & (0 < t < u) \\ R_H Q(1 - e^{-u/\tau})e^{-(t-u)/\tau_H} & (t \geq u) \end{cases} \tag{2.58}$$

Figure 2.20 shows schematically the variation of LD junction temperature under a square wave pulsed current.

When the LD works under a sinusoidal current with modulation frequency of Ω, the generated heat is $\Delta Q \propto \Delta I \sin \Omega t$, the temperature variation is expressed as

$$T(t) = \overline{T} + \Delta T \sin(\Omega t + \varphi), \tag{2.59}$$

where $\Delta T = R_H \Delta Q(1 + \Omega^2 \tau_H^2)^{-1/2}$, $\varphi = \tan^{-1}(\Omega \tau_H)$; $\overline{T}$ is the average temperature. It is seen that the temperature modulation depth is a function of the modulation frequency and the thermal relaxation time; the modulation depth decreases with the frequency increasing.

In order to mitigate the thermal effects, good heatsink and/or cooling is needed, especially for applications requiring high stability and for high power semiconductor lasers. Special technologies have been developed for the purpose, such as optimized design of heatsink, Peltier thermo-electrical cooler, water flow cooling, or cooling by phase transition, micro-channel heatsink, etc. Figure 2.21 depicts a typical LD module with TE cooler, thermometer, PD monitor and fiber pigtail for output.

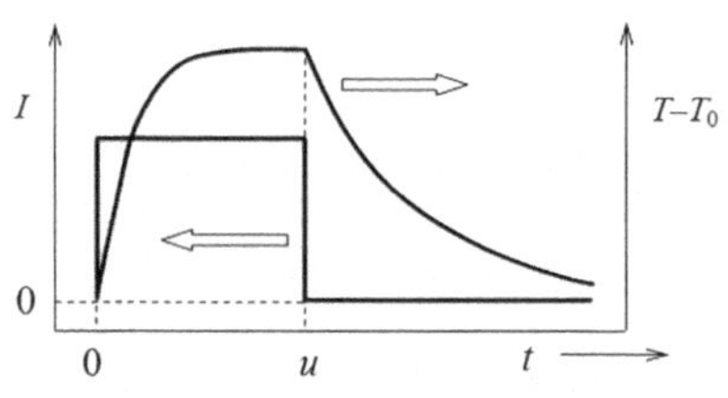

Fig. 2.20 Variation of junction temperature under a square waveform current

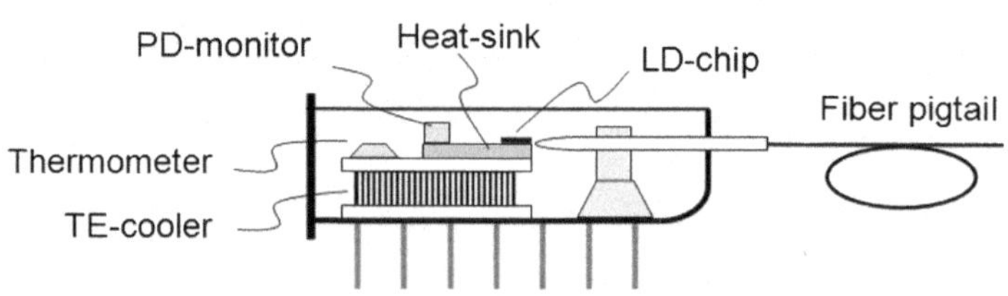

Fig. 2.21 Schematic structure of a LD module with TE cooler and fiber-pigtail output

References

1. Kressel H, Butler JK (1977) Semiconductor lasers and heterojunction LEDs. Academic Press, Cambridge
2. Casey HC, Panish MB (1978) Heterostructure lasers. Academic Press, Cambridge
3. Thompsom GHB (1980) Physics of semiconductor laser devices. Wiley
4. Pertermann K (1991) Laser diode modulation and noise. Kluwer Academic Publishers
5. Agrawal GP, Dutta NK (1993) Semiconductor lasers. Van Nostrand Reinhold
6. Coldren LA, Corzine SW (1995) Diode lasers and photonic integrated circuits. Wiley
7. Yariv A (1997) Optical electronics in modern communications. Oxford University Press
8. Kapon (1999) Semiconductor lasers 1: fundamentals; 2: materials and structures. Academic Press
9. Moll (1964) Physics of semiconductors. McGraw-Hill Book Co
10. Stern F (1976) Calculated spectral dependence of gain in excited GaAs. J Appl Phys 47 (12):5382–5386
11. Saleh BEA, Teich MC (2007) Fundamentals of photonics. Wiley
12. Iga K (2000) Surface-emitting laser—its birth and generation of new optoelectronics field. IEEE J Sel Top Quantum Electron 6(6):1201–1215
13. Yu SF (2003) Analysis and design of vertical-cavity surface-emitting lasers. Wiley
14. Kroemer H (1963) A proposed class of hetero-junction injection laser. Proc IEEE 51 (12):1782–1783
15. Alferov ZI (2000) The double heterostructure: concept and its applications in physics, electronics and technology. http://www.nobelprize.org/mediaplayer/index.php?id=970. Accessed 14 Nov 2016
16. Moss TS (1959) Optical properties of semi-conductors. Butterworths Scientific Publications
17. Holonyak N, Kolbas RM, Dupuis RD et al (1980) Quantum-well heterostructure lasers. IEEE J Quantum Electron 16(2):170–186
18. Adams AR (2011) Strained-layer quantum-well lasers. IEEE J Sel Top Quantum Electron 17(5):1364–1373
19. Capasso F (1987) Band-gap engineering: from physics and materials to new semiconductor devices. Science 235:172–176
20. Faist J, Capasso F, Sivco DL (1994) Quantum cascade laser. Science 264(5158):553–556
21. Nakamura S (1997) The blue laser diode: GaN based light emitters and lasers. Springer
22. Lee TP, Burrus CA, Marcuse D et al (1982) Measurement of beam parameters of index-guided and gain-guided single frequency InGaAsP injection lasers. Electron Lett 18(21):902–904
23. Arnold G, Petermann K, Schlosser E (1983) Spectral characteristics of gain-guided semiconductor lasers. IEEE J Quantum Electron 19(6):974–980
24. Hadley GR, Hohimer JP, Owyoung A (1986) High-order ($\nu > 10$) eigenmodes in ten-stripe gain-guided diode laser arrays. Appl Phys Lett 49(12):684–686
25. Marcuse D (1983) Classical derivation of the laser rate equation. IEEE J Quantum Electron 19(8):1228–1231
26. Mamine T (1983) Astigmatism and spontaneous emission factor of laser diodes with parabolic gain. J Appl Phys 54(4):2103–2105
27. Michalzik R, Ebeling KJ (2003) Operating principles of VCSELs. In: Li H, Iga K (eds) Vertical-cavity surface-emitting laser devices. Springer-Verlag
28. Van Der Ziel JP (1981) Self-focusing effects in pulsating $Al_xGa_{1-x}As$ double-heterostructure lasers. IEEE J Quantum Electron 17(1):60–68
29. Kobayashi S, Yamamoto Y, Ito M et al (1982) Direct frequency modulation in AlGaAs semiconductor lasers. IEEE J Quantum Electron 18(4):582–595
30. Fang Z, Wang S (1984) Longitudinal mode behavior and tunability of separately pumped (GaAl)As lasers. Appl Phys Lett 44(13):13–15
31. Chen C, Xin G, Qu R et al (2006) Measurement of thermal rise-time of a laser diode based on spectrally resolved waveforms. Opt Commun 260:223–226

Chapter 3
Noises and Stability of Semiconductor Lasers

3.1 Characteristics and Inherent Relations of Laser Noises

The linewidth, noises, and stability of a laser are important characteristics for its research and development, and for its practical applications. Their characterization and measurement attract attentions of scientists and engineers in the field. Detailed physical mechanisms and systematic theories on laser noises can be found in monographs [1–5] and references [6–12]. Basically two kinds of mechanisms are involved in laser noises: one is related to the physical processes inside the laser cavity, and the other is induced by external disturbances. According to the basic principles, the laser emission is the amplified spontaneous emission. The spontaneous emission occurs randomly both temporarily and spatially in laser cavity. When the amplified light emits from the mirror, random variations of phases and amplitudes are inevitable. The amplitude and phase of optic field, the electric polarization of the matter, and the inverse population are linked with each other; their fluctuations are also coupled with each other. The parameters of laser structure, e.g., the cavity length and the index of material, will be affected by external disturbances, including temperature fluctuations, mechanical vibrations and acoustic waves, and so on. Besides, the pump's fluctuations, i.e., the injected current noise for LD, will directly cause the laser noises.

3.1.1 *Mathematical Description of Stochastic Variables*

Laser noises include random fluctuations of intensity, phase, and frequency. These noises are regarded as stochastic variables; specifically, as wide-sense stationary stochastic processes for lasers working in continuous wave. Two parameters are generally used to characterize basic properties of a stochastic variable $x(t)$:

Z. Fang et al., *Single Frequency Semiconductor Lasers*, Optical and Fiber Communications Reports 9, DOI 10.1007/978-981-10-5257-6_3

its expectation is denoted by $E[x(t)]$ or $\langle x(t)\rangle$, and its autocorrelation $R_x = E[x(t)x^*(t-\tau)]$, is expressed as:

$$\langle x(t)\rangle = \lim_{T\to\infty}\frac{1}{T}\int_{-T/2}^{T/2} x(t)\mathrm{d}t, \tag{3.1}$$

$$R_x(\tau) = \lim_{T\to\infty}\frac{1}{T}\int_{-T/2}^{T/2} x(t)x^*(t-\tau)\mathrm{d}t. \tag{3.2}$$

The autocorrelation is just the covariance of random process $x(t)$. The temporal average can also be regarded as an ensemble average. In practice, the range of integration cannot be infinitive, but with period T large enough to give a significant value.

The randomness of a stochastic variable is characterized by the power spectral density (PSD), which is the power distribution over different frequency components. PSD is obtained as the Fourier transform of autocorrelation function, i.e., Wiener–Khinchin theorem:

$$S_x(f) = \int_{-\infty}^{\infty} R_x(\tau)\mathrm{e}^{j2\pi f\tau}\mathrm{d}\tau = \lim_{T\to\infty}\frac{1}{T}\left|\int_{-T/2}^{T/2} x(t)\mathrm{e}^{j2\pi ft}\mathrm{d}t\right|^2. \tag{3.3}$$

The last expression is deduced from the autocorrelation property of Fourier transform in its truncated form.

The above formulas are now used to analyze the laser noise. The laser field can generally be written as a sum of its average and the fluctuations in time domain:

$$E(t) = [A + a(t)]\mathrm{e}^{-j[\omega_0 t + \phi(t)]}, \tag{3.4}$$

where $a(t)$ and $\phi(t)$ are the fluctuations of field amplitude and phase, $\omega_0 = m\pi c/(nL)$ is the frequency of m-th longitudinal mode. A single longitudinal mode is considered hereinbelow. The phase fluctuation means the optical frequency noise: $\Delta v(t) = (2\pi)^{-1}\mathrm{d}\phi(t)/\mathrm{d}t$. The distribution of optical frequency around the center frequency forms the spectral shape of laser line characterized mainly by its linewidth, usually measured in full width at half maximum (FWHM).

The effect of laser noises is eventually demonstrated in its intensities, such as the interference intensity of optical waves, which gives the autocorrelation of optical field, written as:

$$
\begin{aligned}
R_{\mathrm{E}}(\tau) &= \lim_{T\to\infty}\frac{1}{T}\int\limits_{-T/2}^{T/2} E(t)E^{*}(t-\tau)\mathrm{d}t \\
&= [A^{2}\langle \mathrm{e}^{j\Delta\phi(t,\tau)}\rangle + \langle a(t)a(t-\tau)\mathrm{e}^{j\Delta\phi(t,\tau)}\rangle + A\langle [a(t)+a(t-\tau)]\mathrm{e}^{j\Delta\phi(t,\tau)}\rangle]\mathrm{e}^{j\omega_{0}\tau},
\end{aligned}
\tag{3.5}
$$

where $\Delta\phi(t,\tau) = \phi(t) - \phi(t-\tau)$. It is shown that the phase noise and the amplitude noise are coupled with each other in calculating their autocorrelation. Under certain conditions, such a coupling may be resolved. When the amplitude noise is much lower than the phase noise, $R_{\mathrm{E}}(\tau) \propto \langle \mathrm{e}^{j\Delta\phi(t,\tau)}\rangle$; contrarily, $R_{\mathrm{E}}(\tau) \approx \langle a(t)a(t-\tau)\rangle$. The PSD of phase noise, amplitude noise, and their coupled noise, S_{ϕ}, S_{a}, and $S_{a\phi}$, can be obtained from the autocorrelation of optical field, respectively, by using certain decoupling methods.

3.1.2 Phase Noise and Frequency Noise

The randomness of optical phase obeys normal distribution with zero mean. Its probability density function (PDF) is a Gaussian written as $P(\Delta\phi) = (\sigma\sqrt{2\pi})^{-1}\exp[-(\Delta\phi)^{2}/2\sigma^{2}]$. It is then deduced that $\langle\exp(j\Delta\phi)\rangle = \exp(-\sigma_{\phi}^{2}/2)$, where $\sigma_{\phi}^{2} = \langle\Delta\phi^{2}\rangle$ is the variance of phase noise. According to Wiener–Khinchin theorem, the PSD of phase noise is expressed as:

$$
S_{\phi}(f) = \int\limits_{-\infty}^{\infty} \mathrm{e}^{-j2\pi f\tau}\langle\phi(t)\phi(t-\tau)\rangle \mathrm{d}\tau. \tag{3.6}
$$

The symbol f here and hereinbelow is the frequency in Fourier transform, i.e., the frequency deviation from the center optical frequency, not the laser frequency itself.

Fluctuation of optical phase is too fast to be measured directly and solitarily. It is analyzed actually by the measured optical frequency spectrum. The PSD of frequency noise $\Delta\nu$ is defined as the spectrum of frequency fluctuation covariance. An intrinsic relation exists between the frequency noise and the phase noise [8, 10]:

$$
S_{\nu}(f) = \int\limits_{-\infty}^{\infty} \mathrm{e}^{j2\pi f\tau}\langle\Delta\nu(t)\Delta\nu(t-\tau)\rangle \mathrm{e}\tau = f^{2}S_{\phi}(f). \tag{3.7}
$$

According to the basic property, Fourier transform of $\Delta\phi(t,\tau) = \phi(t) - \phi(t-\tau)$ is reduced to

$$\widehat{\boldsymbol{F}}[\Delta\phi(t,\tau)] = \widetilde{\Delta\phi}(f,\tau) = (1 - \mathrm{e}^{-j2\pi f\tau})\widehat{\boldsymbol{F}}[\phi(t)] = (1 - \mathrm{e}^{-j2\pi f\tau})\tilde{\phi}(f). \quad (3.8)$$

where "$\sim$" on the top of variables is to denote their Fourier transforms. The inverse transform is $\Delta\phi(t,\tau) = \int_{-\infty}^{\infty} \tilde{\phi}(f)(1 - \mathrm{e}^{-j2\pi f\tau})\mathrm{e}^{j2\pi ft}\mathrm{d}f$. Referring $\delta(f-f') = \int_{-\infty}^{\infty} \exp[j2\pi(f-f')t]\mathrm{d}t$, the variance of phase noise is then deduced as [10, 11]:

$$\begin{aligned}\sigma_\phi^2(\tau) &= \langle|\Delta\phi(t,\tau)|^2\rangle = 2\int_{-\infty}^{\infty} \left|\tilde{\phi}\right|^2 (1 - \cos 2\pi f\tau)\mathrm{d}f \\ &= 4\int_0^{\infty} S_\phi(f)\sin^2(\pi f\tau)\mathrm{d}f = 4\int_0^{\infty} S_v(f)\frac{\sin^2(\pi f\tau)}{f^2}\mathrm{d}f.\end{aligned} \quad (3.9)$$

The spectral line shape is just the power spectral density of optical field. In case the intensity noise can be neglected, $R_\mathrm{E}(\tau) = A^2\mathrm{e}^{-j\omega_0\tau}\exp[-\sigma_\phi^2(\tau)/2]$; PSD of optical field is written as

$$\begin{aligned}S_\mathrm{E}(f) &= A^{-2}\int_{-\infty}^{\infty} R_\mathrm{E}(\tau)\mathrm{e}^{-j2\pi f\tau}\mathrm{d}\tau \\ &= \int_{-\infty}^{\infty} \mathrm{e}^{-j2\pi f\tau}\exp[-2\int_0^{\infty} S_v(f)f^{-2}\sin^2(\pi f\tau)\mathrm{d}f]\mathrm{d}\tau,\end{aligned} \quad (3.10)$$

where PSD of optical field is normalized to the averaged intensity. This is the basic relation between the frequency noise spectrum and the laser line shape. It is indicated that the line shape and linewidth is dependent on the frequency noise spectrum.

The laser frequency noise has different characteristics, depending on its origins. One of the basic types is the white noise, induced by spontaneous emission, which obeys the normal probability distribution with a constant PSD, independent of frequency, $S_v(f) = h$. Assuming such a property covers the whole frequency band, the integral in the exponential term of (3.9) gives an analytic expression:

$$2\int_0^{\infty} S_v(f)f^{-2}\sin^2(\pi f\tau)\mathrm{d}f = 2h\int_0^{\infty} f^{-2}\sin^2(\pi f\tau)\mathrm{d}f = \pi^2 h|\tau|. \quad (3.11)$$

As the result, a normalized Lorentzian line shape is obtained from (3.10):

$$S_{\mathrm{E}}(f) = \frac{h/2}{f^2 + (\pi h/2)^2}. \tag{3.12}$$

Its FWHM linewidth is $\Delta f_{\mathrm{L}} = \pi h$. Usually, the frequency noise has $1/f$ characteristics in low frequency band, similar to the noises of electrical and mechanical systems, as analyzed in Refs. [10, 11]. As discussed in Sect. 3.3 below, the variance of phase fluctuation will be proportional to a power function of τ, τ^{α}. For a typical example, $\sigma_{\phi}^2(\tau) = q_1\tau^2$, thus the normalized PSD of optical field will appear in Gaussian line shape:

$$S_{\mathrm{E}}(f) \approx 2\sqrt{\pi/q_1}\exp(-2\pi f^2/q_1). \tag{3.13}$$

Its linewidth is $\Delta f_{\mathrm{G}} = \sqrt{q_1 \ln 2/2\pi}$. In practical lasers, the power index α may take different figures, the line will then take some shape between Lorentzian and Gaussian, or their convolution, termed Voigt function. The line shape and linewidth should be measured experimentally.

3.1.3 Intensity Noise

The intensity noise is usually measured by the relative intensity noise (RIN), defined as:

$$\mathrm{RIN} = \langle (P - \overline{P})^2 \rangle / \overline{P}^2. \tag{3.14}$$

where $\overline{P}$ is the averaged optical power; $\langle (P - \overline{P})^2 \rangle$ is the power variance, which is proportional to the noise of electric power detected by photodetector. If the amplitude noise and the phase noise are independent of each other, the field autocorrelation will be $R_{\mathrm{E}}(\tau) \propto \langle a(t)a(t-\tau) \rangle \cdot \langle e^{j\Delta\phi(t,\tau)} \rangle$. When the frequency noise is much lower than RIN, or the power is detected by a wide band detector, insensitive to the frequency noise, PSD of RIN is deduced as:

$$\begin{aligned} S_{\mathrm{RIN}}(f) &= 4S_a(f)/\overline{P} \\ S_a(f) &= \int_{-\infty}^{\infty} \langle a(t)a(t-\tau) \rangle e^{-j2\pi f\tau} \mathrm{d}\tau. \end{aligned} \tag{3.15}$$

It is noted that an inherent relation exists between the intensity noise and the phase noise based on the uncertainty principle of quantum mechanics [3]. It is impossible for a laser to have only intensity noise without phase noise, or vice versa. The uncertainty of photon number $\overline{\Delta N} = \langle \Delta N^2 \rangle^{1/2}$ and the uncertainty of field phasor $\overline{\Delta\phi} = \langle \Delta\phi^2 \rangle^{1/2}$ must meet the requirement of

$$2\overline{\Delta N} \cdot \overline{\Delta \phi} \geq 1. \tag{3.16}$$

In most situations of practical lasers, however, the noises are usually much higher than the quantum noise limit.

3.2 Linewidth and Line Shape of Semiconductor Lasers

As stated above, the intrinsic linewidth of a laser originates from the random phase noise of the spontaneous emission. According to the Schawlow–Townes theory [13], the laser's intrinsic linewidth is expressed as

$$\delta v = 4\pi h v (\delta v_c)^2 / P, \tag{3.17}$$

where hv is the photon energy; δv_c is the linewidth of passive cavity, which is inversely proportional to the photon's lifetime; P is laser's output power. It is understandable that the higher the output power, the lower the proportion of random spontaneous emission; and the longer the photon's lifetime, the higher is the amplification of stimulated emission. The linewidth of semiconductor laser obeys the same principle, but with its special features different from other lasers.

3.2.1 Linewidth of Semiconductor Lasers

The linewidth of semiconductor lasers is much wider than that of gas laser and solid state laser. Reference [14] analyzed the broadening mechanism and explained the difference. The related theory is restated briefly here. The random fluctuation of phase in semiconductor lasers is attributed to the spontaneous emission generated by carrier recombination. The process is illustrated as a change of phase vector in the complex plane of Fig. 3.1 with the imaginary and real parts of field as the

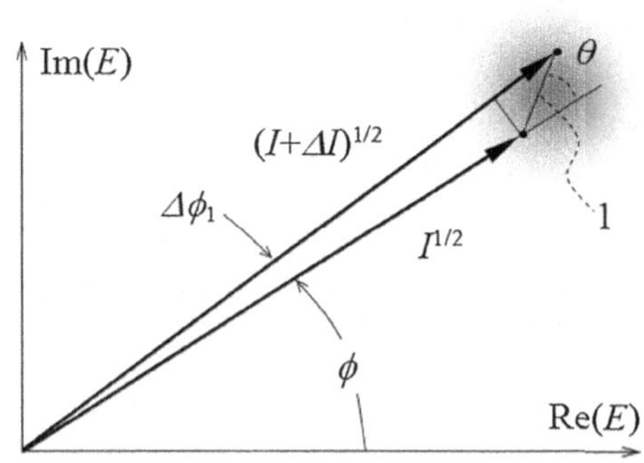

Fig. 3.1 Variation of field amplitude and phase induced by carrier recombination

ordinate and abscissa, where $E(t)e^{-j\omega_0 t} = \sqrt{I(t)}e^{-j[\omega_0 t + \phi(t)]}$ is the laser field, $I = |E|^2$ is the photon's number in cavity, and the change of vector is induced by a spontaneous emission with phase of θ.

In order to deduce the rate equation of complex field in a gain medium, it is necessary to consider the complex index: $\varepsilon/\varepsilon_0 = n^2 = (n_0 + \Delta n_r + j\Delta n_i)^2$, where $\Delta n_i = -g_{\text{net}}/2k_0$ with net gain of $g_{\text{net}} = g - \alpha_c$. It is rewritten as $\varepsilon/\varepsilon_0 \approx n_0^2 + j2n_0\Delta n_i(1 + j\alpha)$, where α is termed the linewidth enhancement factor, expressed as [14, 15]:

$$\alpha = \frac{\Delta n_r}{\Delta n_i} = \frac{\partial n_r/\partial N}{\partial n_i/\partial N}. \tag{3.18}$$

Since the gain in semiconductor laser comes from injected carrier, the second expression is written as the partial derivatives of index with respect to carrier concentration N. Thus the rate equations of laser field is deduced as

$$\frac{\partial E(t)}{\partial t} = \frac{1}{2}v_g g_{\text{net}}(1 + j\alpha)E(t), \tag{3.19}$$

$$\frac{\partial \phi}{\partial t} = \frac{\alpha}{2}v_g g_{\text{net}} = \frac{\alpha}{2I}\frac{\partial I}{\partial t}. \tag{3.20}$$

where v_g is the group velocity in the medium. The phase of spontaneous emission is randomly distributed in range of $[0, 2\pi]$, as shown in Fig. 3.1 by θ. The contribution of a spontaneous emission to the optical intensity is $\Delta I = 1 + 2\sqrt{I}\cos\theta$, and the phase change of laser is $\Delta\phi_1 \approx \sin\theta/\sqrt{I}$. From (3.20), the intensity increment causes also the phase change. The sum of two factors is

$$\Delta\phi_i = (\sin\theta_i - \alpha\cos\theta_i)/\sqrt{I} - \alpha/2I \approx (\sin\theta_i - \alpha\cos\theta_i)/\sqrt{I}. \tag{3.21}$$

It is estimated that the output of 1 mW at 1550 nm corresponds to a photon flow with 10^{16} per second. Therefore, the inequality $I \gg \sqrt{I}$ holds, giving the approximation of last expression of (3.21). For large number of spontaneous emissions, averaging gives $\langle \sin\theta_i \rangle = \langle \cos\theta_i \rangle = 0$ and $\langle \sin^2\theta_i \rangle = \langle \cos^2\theta_i \rangle = 1/2$. The variance of laser phase is thus deduced as

$$\begin{aligned} \langle (\Delta\phi)^2 \rangle &= \frac{1}{I}\left\langle \sum_{i,j}^{M} (\sin\theta_i - \alpha\cos\theta_i)(\sin\theta_j - \alpha\cos\theta_j) \right\rangle \\ &\approx \frac{1}{I}\left\langle \sum_{i}^{M} (\sin^2\theta_i + \alpha^2\cos^2\theta_i) \right\rangle = \frac{1}{2I}(1 + \alpha^2)\gamma_{\text{sp}}R_{\text{sp}}\tau, \end{aligned} \tag{3.22}$$

where R_{sp} is the rate of spontaneous emission; γ_{sp} is the spontaneous emission factor, as introduced in Sect. 2.4; $M = \gamma_{\text{sp}}R_{\text{sp}}\tau$ is the number of spontaneous emission in duration τ. The probability distribution of phase fluctuation caused by spontaneous emission is in Gaussian type, showing a white frequency noise; its variance is expressed as [14]:

$$\langle \exp j\Delta\phi(t,\tau)\rangle = \exp\frac{-\langle \Delta\phi^2(t,\tau)\rangle}{2} = \exp\frac{-\tau}{\tau_c}, \tag{3.23}$$

where $\tau_c = 1/(2\pi\delta\nu)$ is the coherence time. By referring the relation between photon number and output power, $P = h\nu v_g \alpha_m I/2$ with cavity mirror loss of $\alpha_m = L^{-1}\ln(1/R)$, and the relation between spontaneous emission rate and gain coefficient, i.e., formula (2.8), the FWHM linewidth is reduced to [14]:

$$\delta\nu = \frac{v_g^2 h\nu\gamma_{sp} g\alpha_m}{8\pi P}(1+\alpha^2) = \delta\nu_{ST}(1+\alpha^2) \tag{3.24}$$

where $\delta\nu_{ST}$ is the linewidth given by Schawlow–Townes theory, i.e., the linewidth without linewidth enhancement factor. According to (3.12), the laser line has Lorenzian shape:

$$S_E(\nu) = \frac{2\tau_c}{1+[\pi(\nu-\nu_0)\tau_c]^2} \tag{3.25}$$

The linewidth enhancement factor of semiconductor laser is quite large. It is measured that α of F–P cavity LD is basically in the range of 5–7, and $1+\alpha^2$ reaches 30–50. For gas lasers and solid state lasers, such a factor is negligible small, because the index of gas and solid state medium is hardly changed by the population variation, whereas the index of semiconductor varies with injected carrier concentration directly and greatly.

It should be noticed that the value of α is not a constant, but varies with wavelength and pump current level [16]. As the pump level increases, the gain peak moves toward shorter wavelength, as shown in Fig. 2.2. It is seen that the value of $|\partial n_i/\partial N|$ at short wavelength side is much larger than that at longer wavelength side; therefore, α takes a much smaller value at the short wavelength side of gain spectrum.

The spontaneous emission factor γ_{sp} of LD is much larger than that of other lasers. It is another factor causing a much wider linewidth. γ_{sp} is dependent on the laser structure in a certain degree, as discussed in Chap. 2. The line of index-guided laser is narrower than that of gain-guided laser under the same total output power, because the former has fewer modes and higher mode power than that of the latter, resulting in smaller linewidth of the former, according to Schawlow–Townes formula. The linewidth can be greatly reduced by special laser structures and measures of frequency stabilization as discussed in Chaps. 4 and 5.

3.2.2 *Effect of Photon-Carrier Coupling on Line Shape and Noise*

The noises coming from different origins are coupled with each other through the rate equations. Such a coupling is much stronger in semiconductor laser than other lasers due to its particular pumping mechanism, the direct electric current injection. The transient and modulation characteristics are analyzed in Sect. 2.6 by the rate equations. The noises of related valuables are perturbations of the equation [4, 17, 18]. However, the equation of photon density has to be decomposed into two for amplitude a and phase ϕ, respectively. The laser is assumed working at a stationary point with a constant DC bias current. As small quantities, the spontaneous emission factor and other multiplied perturbations can be neglected in the small signal analysis. The rate equations of perturbations are thus obtained [4]:

$$\tfrac{\partial N}{\partial t} = -\left(\tfrac{1}{\tau_{sp}} + v_g g_N A^2\right)\Delta N - \tfrac{2}{\tau_p} A a + F_N(t) \tag{3.26a}$$

$$\tfrac{\partial a}{\partial t} = \tfrac{1}{2} v_g g_N A \Delta N + F_a(t) \tag{3.26b}$$

$$\tfrac{\partial \phi}{\partial t} = \delta\omega = \tfrac{\alpha}{2} v_g g_N \Delta N + F_\phi(t). \tag{3.26c}$$

where $F_{N,a,\phi}$ are the Langevin forces, describing the origins of related noises. The noise spectra can be deduced by Fourier transform of the equations. If the injection current noise δI is taken into consideration, $F_N = \delta I/ed$, while F_a and F_ϕ are omitted, the Fourier frequency spectra of carrier concentration, field amplitude, and optical frequency are deduced as:

$$\widetilde{N}(f) = \frac{jfF_N}{f_R^2 - f^2 - j\gamma_R f}, \tag{3.27a}$$

$$\tilde{a}(f) = \frac{v_g g_N A}{2} \frac{F_N}{f_R^2 - f^2 - j\gamma_R f}, \tag{3.27b}$$

$$\tilde{v}(f) = \frac{\alpha v_g g_N}{4\pi} \frac{F_N}{f_R^2 - f^2 - j\gamma_R f}. \tag{3.27c}$$

where f_R and γ_R are the resonant frequency and the damp factor, as introduced in Sect. 2.6.

If the injection current noise is suppressed, the amplitude noise and phase noise induced by spontaneous emission are inevitable. From expression (3.22), we have $F_a \propto \sum_i \cos\theta_i \delta(t - t_i)$ and $F_\phi \propto A^{-1} \sum_i \sin\theta_i \delta(t - t_i)$, where $\delta(t - t_i)$ stands for the moment when a single spontaneous emission occurs. The randomness of θ_i gives relation of $|F_\phi| = |F_a|/A$. PSD of intensity noise and frequency noise are obtained from rate Eq. (3.26), expressed by the square of Fourier transforms [4]:

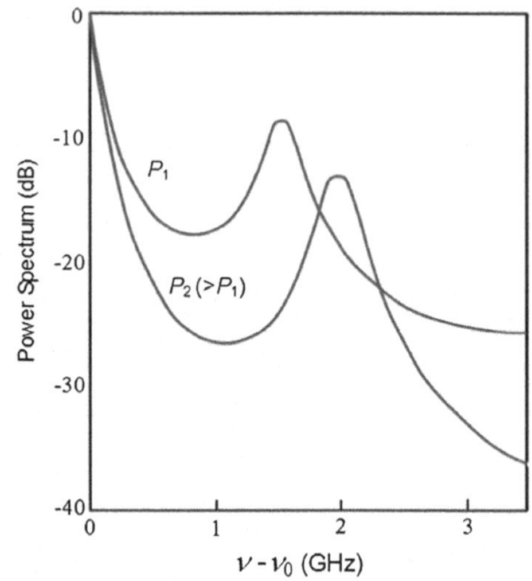

Fig. 3.2 Schematic line profiles of semiconductor laser, P_1, P_2: output powers

$$S_a(f) = |\tilde{a}(f)|^2 = |F_a|^2 \frac{f^2 + \gamma^2}{(f_R^2 - f^2)^2 + (\gamma f)^2}, \tag{3.28a}$$

$$S_\nu(f) = |\tilde{\nu}(f)|^2 = \frac{\delta\nu_{ST}}{\pi}\left[1 + \frac{\alpha^2 f_R^4}{(f_R^2 - f^2)^2 + (\gamma f)^2}\right]. \tag{3.28b}$$

Formulas (3.26–3.28b) describe the effect of coupling between photons and electrons on intensity noise spectrum and line shape, showing behaviors similar to relaxation oscillation. Figure 3.2 gives line shapes of semiconductor lasers at different output powers [17, 18].

3.2.3 *Measurement of Linewidth*

The first step of laser spectrum measurement is to find out its longitudinal mode characteristics, whether single mode or multimode, how many mode numbers, what about its envelop profile and side mode suppression ratio, and so on. The commonly used optical spectrum analyzer is usually based on an optical grating with wavelength resolution in the range of 0.1–0.01 nm, corresponding to frequency resolution of around 10–1 GHz in the near infrared band. The linewidth larger than such a resolution can be obtained. For narrower linewidths and smaller longitudinal mode spacing, the Fabry–Perot etalon and the scanned F–P analyzer are often used. The F–P characteristics are explained in details in many textbooks, e.g., Ref. [19], and will be restated in Sect. 5.4.

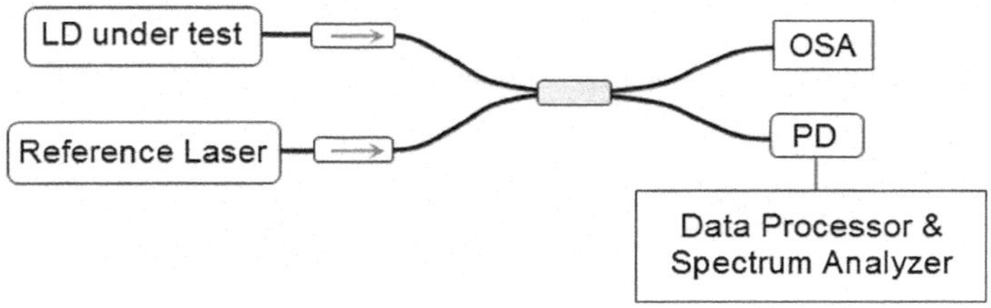

Fig. 3.3 Heterodyne with a narrow line reference laser

For single-frequency lasers with higher coherences, e.g., $\delta\nu \leq 100\,\text{kHz}$, the required resolution exceeds the performance of conventional optical spectrum analyzer. In such cases, the heterodyne method is often used, that is, to get interference of the laser under test and another reference laser with much narrower linewidth, and the linewidth can be obtained from the beat spectrum given by an electrical frequency analyzer [20], as illustrated in Fig. 3.3, where a 2 × 2 fiber coupler is used with field split ratio of

$$\begin{pmatrix} A \\ B \end{pmatrix} = \begin{pmatrix} \sqrt{\kappa} & j\sqrt{1-\kappa} \\ j\sqrt{1-\kappa} & \sqrt{\kappa} \end{pmatrix} \begin{pmatrix} A_0 \\ B_0 \end{pmatrix}. \tag{3.29}$$

Two optical isolators are inserted in the output paths of the two lasers to avoid influences of returned waves. If the phase noise of narrow line reference laser can be neglected, its field is written as $E_{\text{ref}} = E_0 e^{j\omega_0 t}$, and the field of LD under test is $E_{\text{LD}} = E_1 e^{-j[\omega_1 t + \phi(t)]}$, the interference signal detected by PD is expressed as

$$I_{\text{LD}}(t) \propto E_0^2 + E_1^2 + 2E_0 E_1 \cos[(\omega_1 - \omega_0)t + \phi(t)], \tag{3.30}$$

where the split ratio of fiber coupler is supposed to be 50:50, and the polarizations of two sources have been adjusted parallel. The variance of phase noise $\sigma_\phi^2 = \langle \phi^2(t) \rangle$ and the line shape can then be obtained by the electrical spectrum analyzer and data processor. Advanced analyzers based on the method are available commercially with resolution of 1 pm (~0.1 GHz) or better, as introduced in Sect. 7.1.

If the laser with linewidth much narrower than that of LD under test is not available, self-heterodyne and self-homodyne delayed autocorrelations can be used; that is, to get the interference of laser wave with itself delayed long enough; the linewidth can then be calculated from the autocorrelation spectrum [7, 21, 22]. The basic configuration of self-heterodyne and self-homodyne is a fiber Mach–Zehnder interferometer (MZI) or a fiber Michelson interferometer (MI) with path difference of the two arms large enough, as shown in Fig. 3.4a, b. In the setup, an acousto-optic modulator (AOM), modulated by a certain RF frequency, is inserted in one of the arms for the heterodyne interferometers, as shown in the figures by dotted frames. When it is removed, the system gives homodyne measurement. The signal of self-heterodyne is biased at the AOM modulation frequency, beneficial for

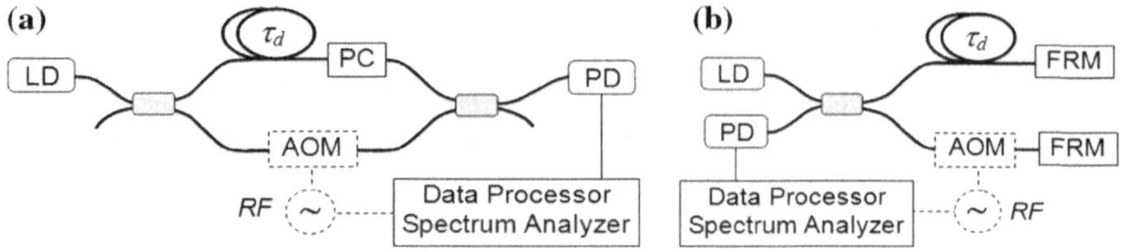

Fig. 3.4 Delayed autocorrelation interferometer, **a** MZI; **b** MI

mitigating low frequency disturbances and DC drift, which exist often in homodyne interferometers.

The quantitative relation between the beat signal and the laser's line shape is derived as follows, with the MZI scheme taken as an example. With the laser field denoted as $E(t) = E_0 e^{-j[\omega_0 t + \phi(t)]}$, the combined field of two returned waves from the MZI is written as

$$\begin{aligned} E_{\mathrm{T}}(t) &= \kappa E_0 e^{-j[(\omega_0 + \omega_{\mathrm{m}})t + \phi(t)]} + (1-\kappa) E_0 e^{-\alpha_{\mathrm{F}} \Delta L/2} e^{-j[\omega_0 (t-\tau_{\mathrm{d}}) + \phi(t-\tau_{\mathrm{d}})]} \\ &= \kappa E_0 e^{-j[(\omega_0 + \omega_{\mathrm{m}})t + \phi(t)]} \left[1 + b e^{j\Delta\phi(t,\tau_{\mathrm{d}})} e^{j(\omega\tau_{\mathrm{d}} + \omega_{\mathrm{m}} t)}\right], \end{aligned} \tag{3.31}$$

where $b = [(1-\kappa)/\kappa] \exp(-\alpha_{\mathrm{F}} z/2)$ with fiber loss coefficient of α_{F}, ω_{m} is the frequency shift of AOM; $\tau_{\mathrm{d}} = n\Delta L/c$ is the delay with fiber length difference of ΔL between the two arms. The interference signal can also be obtained from the coupler port other than that depicted in the figure, giving the same characteristics.

The beat signal is detected by the photodiode (PD) and analyzed by the spectrum analyzer (SA). The spectrum given by SA is proportional to the autocorrelation of optical intensity:

$$\begin{aligned} R_{\mathrm{I}}(\tau) &= \langle E_{\mathrm{T}}(t) E_{\mathrm{T}}^*(t) E_{\mathrm{T}}(t-\tau) E_{\mathrm{T}}^*(t-\tau) \rangle \\ &= \kappa^4 E_0^4 \langle \{1 + b^2 + b[e^{j\Delta\phi(t,\tau_{\mathrm{d}})} e^{j(\omega\tau_{\mathrm{d}} + \omega_m t)} + c.c.]\} \\ &\quad \cdot \{1 + b^2 + b[e^{j\Delta\phi(t-\tau,\tau_{\mathrm{d}})} e^{j[\omega\tau_{\mathrm{d}} + \omega_{\mathrm{m}}(t-\tau)} + c.c.]\} \rangle. \end{aligned} \tag{3.32}$$

For simplicity, the autocorrelation for homodyne with $\omega_m = 0$ is given as [21, 22]

$$R_{\mathrm{I}}(\tau) = \kappa^4 E_0^4 [(1+b^2)^2 + 4b(1+b^2) e^{-U} \cos\omega\tau_{\mathrm{d}} + 2b^2 (e^{-V} + e^{-W} \cos 2\omega\tau_{\mathrm{d}})],$$

where

$$\begin{aligned} U &= \langle [\phi(t) - \phi(t-\tau_{\mathrm{d}})]^2 \rangle / 2 \\ V &= \langle [\phi(t) - \phi(t-\tau_{\mathrm{d}}) - \phi(t-\tau) + \phi(t-\tau-\tau_{\mathrm{d}})]^2 \rangle / 2 \, . \\ W &= \langle [\phi(t) - \phi(t-\tau_{\mathrm{d}}) + \phi(t-\tau) - \phi(t-\tau-\tau_{\mathrm{d}})]^2 \rangle / 2 \end{aligned} \tag{3.33}$$

The time averages of $\Delta\phi$ with different delays in (3.33) can be reduced by using Fourier transform relation of $\widetilde{\Delta\phi}(f,\tau) = \widetilde{\phi}(f)(1 - e^{-j2\pi f\tau})$, resulting in

$$\begin{aligned}
U &= 2\int_0^\infty S_\nu(f)\sin^2(\pi f\tau_d)f^{-2}df \\
V &= 8\int_0^\infty S_\nu(f)\sin^2(\pi f\tau_d)\sin^2(\pi f\tau)f^{-2}df \\
W &= 8\int_0^\infty S_\nu(f)\sin^2(\pi f\tau_d)\cos^2(\pi f\tau)f^{-2}df.
\end{aligned} \tag{3.34}$$

For white noise, the three expressions are reduced to

$$\begin{aligned}
U &= \tau_d/2\tau_c \\
V &= (2\tau_d + 2|\tau| - |\tau_d - \tau| - |\tau_d + \tau|)/2\tau_c \\
W &= (2\tau_d - 2|\tau| + |\tau_d - \tau| + |\tau_d + \tau|)/2\tau_c.
\end{aligned} \tag{3.35}$$

Substituting (3.35)–(3.9) for the PSD of optical field, taking frequency shift of heterodyne into account, the normalized spectrum for white noise is deduced as [21]:

$$\begin{aligned}
S(f) = {} & e^{-\tau_d/\tau_c}\delta(f - f_m) + \frac{\tau_c/2}{1 + 4\pi^2(f - f_m)^2\tau_c^2} \\
& \cdot \left\{1 - \left[\cos 2\pi(f - f_m)\tau_d + \frac{\sin 2\pi(f - f_m)\tau_d}{2\pi(f - f_m)\tau_c}\right]e^{-\tau_d/\tau_c}\right\},
\end{aligned} \tag{3.36}$$

where $b = 1$ is assumed for simplicity. The term with $\cos 2\omega\tau_d$ in (3.33) gives smaller contributions to the PSD, being omitted in the formula. With $f_m = 0$, (3.36) gives the self-homodyne measured spectrum. The term in the square brackets multiplied by the visibility, $\exp(-\tau_d/\tau_c)$, in (3.36) is the results of interference, which will vanish when the delay line is much longer than the laser's coherence length with $\tau_d \gg \tau_c$. The resultant autocorrelation spectrum is thus with a Lorenzian shape. The FWHM linewidth of autocorrelation $\delta\nu = 1/(2\pi\tau_c)$ can then be obtained from the spectrum from SA; which is twice the laser's Lorentzian linewidth. Figure 3.5 shows a calculated self-homodyne spectrum as an example for illustration.

If the frequency noise is not an exact white noise, especially when the frequency noise in low frequency band is taken into account, analytic expressions of U, V, W in (3.35) may not be obtained. Digital calculations are then needed. If laser's line shape is in Gaussian, its autocorrelation will show Gaussian type also.

In order to measure the linewidth precisely several technical issues in the measurement have to be considered. For high coherence lasers, the fiber length used in the interferometer is so long that the returned wave would be very weak due to

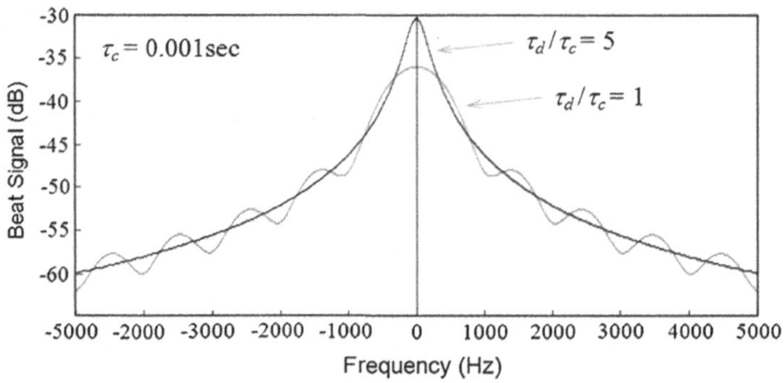

Fig. 3.5 Simulated curves of delayed homodyne signals

fiber loss. Apart from the ordinary loss, the stimulated Brillouin scattering will induce an additional loss for narrow line optical wave. In practice, an interferometer with OPD less than coherence length may be used, then the linewidth has to be calculated by fitting the measured data with formula (3.36) [21, 22]. A method of recirculating delayed self-heterodyne was proposed to enhance the measurement accuracy [23]. An extremely narrow linewidth may be obtained by digital calculations from the frequency noise spectrum, as introduced in the next subsection.

It is also needed to adjust the effective split ratio b to be near unity to get higher extinction interference signal. The stability of MZI or MI is another factor, which may induce errors to the measurement. Environmentally induced mechanical vibrations and temperature fluctuations should be reduced, especially for the long fiber. The polarization fading induced by polarization evolution in fiber has to be avoided, for example, by using Faraday rotation mirrors (FRM) in MI configuration as depicted in Fig. 3.4b. The action of FRM is to rotate the polarization of reflected beam 90° to the incident beam polarization, so that the effect of birefringence induced externally can be canceled out. For MZI system, a polarization controller may be inserted in one of its arms. The reliability of measured linewidth is dependent on the precision of electrical frequency spectrum analyzer (SA), including its frequency resolution and amplitude resolution. When the beat signal is very weak with low SNR, the 3 dB down frequency may not be read precisely. In such a case, the FWHM spectral width may be calculated by reading the frequencies at a lower level, e.g., 20 dB below, and by fitting the Lorenzian function.

3.3 Noises of Semiconductor Laser

The intrinsic linewidth discussed above is a reflection of frequency noise induced by the random spontaneous emission. The laser noises are generated also by external disturbances, such as temperature fluctuations, mechanical vibrations (including acoustic waves), and pump current noise. The external electromagnetic interferences (EMI), the external feedback of laser's output via cavity mirrors, and even the artificial factors are also noise origins. The different noises are coupled with each other by the rate equations. Some kinds of the externally induced noises may be traceable; some others may come from factors people have not yet understood and cannot be controlled.

The noise induced by the spontaneous emission is a white noise in a broad frequency band; its upper limit is beyond the reciprocal of photon lifetime in cavity. Contrarily, the noise induced by external disturbances is in low frequency bands. Similar to the noises of electrical and mechanical devices, the laser noises in low frequency bands have characteristics of 1/*f*, i.e., the power spectral density (PSD) is proportional to f^{α}. The noise behaviors in low frequency bands are described by some models, such as random walking (similar to Brownian motion) and flicker. Index α will take different values: $\alpha \approx -1$ for flicker noises; $\alpha \approx -2$ for random walking noises; generally α may take a figure in a wider range [24].

3.3.1 Characteristics of Frequency Noise in Low Frequency Band

The external disturbances and pump current variation cause directly fluctuations of the mode frequency $\omega_0 = m\pi c/(nL)$, because both of the cavity length and the refractive index are functions of temperature and strains, and the index varies also with injected carrier concentration. It is known that the laser oscillation can be set up in a very short period corresponding to several round-trips of light propagation in the cavity, which is less than about 0.1 ns for semiconductor lasers. The noises caused by external disturbances are in a relatively low frequency band; such noises can be regarded as drifting of narrow linewidth laser oscillation in space of $\mathrm{Im}(\boldsymbol{E})$ versus $\mathrm{Re}(\boldsymbol{E})$, as shown in Fig. 3.6a. The scope of such drifting is much larger than randomness of spontaneous emission. Therefore, the line shape measured in a

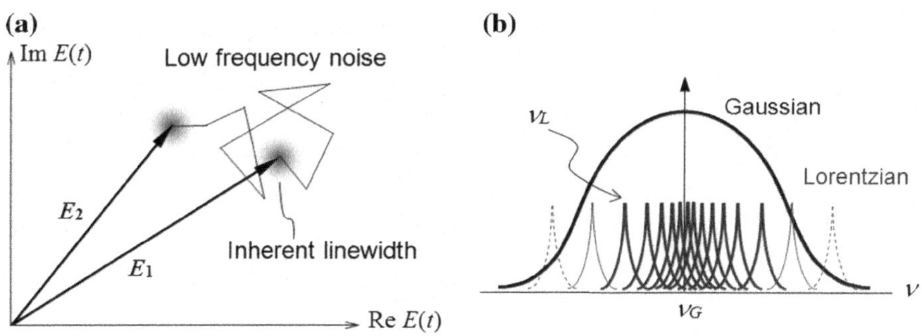

Fig. 3.6 a Random drifts of field amplitude and frequency; **b** Synthesized line of drifted frequency

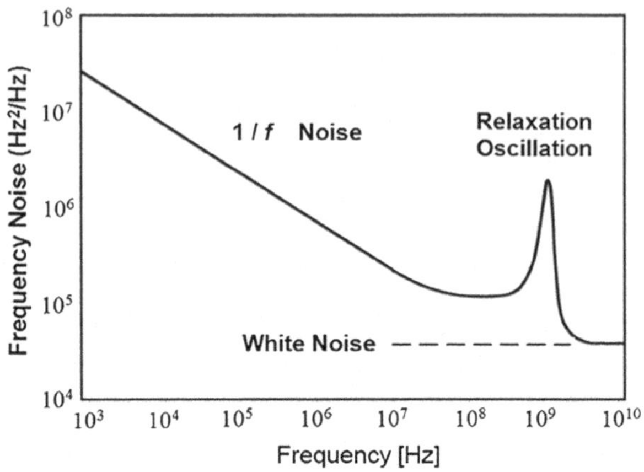

Fig. 3.7 Schematic of LD frequency noise PSD

certain period is actually a synthesis of drifted narrow linewidth laser wave, as shown in Fig. 3.6b [20, 25]. With the $1/f$ frequency noise taken into consideration, the frequency noise spectrum in the whole range of frequency band is no longer a white noise, but a sum of several terms, expressed as [10–12]:

$$S_v(f) = h_0 + h_R(f) + h_\alpha f^\alpha, \tag{3.37}$$

where h_0 is the white noise, h_R stands for the contribution of relaxation oscillation. Figure 3.7 shows the frequency noise PSD schematically.

According to the central limitation theorem, when a great number of laser frequency fluctuations exist, the probability of randomness obeys the normal distribution; thus a Gaussian type line profile results, as demonstrated experimentally. For the theoretical analysis of $1/f$ noise, the variance of phase fluctuation is written as

$$\begin{aligned} \sigma_\phi^2(\tau) &= 4h_\alpha \int_0^\infty f^{\alpha-2} \sin^2(\pi f \tau) \mathrm{d}f = q(\alpha)\tau^{1-\alpha} \\ q(\alpha) &= 4h_\alpha \pi^{1-\alpha} \int_0^\infty x^{\alpha-2} \sin^2 x \mathrm{d}x, \end{aligned} \tag{3.38}$$

For $\alpha = -1$, $\sigma_\phi^2 = q_1 \tau^2$, a Gaussian line shape is then given by integral (3.10) and written as (3.13). In case $\alpha < -1$ and is not a negative integer, coefficient $q(\alpha)$ can be expressed with Gamma function of $\Gamma(x) = \int_0^\infty u^{x-1} \mathrm{e}^{-u} \mathrm{d}u$ [26, 27]:

$$q(\alpha) = \frac{2^{2-\alpha}\pi^{1-\alpha}h_\alpha}{\alpha - 1} \sin\frac{\alpha\pi}{2} \Gamma(\alpha). \tag{3.39}$$

It is noticed that Gamma function is divergent if x is a negative integer. In practice, the power index α is usually not an exact negative integer; then the divergence of Gamma function there can be avoided. Nevertheless, expression (3.38) gives an explanation of the Gaussian line shape, observed mostly in experiments.

Actually, the noise spectrum cannot expand to the whole range from zero to infinitive. Instead, the range should be between a finite lowest frequency f_L and a finite upper frequency f_H. In practice, f_L corresponds to the reciprocal of observation duration; f_H is determined by the frequency response of measurement setup. The integral in (3.38) should then be rewritten as

$$q(\alpha) = 4h_\alpha \pi^{1-\alpha} \int\limits_{x_1}^{x_2} x^{\alpha-2} \sin^2 x \mathrm{d}x, \tag{3.40}$$

where $x_{1,2} = \pi\tau f_{L,H}$. The integral can be expressed with cosine integral functions [27]. In such a case, the integral will not be divergent even for an exact negative integer α. For flicker noises, $\alpha = -1$, the integral is reduced to

$$\int\limits_{x_1}^{x_2} x^{-3} \sin^2 x \mathrm{d}x = 3/2 + Ci(x_2) - Ci(x_1), \tag{3.41}$$

where $Ci(x) = -\int_x^\infty t^{-1} \cos t \mathrm{d}t$ is the cosine integral function. For random walking noises, $\alpha = -2$, $\int_{x1}^{x2} x^{-4} \sin^2 x \mathrm{d}x \approx 1/x_1 - \pi/3$ is obtained [28], where approximation of $x_2 \gg 1$ is taken. Formulas (3.38)–(3.41) show that the phase variance is proportional to a power function of τ, which gives basically a Gaussian-like line shape.

The $1/f$ noise and white noise appear in the integral (3.9) as a product of two exponential functions:

$$S_E(f) = \int\limits_{-\infty}^{\infty} \left[\exp\left(-2\pi^2 h_0 |\tau|\right) \cdot \exp\left(-2q_1\tau^2\right)\right] \mathrm{e}^{-j2\pi f\tau} \mathrm{d}\tau. \tag{3.42}$$

According to the convolution theorem of Fourier transform, their composed line is the convolution of their respective Fourier transforms, Lorenzian and Gaussian, i.e., Voigt waveform [10, 11]:

$$S_V(f) = S_0 \int\limits_{-\infty}^{\infty} \frac{\exp(-t^2/b^2)}{(f-t)^2 + a^2} \mathrm{d}t, \tag{3.43}$$

where $a = \Delta f_L/2$, $b = \Delta f_G/2\sqrt{\ln 2}$, and $S_0 = 2\sqrt{\ln 2}\Delta f_L/(\pi^{3/2}\Delta f_G)$. The Voigt linewidth can be approximated by expression of $\Delta f_V = \sqrt{\Delta f_G^2 + 0.21666\Delta f_L^2} + 0.5346\Delta f_L$ [10].

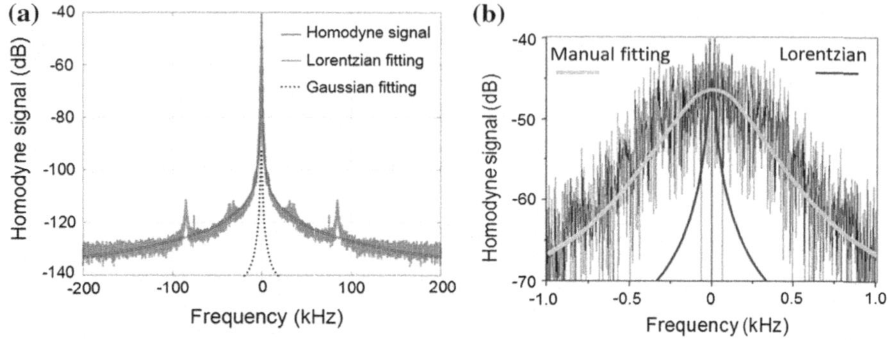

Fig. 3.8 **a** Measured self-homodyne signal and its fittings; **b** In an expanded abscissa

Generally, the intrinsic linewidth is much smaller than the Gaussian linewidth caused by external disturbances; therefore, the composite linewidth is mainly determined by the latter. However, the two wings of line shape are mainly attributed to the Lorenzian line. Figure 3.8 shows an example of self-homodyne signals, measured with a setup shown in Fig. 3.4 [29]. Figure 3.8a is with a larger scale of coordinate, and Fig. 3.8b is magnified curves. It is shown that the measured FWHM linewidth is about 150 Hz by the light gray fitting curve, whereas the Lorentzian line shape fitting the two wings corresponds to an intrinsic linewidth of 10 Hz.

3.3.2 *Measurement of Laser Noises*

Measurement of intensity noise and frequency noise is an important subject both for research and development of single-frequency lasers and for their applications. In the interferometry, the linewidth of laser source determines the visibility of interference fringes, while the laser noise, especially the noise in low frequency band, degrades SNR of the interference signal. The method of noise measurement is based on interferometers [20, 25–28]; but its focus is on the frequency spectrum of interference signal, instead of the spectral shape of autocorrelation in the linewidth measurement.

(1) **Basic Principle**.

In case a low noise reference laser is available, and the difference between the central frequencies of reference laser and laser under test is within the range of electrical spectrum analyzer, the phase noise can be measured by heterodyne of the two lasers [20], as shown in Fig. 3.9, where two laser beams combined at the 3 dB coupler are detected by a double-balanced photodiode (DB-PD), and the signal is processed in the spectrum analyzer (SA). DB-PD is composed of two photodiodes connected in series, as depicted in the figure, and biased reversely. The output is acquired from their connection point, proportional to the difference of the two PD

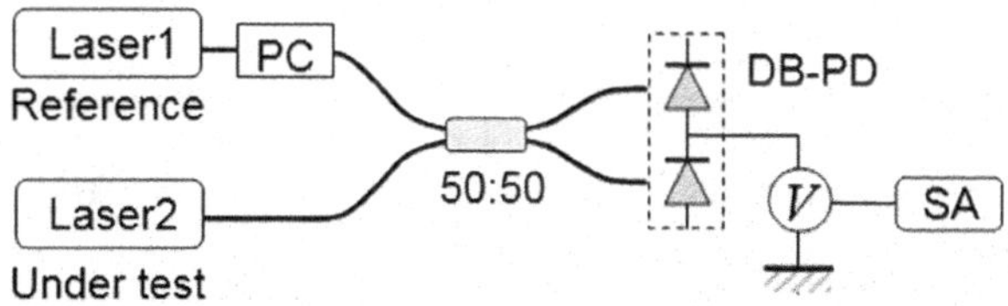

Fig. 3.9 Frequency noise measurement with heterodyne method; PC: polarization controller

signals, so that the DC component will be canceled, whereas the AC components with opposite phases will be detected.

The optical fields at the output ports of 3 dB coupler are written as

$$E_a = j[E_1 e^{-j\omega_{ret}t} + E_2 e^{-j[\omega t + \phi(t)]}]/\sqrt{2}$$
$$E_b = [E_1 e^{-j\omega_{ret}t} - E_2 e^{-j[\omega t + \phi(t)]}]/\sqrt{2}. \tag{3.44}$$

The output signal of DB-PD is then obtained as

$$V(t) = \eta(|E_a|^2 - |E_b|^2) = 4\eta E_1 E_2 \cos[(\omega_1 - \omega_2)t - \phi(t)], \tag{3.45}$$

where η stands for the responsivity of PD and amplification in electronics. From its in-phase (I) and quadrature (Q) components interrogated in the spectrum analyzer, the phase noise is detected as

$$\tan[\phi(t)] = \frac{\langle V(t)\ \sin[(\omega_1 - \omega_2)t\rangle}{\langle V(t)\ \cos[(\omega_1 - \omega_2)t\rangle}. \tag{3.46}$$

When the proper reference laser is not available, delayed self-heterodyne (DSH) or delayed self-homodyne (DSO) are usually used. The delayed MZI and MI are the basic setup, similar to the linewidth measurement; but the OPD is short than the coherence length of laser under test, so that the low frequency band noise can be manifested clearer. Figure 3.10 shows a typical setup [30], where the laser output is

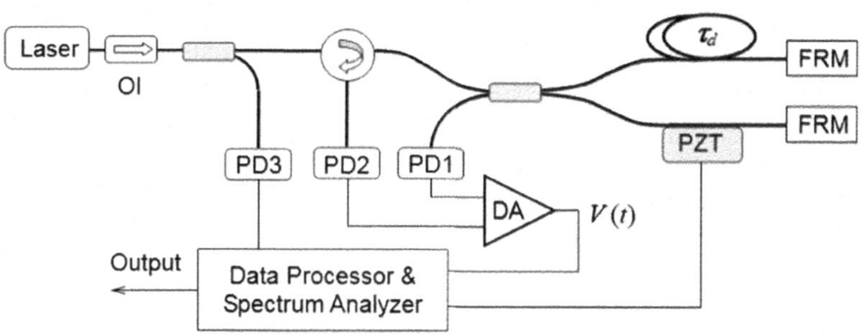

Fig. 3.10 Laser noise measurement setup

divided into two by the fiber coupler, one is detected by a photodiode PD3, giving information of output power and intensity noise; the other is analyzed by a fiber Michelson interferometer (MI) with delay of $\tau_d = 2nk\Delta L$, which can be adjusted by PZT. The two returned waves from the two arms of interferometer are detected by PD1 and PD2. With the similar function of DB-PD, the signals are amplified in a differential amplifier (DA); and converted to digital data by a data acquisition, and processed with fast Fourier transform (FFT). An optical isolator (OI) is inserted in the LD output path to avoid influence of returned wave to the laser's status. Two Faraday rotation mirrors (FRM) are used to remove the polarization fading due to polarization change induced by external disturbances. The Mach–Zehnder interferometer (MZI) can also be used in the setup to get the same function.

The signal of interferometer contains information of frequency fluctuation of the laser. Since the RIN is measured by PD1, the optical field to be measured for the phase noise is regarded as with a constant amplitude: $E(t) = E_0 \exp[-j\omega t - j\phi(t)]$. The fiber coupler with split ratio $\kappa : (1-\kappa)$ is assumed lossless. The returned wave fields and intensities from the coupler are written as:

$$\begin{aligned} E_1 &= j\sqrt{\kappa(1-\kappa)}E_0\left[e^{-j[\omega t + \phi(t)]} + \rho e^{-j[\omega(t-\tau_d) + \phi(t-\tau_d)]}\right] \\ E_2 &= E_0\left[\kappa e^{-j[\omega t + \phi(t)]} - \rho(1-\kappa)e^{-j[\omega(t-\tau_d) + \phi(t-\tau_d)]}\right], \end{aligned} \tag{3.47}$$

$$\begin{aligned} I_1 &= \kappa(1-\kappa)I_0[1 + \rho^2 + 2\rho\ \cos(\omega\tau_d + \Delta\phi)] \\ I_2 &= I_0[\kappa^2 + \rho^2(1-\kappa)^2 - 2\rho\kappa(1-\kappa)\ \cos(\omega\tau_d + \Delta\phi)], \end{aligned} \tag{3.48}$$

where $\Delta\phi = \phi(t) - \phi(t-\tau_d)$ is the differential phase noise, from which the frequency noise and its spectrum will be obtained; $\rho = \exp(-\alpha_F\Delta L/2)$ is the fiber loss factor. If the coupler with $\kappa = 0.5$ is used, the DC component of interference signals can be canceled in the differential amplifier, the AC component of interferometer output is obtained as

$$V(t) = \eta(I_1 - I_2) = \eta\rho I_0 \cos(\omega\tau_d + \Delta\phi) \to \eta\rho I_0 \sin\Delta\phi, \tag{3.49}$$

where η stands for the responsivity of PD and the amplification of DA. In order to obtain the quantitative value of phase noise variance, it is needed to calibrate the ratio of optical signal to electrical signal. By modulating the PZT or by tuning the laser under test to adjust the interference phase $\omega\tau_d$, the maximum and minimum values of interferometer signal, V_{max} and V_{min} are measured, giving $\eta\rho I_0 = (|V_{max}| + |V_{min}|)/2$. Furthermore, to maximize the sensitivity of interference to the phase fluctuation $\Delta\phi$, the interferometer should be adjusted to work at quadrature point $\omega\tau_d = (m + 1/2)\pi$. Signal $V(t)$ is then averaged to zero, as shown by the last expression of formula (3.49), and its variance is reduced to

$$\langle\sin^2\Delta\phi\rangle = \langle V^2(t)\rangle/(\eta\rho I_0)^2 = \sigma_V^2/(\eta\rho I_0)^2. \tag{3.50}$$

If the fluctuation of differential phase is small enough to meet condition of $\Delta\phi \ll 1$, the variance of differential phase will be proportional to the variance of detected signal. Based on the relation of Fourier transforms $\widetilde{\Delta\phi}(f,\tau_d) = (1 - e^{-j2\pi f\tau_d})\widetilde{\phi}(f)$, PSD of laser's phase noise is obtained as

$$S_\phi(f) = \frac{S_{\Delta\phi}(f)}{4\sin^2(\pi f\tau_d)}. \tag{3.51}$$

It is seen that the measured data of phase noise are related to the OPD of interferometer. The relation between the differential phase and the frequency noise δv can be expressed as $\Delta\phi = 2\pi\int_{t-\tau_d}^{t}\delta v(t)dt = 2\pi\tau_d\langle\delta v\rangle$, meaning that the average of frequency fluctuation is obtained in range of τ_d, and the measured spectrum is reliable in the range lower than $1/\tau_d$. The power spectral density $S_v(f)$ of frequency noise in the range can then be obtained by Fourier transform of measured frequency noise variance, $\sigma_v^2 = \sigma_{\Delta\phi}^2/(4\pi^2\tau_d^2)$. For comparable information, the reported measured data in the literature are usually provided with an annotation about the delay.

Although a larger delay is beneficial for a higher signal of $\Delta\phi$ under certain frequency noise levels, it should be limited to meet the requirement of $\sin\Delta\phi \approx \Delta\phi$, especially in the range of very low frequency, where the noise level will increase greatly for $1/f$ noise.

The paths of fiber interferometer are usually susceptible to external disturbances, which will cause errors of laser noise measurement. Therefore, the interferometer has to be put into a temperature-stabilized and vibration-isolated box. It is necessary to use low noise detectors and electronics to decrease their influence on the measured data. Practically such noises exist inevitably; the noise spectrum of measurement setup itself has to be measured separately and used to calibrate the measured data of laser noise. On the other hand, the frequency responses of the detectors and electronics, including the data acquisition device, should be taken into account also. Their response will determine the upper limit of measured laser noise spectrum.

The methods of laser noise measurement are being developed for higher reliability and for lower cost. References [20, 32] proposed a scheme based on the technology of optical phase-locked loops (OPLL), which is the basic technique used in coherent optical communications. Instead of frequency shift, the frequency modulation is used in self-heterodyne [28, 33].

(2) **Improvement of Interferometer with 3 × 3 Coupler.**

The MZI or MI composed of a 2 × 2 fiber coupler has essential shortcomings; that is, dead regions exist at positions with phase factor of $m\pi$, where the sensitivity to interference signal is toward zero. In addition, the direction of signal's variation cannot be distinguished there, and the accumulative calculation of phase changes over 2π becomes impossible, i.e., the phase unwrapping there has to rely on some indirect methods. Therefore the working point at quadrature should be adjusted and

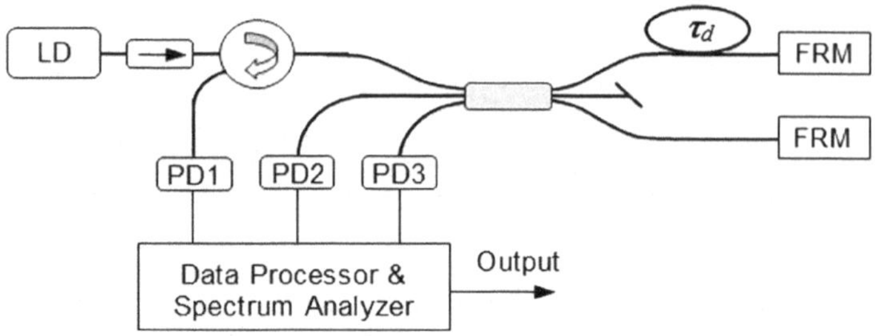

Fig. 3.11 Frequency noise measurement setup with a 3 × 3 coupler

stabilized, as discussed above. To avoid such shortcomings, an interferometer composed of a symmetric 3 × 3 coupler was proposed [34, 35], as shown in Fig. 3.11. The transfer matrix of a 3 × 3 lossless coupler is written as [36]:

$$\begin{pmatrix} A \\ B \\ C \end{pmatrix} = \frac{j}{\sqrt{3}} \begin{pmatrix} s & 1 & 1 \\ 1 & s & 1 \\ 1 & 1 & s \end{pmatrix} \begin{pmatrix} A_0 \\ B_0 \\ C_0 \end{pmatrix}, \tag{3.52}$$

where A_0, B_0, C_0 and A, B, C are the input and output optical fields of the three ports, respectively. A phase shift between adjacent ports exists as expressed in $s = |s|\exp(j\varphi)$. For an ideal 3 × 3 coupler with split ratio of 1:1:1, $|s| = 1$, $\varphi = 2\pi/3$. The interferometer in Fig. 3.11 uses two output ports of the 3 × 3 coupler, with the third port made reflection-free.

The three returned waves of the MI are obtained as

$$\begin{pmatrix} E_1 \\ E_2 \\ E_3 \end{pmatrix} = \frac{-1}{3} \begin{pmatrix} s & 1 & 1 \\ 1 & s & 1 \\ 1 & 1 & s \end{pmatrix} \begin{pmatrix} e^{j\theta_1} & 0 & 0 \\ 0 & 0 & 0 \\ 0 & 0 & e^{j\theta_2} \end{pmatrix} \begin{pmatrix} s & 1 & 1 \\ 1 & s & 1 \\ 1 & 1 & s \end{pmatrix} \begin{pmatrix} E_0 \\ 0 \\ 0 \end{pmatrix}, \tag{3.53}$$

$$\begin{pmatrix} I_1 \\ I_2 \\ I_3 \end{pmatrix} = \frac{2}{9} \begin{pmatrix} 1 - \cos(\Delta\theta + \pi/3) \\ 1 - \cos(\Delta\theta - \pi/3) \\ 1 + \cos\Delta\theta \end{pmatrix} E_0^2, \tag{3.54}$$

where $\Delta\theta = \theta_1 - \theta_2 = \Delta\phi + \omega\tau_d$ is the differential phase of MI. Figure 3.12 shows a measured interference signals with a tunable source. It is seen that no matter how much the bias phase is, at least two of the three outputs are given with definite values and sensitivities; and when one of them reaches its maximum, the other two show opposite slopes, which can be used to judge the direction of phase change. The interference phase can be demodulated by three detected signals:

$$\tan\Delta\theta = \frac{1}{\sqrt{3}} \frac{I_2 - I_1}{2I_3 - I_1 - I_2} \tag{3.55}$$

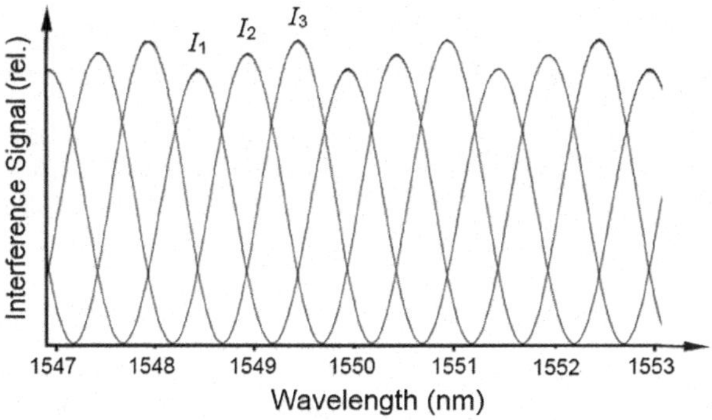

Fig. 3.12 Signals of MI composed of a 3 × 3 coupler

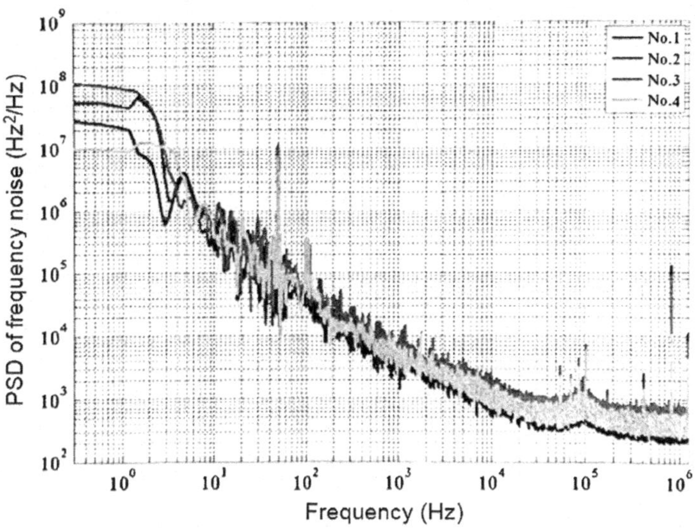

Fig. 3.13 An example of measured PSD of frequency noise, with the setup of Fig. 3.11

To reveal the information of laser noise, the differential phase $\omega\tau_d$ is demodulated first by using average signals of I_1, I_2, and I_3, since the average of laser phase noise is zero. Then, the phase noise can be obtained from their dynamic values. Mach–Zehnder interferometer with a 3 × 3 coupler has the similar property. Figure 3.13 shows an example of frequency noise PSD of a waveguide grating external cavity semiconductor laser, measured by the setup shown in Fig. 3.11 with a 50 m long delay line fiber, where four traces are the results of successive measurements [35]. In practice, the 3 × 3 coupler may deviate from the ideal case; generally, the transfer matrix should be written as

$$\begin{aligned} I_1/I_0 &= A - A_1 \sin\Delta\phi - A_2 \cos\Delta\phi \\ I_2/I_0 &= B - B_1 \sin\Delta\phi + B_2 \cos\Delta\phi\,. \\ I_3/I_0 &= C + C_1 \sin\Delta\phi \end{aligned} \tag{3.56}$$

The coefficients in (3.56) have to be calibrated experimentally. The responsivities of the three photodetector and amplifications of electronic circuits have to be calibrated also.

(3) **Calculation of Dynamic Linewidth with Frequency Noise PSD.**

For an extremely narrow line laser, the self-correlation methods introduced in Sec. 3.2 may not be effective, because the OPD of interferometer matching with the coherence length is so large that the fiber loss is too high for returned wave to be received. In such a case, the line shape and linewidth can be obtained by calculating integral (3.10) with the measured PSD of frequency noise. On the other hand, the variation of linewidth with the observation time is an important parameter both for researches and applications of the laser.

It is imaged from Fig. 3.6 that the composite linewidth increases with the observation time since the drifting of peak frequency will give more contribution. The effect of finite observation time T can be regarded as a filter in frequency domain applied on PSD of frequency noise with cutoff frequency of $1/T$ [37]:

$$S_\nu(f,T) = S_\nu(f)\left[1 - \frac{\sin^2(\pi fT)}{(\pi fT)^2}\right]. \tag{3.57}$$

The autocorrelation of differential phase is given by the term with e^{-V} in formula (3.33) and (3.34), which is now expressed as

$$\sigma_{\Delta\phi}^2 = 16\int_0^\infty S_\nu(f)\frac{\sin^2(\pi f\tau_\mathrm{d})\sin^2(\pi f\tau)}{f^2}\left[1 - \frac{\sin^2(\pi fT)}{(\pi fT)^2}\right]\mathrm{d}f. \tag{3.58}$$

The line shape is then written as

$$S_\mathrm{E}(f,T) = \int_{-\infty}^{\infty} \exp[-\sigma_{\Delta\phi}^2(\tau,T)/2]e^{-j2\pi f\tau}d\tau. \tag{3.59}$$

For an arbitrary line shape other than Lorentzian and Gaussian, it is needed to define an effective linewidth. One of the definitions is based on the equality of integrated areas of line shape, expressed as

$$\Delta f = \int_0^\infty S_\mathrm{E}(f)\mathrm{d}f/S_\mathrm{E}(0) = 1/S_\mathrm{E}(0). \tag{3.60}$$

The last expression of (3.60) is for a normalized line shape, and its peak can be written as $S_E(0) = \int_{-\infty}^{\infty} \exp[-\sigma_\phi^2(\tau)/2]d\tau$. For an exact Gaussian line, the ratio of (3.60) to the standard Δf_G is $\sqrt{\pi/\ln 2}/2 \approx 1.064$; it means (3.60) gives a good fitting for Gaussian-like lines. For Lorenzian line, the ratio is $\pi/2$; the big deviation is due to its higher wings. Another definition of effective linewidth is also used in some literature, expressed as [38]:

$$\Delta f = \left[\int_0^{\infty} S_E(f)df\right]^2 / \int_0^{\infty} S_E^2(f)df. \tag{3.61}$$

Difference between the contributions of PSD spectral components of low and high frequency bands to the linewidth is discussed in Ref. [39]. It is shown that the integral over lower frequency range gives the main contribution to the linewidth, with the boundary of two ranges set at frequency of $S_\nu(f) \approx f$, called β-line. The empirical formula gives a simple method to estimate the linewidth to avoid the calculation of PSD integral.

3.3.3 Intensity Noise and Its Suppression

Direct current pumping is a remarkable advantage of semiconductor lasers; however, the current noise will become an origin of laser's noises, especially the intensity noise. Owing to high gain coefficient, low reflectivity mirrors can be used to build the laser cavity. It is another advantage of semiconductor laser. However, it makes the laser susceptible to external optical feedback; it is another origin of phase noise and intensity noise.

The PSD of intensity noise induced by pumping noise can be derived from (3.27b), as

$$S_{RIN}(f) = \frac{v_g^2 g_N^2}{4} \frac{|F_N|^2}{(f^2 - f_R^2)^2 + (\gamma_1 f)^2}, \tag{3.62}$$

which has similar behavior to current modulation spectrum. Taking the $1/f$ noise and the white noise into account, the PSD of RIN is shown schematically in Fig. 3.14a. The G–R noise in the figure is so-called generation–recombination noise, which often exists in semiconductor devices, such as photodetector, electrical amplifier and in laser driver source [40].

The intensity noise can be suppressed greatly by controlling the pumping current with feedback error signals. The method is called the automatic power control (APC). Its usual procedure includes the following steps: (1) Detect the output power

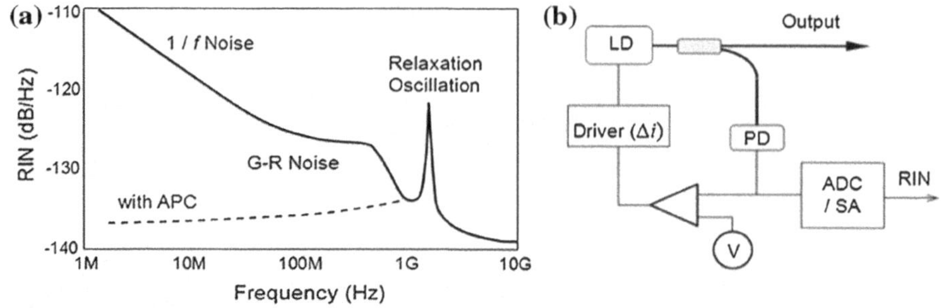

Fig. 3.14 **a** Typical PSD curve of intensity noise. **b** Schematic diagram of LD intensity noise measurement with APC

by a photodiode (PD); (2) Get the difference between the detected signal and a stabilized reference voltage as error signals; (3) Adjust the laser driver with the error as feedback signal. Figure 3.14b shows a schematic frame of APC, where an analog-to-digital converter (ADC) and a spectrum analyzer (SA) are used to measure the noise characteristics of power stabilized output. The dashed line in Fig. 3.14a is the result with APC applied, showing a great suppression of RIN in the low frequency band. The performance of APC is dependent on the response speed of the detector and the feedback circuit; it is more effective in the low frequency band.

The relaxation oscillation behavior is not welcome in most applications. A number of efforts have been made to suppress the resonance peak. The basic idea is to increase the photon lifetime and thus to make the semiconductor laser transferred from so-called B-class to A-class. (The laser classifications on their stabilities will be stated later in Sect. 3.5.1.) The relaxation oscillation amplitude goes up with output power increasing firstly, and then goes down after a maximum, as shown in Fig. 2.16. It is derived from (2.43) that the relaxation oscillation peak will disappear if the $\omega_R^4 \leq (\omega_R^2 - \gamma_R^2/4)\gamma_R^2$, or equivalently the photon lifetime τ_p meets the following condition:

$$\tau_p \geq \frac{2v_g\Gamma g_N p_0 \tau_{sp}^2}{(1 + v_g\Gamma g_N p_0 \tau_{sp})^2}. \tag{3.63}$$

In such a case a flat frequency spectrum of relaxation oscillation can be obtained. Several schemes for the purpose have been proposed and demonstrated by designing or adjusting the photon's lifetime τ_p or the working point p_0, such as by using a long extended cavity length [41] and by inserting an optical element with high dispersion index, e.g., with property of electromagnetically induced transparency (EIT) [42]. Other schemes use nonlinear damp elements, e.g., a two-photon absorber or a SHG-based buffer reservoir [43], which can be regarded as a gain saturation element to damp the power fluctuation.

3.4 Frequency Stability and Allan Variance

3.4.1 Characterization of Frequency Stability

The single frequency laser is widely used as the source to provide a frequency reference in applications of frequency standard, time services, precise metrology, and others. The frequency stability in long terms is one of the most important concerns for such applications. The stability has intrinsic relations with frequency noise, or more exactly with frequency drift. The stability involves characterizations in a long period operation, from the order of milliseconds through days, months, and even longer. In view of frequency noise spectrum, it corresponds to the noise in very low frequency bands toward zero. It has been found in R&D of electronic frequency standard technologies that the conventional method of variance analysis of random variables will result in divergence, as shown in Fig. 3.15 [44]. It may be understood by the argument that the frequency noise itself, as a random variable, has $1/f$ noise characteristics in the low frequency bands. It is meant that the longer the period of inspection, the lower the frequency band is concerned, and thus the stronger the effect of $1/f$ noise exists.

To overcome such a difficult, Allan variance and related modified schemes were proposed as commonly recognized methods for characterizing the frequency stability [44]. The basic definition of Allan variance is written as [24, 44–46]

$$\sigma_y^2(\tau) = \frac{1}{2}\langle (y_{i+1} - y_i)^2 \rangle = \frac{1}{2(N-1)}\sum_{i=1}^{N}(y_{i+1} - y_i)^2, \tag{3.64}$$

where $y_i = \nu_i/\nu_0$, ν_0 is the nominal frequency, ν_i is the i-th frequency measured in the observation time (or integration time) of τ. The period of repeated measurements is T, which may be larger than the observation time, and depends mainly on the availability of measurement apparatus. Symbol of $\langle \cdots \rangle$ stands for the ensemble

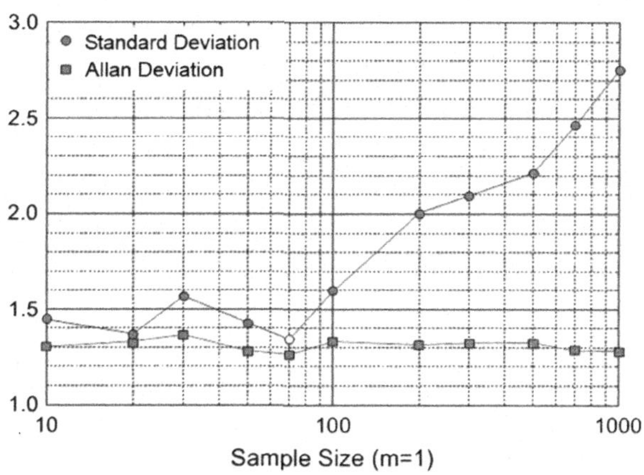

Fig. 3.15 Convergence of standard deviation and Allan deviation. Reprinted from [44] with permission

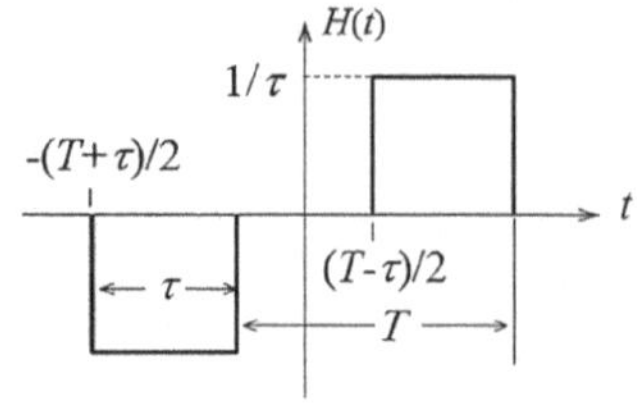

Fig. 3.16 Sampling function of y_Δ

average or time average; the last expression of (3.64) is the estimate of variance, where N is the number of samples. Allan variance is actually the variance of frequency difference between successive measurements. The frequency stability is usually displayed by a curve of $\log \sigma_y \sim \log \tau$, where $\sigma_y = \sqrt{\sigma_y^2}$ is Allan deviation (ADEV). The divergence of conventional variance is then avoided, as shown in Fig. 3.15.

Allan variance is interrelated directly with the frequency noise. The frequency difference of successive nominal frequencies $y_\Delta = y_{i+1} - y_i$ can be written as a convolution:

$$y_\Delta = \int_{-\infty}^{\infty} y(t)H(t - t_i)\mathrm{d}t, \tag{3.65}$$

where $y(t) = v(t)/v_0$. The sampling function $H(t)$ shown in Fig. 3.16, is expressed as

$$H(t) = \begin{cases} -1/\tau & -(T+\tau)/2 \le t < -(T-\tau)/2 \\ 1/\tau & (T-\tau)/2 \le t < (T+\tau)/2 \\ 0 & \text{else} \end{cases} \tag{3.66}$$

Allan variance is then deduced as [44, 46]:

$$\begin{aligned} \sigma_y^2(\tau) &= \frac{1}{2}\langle (y_\Delta)^2 \rangle = \frac{1}{2}\langle |\int_0^\infty \tilde{y}_\Delta e^{j2\pi f t}\mathrm{d}f|^2 \rangle = \frac{1}{2}\langle |\int_0^\infty \tilde{y} \cdot \widetilde{H} e^{j2\pi f t}\mathrm{d}f|^2 \rangle \\ &= \frac{1}{2}\int_0^\infty \int_0^\infty |\tilde{y} \cdot \widetilde{H}|^2 \langle e^{j2\pi(f-f')t} \rangle \mathrm{d}f\mathrm{d}f' = \frac{1}{2}\int_0^\infty S_y |\widetilde{H}|^2 \mathrm{d}f \\ &= 2\int_0^\infty S_y(f)\frac{\sin^2(\pi f\tau)\sin^2(\pi f T)}{(\pi f \tau)^2}\mathrm{d}f \to 2\int_0^\infty S_y(f)\frac{\sin^4(\pi f \tau)}{(\pi f \tau)^2}\mathrm{d}f. \end{aligned} \tag{3.67}$$

where $S_y(f) = S_v(f)/v_0^2 = f^2 S_\phi(f)/v_0^2$ is the power spectral density of $y(t)$, and the Fourier transform of $H(t)$, $\widetilde{H} = -j2\sin(\pi f\tau)\sin(\pi fT)/(\pi f\tau)$, is used. The last expression is for $T = \tau$.

If the PSD of frequency noise is dominated by the power law, written as $S(f) = h_\alpha f^\alpha$, Allan variance is reduced to

$$\sigma_y^2(\tau) = 2h_\alpha(\pi\tau)^{-\alpha-1}\int_0^\infty \frac{\sin^4 x}{x^{2-\alpha}}\mathrm{d}x \propto \tau^\mu, \tag{3.68}$$

showing different behaviors for different indexes of α. The noises are classified as white frequency (FM) noise ($\alpha = 0$), flicker FM noise ($\alpha = -1$), random walk FM noise ($\alpha = -2$), white phase modulation (PM) noise ($\alpha = 2$), and flicker PM noise ($\alpha = 1$). For the white and flicker PM noise, integral (3.68) will be divergent. However, in practical cases an infinite band of frequency measurement is impossible; its low and high edges, f_L and f_H, should be considered, and the integral will result in a finite coefficient. Table 3.1 gives the power spectral density and Allan variance of different noise types; c_1 and c_2 in Table 3.1 are functions of f_H [24, 44].

Figure 3.17a gives a schematic illustration of the behaviors of $\sigma_y \sim \tau$ diagram, from which different noise types can be distinguished. It is seen from Table 3.1 that Allan variances of white and flicker PM noise have similar power law of τ.

Table 3.1 Allan variances for different power law noises [24, 44]

	White PM	Flicker PM	White FM	Flicker FM	Random walk FM
$S_y(f)$	$h_2 f^2$	$h_1 f$	h_0	$h_{-1}f^{-1}$	$h_{-2}f^{-2}$
$\sigma_y^2(\tau)$	$c_2 h_2/\tau^2$	$c_1 h_1/\tau^2$	$h_0/(2\tau)$	$2h_{-1}\ln 2$	$2\pi^2 h_{-2}\tau/3$

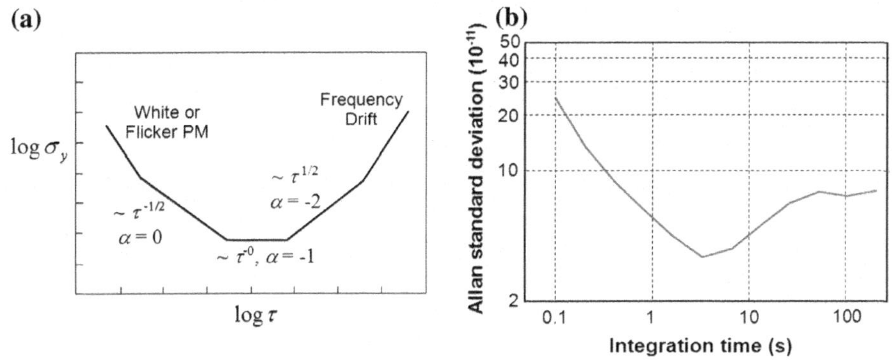

Fig. 3.17 **a** Allan variance versus integration time for different noise types. **b** Measured Allan deviation versus integration time

To distinguish the two noise types, the modified Allan variance was proposed [46], which is defined as:

$$\mathrm{mod}\,\sigma_y^2(\tau) = \frac{1}{2}\left\langle \left[\frac{1}{n}\sum_{i=1}^{n}(\bar{y}_{i+n} - \bar{y}_i)\right]^2 \right\rangle, \tag{3.69}$$

where $\tau = n\tau_0$, τ_0 is the time needed for the used frequency counter to read out; $\bar{y}_i$ is the average of n measurements. The modified Allan variance of white PM will display a behavior of $\propto \tau^3$, thus distinguishable from flicker PM with $\propto \tau^2$.

3.4.2 Measurement of Allan Variance

The frequency can be measured by a frequency counter, which is a commercially available apparatus. For the laser frequency measurement, it is necessary to use the method of correlation, i.e., mixing the laser under test with a reference laser with stable and precision frequency to convert the high frequency down to low frequency in the band suitable for the frequency counter.

Figure 3.18a is a schematic diagram; Fig. 3.18b shows an example of measurement setup, which uses a high stable laser as the reference frequency; the error signal provided by the PD is used to lock the single frequency laser (SFLD), and also to calculate Allan variance. Figure 3.17b gives a measured curve of Allan deviation versus integration time of a frequency-stabilized semiconductor laser, showing the instability of 5×10^{-11} in second scale [47].

In practice, the integration time τ may take different ranges for short-term, medium-term, or long-term applications, and depends on different concerns of frequency stability. The analysis of Allan variance can be carried on either in real time or by post-processing. For a certain frequency counter with readout time τ_0 and total data number M, the sample's number is $N = M\tau_0/\tau$. Therefore, the confidence interval of Allan variance varies with the integration time; the $\sigma_y \sim \tau$

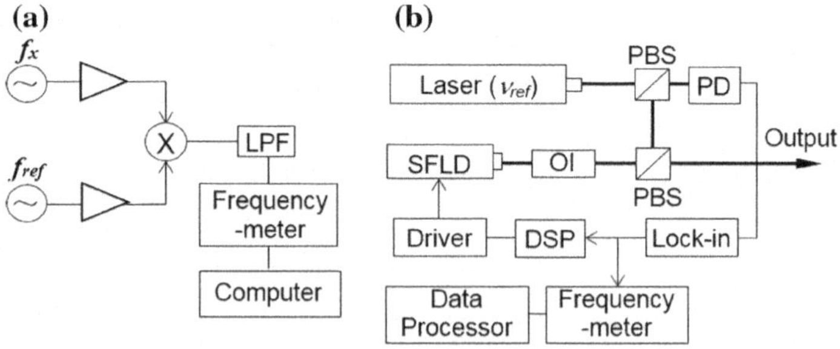

Fig. 3.18 Setup of Allan variance measurement. **a** Conceptual diagram; **b** With a reference laser. *OI* Optical Isolator; *FD* Frequency Discriminator; *PBS* Polarization Beam Split; *WP* Wave Plate

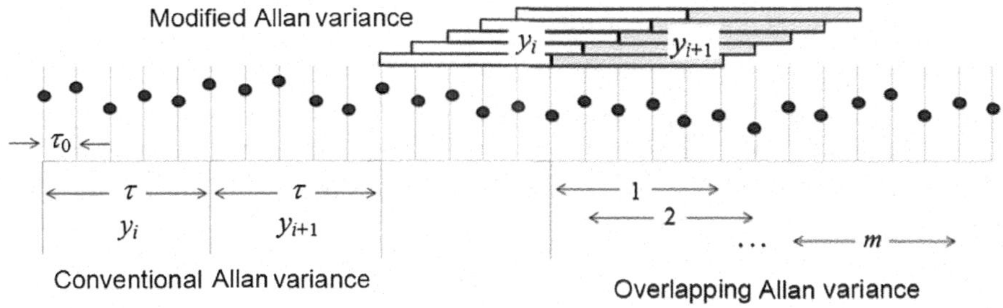

Fig. 3.19 Illustration of different statistic methods of Allan variance

curve may display dispersions for larger τ. It is shown by analysis of independent random variables that the probability of multiply measured Allan variance obeys χ^2 distribution, expressed as [48]

$$\chi_N^2(x) = x^{N-1}\mathrm{e}^{-Nx/\bar{x}}/[2^N(N-1)!], \tag{3.70}$$

where $\bar{x} = \int_0^\infty xp(x)\mathrm{d}x$ with $p(x)$ as the probability of single measurement. It is deduced that the variance of N-time measured variable decreases with the number of samples: $\sigma^2 = \bar{x}^2/N$.

For better confidence, with shorter error bars in $\sigma_y \sim \tau$ curve, the overlapping method used in sample's averaging can increase the sample's number, which is expressed as [49]:

$$\sigma_y^2(\tau) = \frac{1}{2m^2(N-2m+1)} \sum_{i=1}^{N-2m+1} \sum_{j=i}^{i+m-1} (y_{j+m} - y_j)^2. \tag{3.71}$$

Figure 3.19 gives an illustration of statistic method for conventional and modified Allan variance, and its overlapping calculation. Other modified methods, such as Hadamard variance, are also developed to conform with the complicated situations in practice, and to give results with higher accuracy and stronger capability of tracking noise origins [49, 50].

Many theoretical problems and technical issues have been studied for characterization of frequency stabilities in practice; the results are published in journals and monographs. Interested readers can find them in the literature.

3.5 Other Effects and Manifestations of Laser Noises

Apart from linewidth, frequency noise, and frequency stability described above, laser will show other appearances of imperfections. Three of them are introduced very briefly in this section: chaos, jitter and multimode noise, and those related to internal defects.

3.5.1 Chaos of Semiconductor Lasers

Chaos was discovered first in investigations of weather and climate [51]; it was also observed in many fields, including physics. Chaos looks random, but actually is deterministic. Chaos is a nonlinear process, described by nonlinear equations. The evolution of concerned parameters is determined by definite equations, different from the random noise described by Langevin forces. Chaos features that primary conditions with a minor difference will result in outputs with very different characteristics; and a minor disturbance will lead to big instabilities. Chaos displays often behaviors of bifurcation and strange attractors, and so on [52].

The origin of chaos in lasers can be understood and described as follows. According to Maxwell electromagnetic wave theory, the optical field $\boldsymbol{E}(z,t)$ is induced by the electric polarization $\boldsymbol{P}(z,t)$; the latter is related to transitions of active atoms, which is determined by the population inversion $W(z,t)$. In the semiclassical theory of laser physics, the three quantities are coupled with each other in the rate equations [5], written as:

$$\begin{aligned}&\tfrac{\partial \boldsymbol{E}}{\partial z}+\tfrac{n}{c}\tfrac{\partial \boldsymbol{E}}{\partial t}=j\tfrac{k}{\varepsilon}\boldsymbol{P}-\tfrac{n}{2c\tau_{\mathrm{p}}}\boldsymbol{E}\\&\tfrac{\partial \boldsymbol{P}}{\partial t}=j(\omega-\omega_0)\boldsymbol{P}+j\tfrac{\mu^2}{2\hbar}W\boldsymbol{E}-\tfrac{\boldsymbol{P}}{T_2}\\&\tfrac{\partial W}{\partial t}=\tfrac{j}{\hbar}(\boldsymbol{P}\cdot\boldsymbol{E}^*-\boldsymbol{E}\cdot\boldsymbol{P}^*)-\tfrac{W-W_0}{T_1},\end{aligned} \tag{3.72}$$

where $\mu=\langle e\boldsymbol{r}\rangle$ is the electric dipole, $\omega_0=(E_2-E_1)/\hbar$ is the transition frequency, W_0 is the biased population inversion, τ_{p} is the photon's lifetime, T_2 is the relaxation time of electric polarization, termed the transverse relaxation time, and T_1 is the lifetime of inversed population, termed the longitudinal relaxation time. The rate equations contain nonlinear terms of $\boldsymbol{E}\cdot\boldsymbol{P}$ and $W\boldsymbol{E}$; they can be reduced to equations similar to the famous Lorenz equation in the chaos theory, typically expressed as

$$\begin{aligned}\mathrm{d}x/\mathrm{d}t&=\sigma(y-x)\\\mathrm{d}y/\mathrm{d}t&=rx-xz-y\\\mathrm{d}z/\mathrm{d}t&=xy-bz.\end{aligned} \tag{3.73}$$

It is indicated by theoretical study that the stability of laser operation depends on relative values of the three time constants. Correspondingly the lasers are classified into three classes. When the time constants of the relaxations are of the same order, $\tau_{\mathrm{p}}\sim T_1\sim T_2$, the laser is classified to Class C, which works usually in chaos state, showing unstable oscillation like irregular pulsations. When the transverse relaxation time is much smaller than the photon lifetime and the population lifetime, $T_2\ll\tau_{\mathrm{p}},T_1$, the laser is classified to Class B, which works usually stably, but is susceptible to external disturbances and goes into chaos under certain conditions. The laser with $T_1,T_2\ll\tau_{\mathrm{p}}$ is called Class A laser, which works stably.

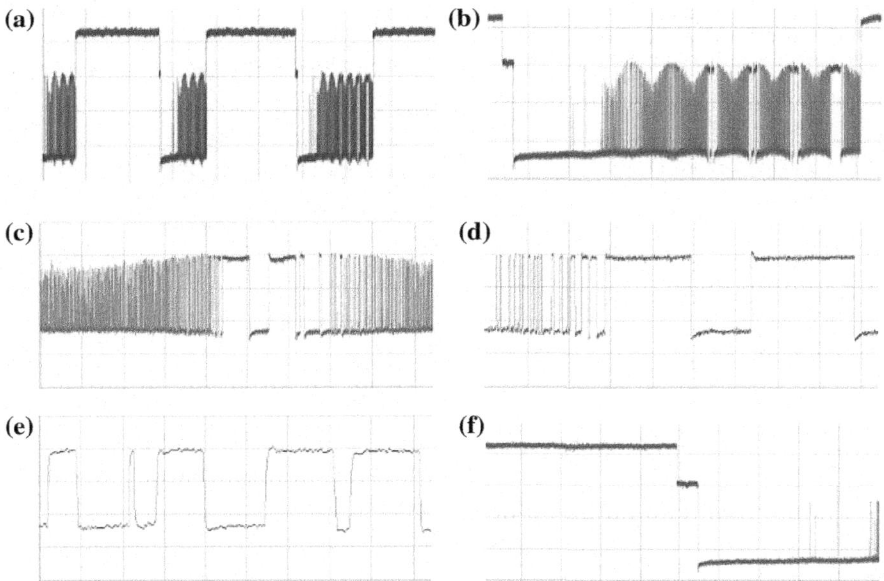

Fig. 3.20 Self-pulsation waveforms with time scales of **a** 100 ms/div.; **b** 20 ms/div.; **c** 2 ms/div.; **d** 500 μs/div.; **e** 50 μs/div.; **f** 10 ms/div

The semiconductor laser is classified to Class B, because its transverse relaxation time is in the order of 10^{-13} s, determined by the carrier relaxation inside the bands [5], much smaller than the photon lifetime and carrier recombination lifetime. The mutual reaction between photons and carriers in semiconductor lasers is the strongest and the most direct among various lasers. Because of the high gain coefficient and low reflectivity of cavity mirror, the output optical wave is easier to reenter the LD cavity, resulting in instability higher than other lasers. The direct electric current pump makes the LD susceptible to the noise of driver and electromagnetic interferences. The chaos of laser operation state displays not only in instability of its intensity, but also in its longitudinal mode characteristics. Detailed theoretical analyses and experimental studies can be read in journals and books [5, 53–56].

As discussed in Sect. 2.4, the output power of LD shows a linear L-I curve. However, kink of L-I curve is often observed, which is commonly attributed to mode competitions. Kinks can be described by rate equations without noise factors, as will be discussed in Chap. 5. At the kinks, self-pulsation is usually taking place, and chaos occurs often. Figure 3.20 shows the pulsation waveforms of an external cavity semiconductor laser recorded at kink points of L-I curve when the pumping current injected to LD chip is kept constant [57]. Note that the scales of abscissa are different in the figure, showing bifurcation of self-pulsation period, which is one of the typical behaviors of chaos.

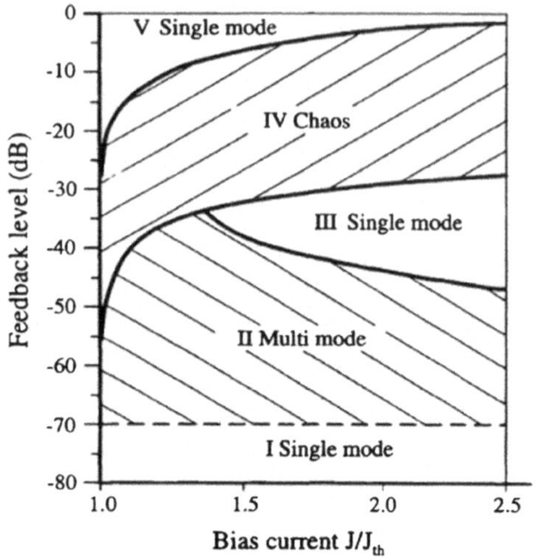

Fig. 3.21 Regions of laser stability states with different external feedback levels. Reprinted from [59] with permission

The stability is one of the most important performances of semiconductor lasers, especially in the optical communication applications. In optical fiber systems, forward and backward scatterings occur frequently at the connections between optical components and fiber sections. If such returned waves enter LD cavity, the laser operation will be affected greatly.

The instability caused by such an optical feedback has been studied in details [58–60]. It is indicated by experimental studies and theoretical analyses that the stabilities of LD operations show different appearances in five regions according to feedback strength and bias level, as shown in Fig. 3.21 [59], where the feedback is measured as the fraction of output power, and the bias is normalized to the threshold. Due to its randomness, the feedback may generate different influences. When the optical feedback is lower than some critical level, as marked by (I) in the figure, the laser works in a single longitudinal mode. As the feedback increases it will work in multiple longitudinal modes, and the noise caused by mode competition will occur in region II. If the bias increases to a certain degree for the case of same feedback, the main mode may dominate; the laser will go back to the single mode operation, as marked by (III). As the feedback increases further, the laser operation enters into chaos state in region (IV), where great instabilities occur with big intensity fluctuations and linewidth broadening. It is called the state of coherence collapse. The scope of chaos region is dependent on the bias level as shown in the figure. When the feedback increases to a higher level, the laser state will go back to a stable single mode state in region (V), where the optical feedback plays a role of noise suppression and linewidth reduction. Lasers with feedback from a specially designed external cavity are classified to this case, which will be discussed in Chap. 5.

The above-mentioned characteristics are only a qualitative description. Complicated phenomena with big differences may be observed for lasers with different structures and under different conditions. The laser chaos is a topic involving deep physical mechanisms and mathematical connotation, beyond the scope of this book. Its detailed analyses and discussions may be read in monograph [5, 52] and references [53–56].

3.5.2 Jitter of Laser Pulses, Mode Partition Noise, and Speckles

Noises of semiconductor lasers display in different appearances. Among them jitter of laser pulses and mode partition noise are widely concerned for some important applications, especially for optical communication and optical storage.

(1) **Turn-on Jitter**

It is evident that the fluctuation of pump current pulse front $\delta\, t$ and the fluctuations of its amplitude $\delta\, I$ will cause the turn-on jitter of laser pulses in time domain. From the delay time given in Sect. 2.6, $t_d \approx \tau_{\rm sp} \ln[(I - I_b)/(I - I_{\rm th})]$, the turn-on jitter can be estimated as

$$\delta t_d \approx \left[\delta t^2 + \frac{\tau_{\rm sp}^2 (I_{\rm th} - I_b)^2}{(I - I_b)^2 (I - I_{\rm th})^2} \delta I^2 \right]^{1/2}. \tag{3.74}$$

It is seen that the second term declines as the pulse current amplitude increases. Other factors of noise, such as temperature fluctuations and vibrations, may also induce the jitter, because these factors will cause noise of carrier concentration and field amplitude by their mutual coupling, as described by the rate Eqs. (3.26–3.27c).

(2) **Mode Partition Noise**

Mode partition noises are attributed to random launching of different longitudinal modes. Their interferences will result in random fluctuations of output power due to the random phase difference between the modes. References [6, 61, 62] gave detailed descriptions and analyses. No mode partition noise would occur for a pure single longitudinal mode laser. It is noticeable that the amplitude of mode partition noise declines with increasing of the mode number due to the averaging effect over the modes. When only two or a few modes exist in the laser beam, the amplitude of mode partition noise will be high. Mode partition noise is harmful to the performance of laser application system. For example, a high bit error rate will be induced in writing and reading data of optical storage disks. Lasers with many longitudinal modes are preferable in such applications.

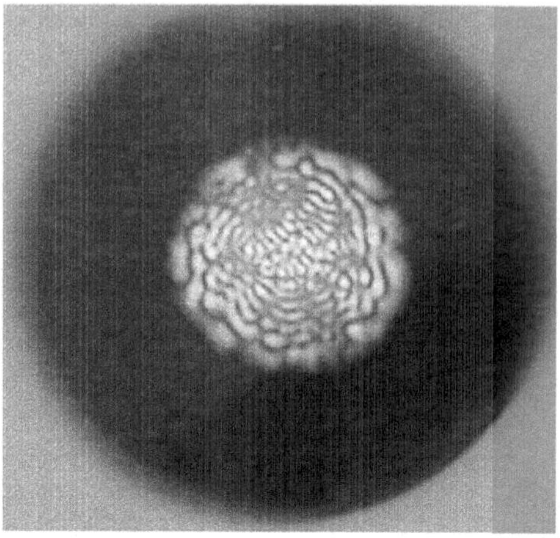

Fig. 3.22 Speckle pattern on facet of multimode fiber

(3) **Speckles**

Another effect of the multi-longitudinal mode laser is the specklegram, which is attributed to interference of longitudinal modes in space, and interference of transverse modes in waveguide [48, 63]. Figure 3.22 displays a photo of speckle pattern observed in a multimode fiber [64]. The pattern is formed by constructive and destructive interferences due to different propagation phase shifts of the modes. The speckle will induce intensity noise in applications of laser far-field; on the other hand, it can be used as a sensor for detecting variations of propagation phases. Detailed theory of laser specklegram and its applications can be found in references and books [63–65].

3.5.3 Noise and Instability Related to Internal Defects

Material defect in the semiconductor is one of the main factors affecting laser's performances, such as its threshold and output power, and also inducing its noises. The defects may play a role of recombination centers, causing non-radiative transitions and reducing the internal quantum efficiency. The defects may absorb photons, increasing the loss coefficient in the laser cavity. The effect of defects is also reflected in the characteristics of P-N diode, such as the turn-on voltage, the derivative conductivity, junction capacity, and its electrical noises. Defects will generate during long-term operations as a symbol of LD degradation. Therefore, the measurement and inspection of LD noises and I–V characteristics can be utilized to predict its reliability [66–68].

The self-sustained pulsation is a typical instability observed in semiconductor lasers; that is, even under a DC current injected, the laser can oscillate without stop [69]. It originates from some kind of defects and imperfections, including existence of regions with nonuniform and low pump levels in the cavity. For example, the carrier's recombination by surface states at the cleaved facets will generate absorption regions in longitudinal direction of the cavity. The injected current may expand laterally, generating lateral absorption. The defects induced during material growth and during long-term operations will also cause such a phenomenon. It is indicated theoretically that such defects and imperfections behave somewhat like saturated absorbers [70, 71], i.e., its absorption will decrease with increasing of optical intensity, and the effect is reversible. The oscillation can be described by rate equations modified with such absorbers involved; the oscillation frequency of self-sustained pulsation is basically equal to the frequency of relaxation oscillation, as described in Sect. 2.6.

References

1. Pertermann K (1991) Laser diode modulation and noise. Kluwer Academic Publishers, Dordrecht
2. Ohtsu M (1992) Highly coherent semiconductor lasers. Artech House
3. Yariv A (1997) Optical electronics in modern communications. Oxford University Press, Oxford
4. Okoshi T, Kikuchi K (1988) Choherent optical fiber communications. KTK Scientific Publisher, Tokyo
5. Ohtsubo J (2013) Semiconductor lasers—stability, Instability and Chaos. 3rd edn. Springer, Berlin
6. Petermann K, Weidel E (1981) Semiconductor laser noise in an interferometer system. IEEE J Quantum Electron 17(7):1251–1256
7. Okoshi T, Kikuchi K, Nakayama A (1980) Novel method for high-resolution measurement of laser output spectrum. Electron Lett 16(16):630–631
8. Spano P, Piazzolla S, Tamburrini M (1984) Theory of noise in semiconductor lasers in the presence of optical feedback. IEEE J Quantum Electron 20:350–357
9. Gallion PB, Debarge G (1984) Quantum phase noise and field correlation in single frequency semiconductor laser systems. IEEE J Quantum Electron 20(4):343–349
10. Mercer LB (1991) 1/f frequency noise effects on self-heterodyne linewidth measurements. J Lightwave Technol 9(4):485–493
11. Stéphan GM, Tam TT, Blin S et al (2005) Laser line shape and spectral density of frequency noise. Phys Rev A 71:043809(1–9)
12. Kikuchi K (1989) Effect of l/*f*-type FM noise on semiconductor-laser linewidth residual in high-power limit. J Quantum Electron 25(4):684–688
13. Schawlow AL, Townes CH (1958) Infrared and optical masers. Phys Rev 112:1940–1949
14. Henry CH (1982) Theory of the linewidth of semiconductor lasers. IEEE J Quantum Electron 18(2):259–264
15. Osinski M, Buus J (1987) Linewidth broadening factor in semiconductor lasers-An overview. IEEE J Quantum Electron 23(1):9–29
16. Henry CH (1986) Phase noise in semiconductor lasers. J Lightwave Technol 4(3):298–311
17. Henry CH (1983) Theory of the phase noise and power spectrum of a single mode injection laser. IEEE J Quantum Electron 19(9):1391–1397
18. Daino B, Spano P, Tamburrini M et al (1983) Phase noise and spectral line shape in semiconductor lasers. IEEE J Quantum Electron 19(3):266–270
19. Born M, Wolf E (1999) Principles of optics, 7th edn. Cambridge University Press, Cambridge

20. Camatel S, Member Ferrero V (2008) Narrow linewidth CW laser phase noise characterization methods for coherent transmission system applications. J Lightwave Technol 26(17):3048–3055
21. Richter LE, Memdelberg HI, Kruger MS et al (1986) Linewidth determination from self-heterodyne measurements with subcoherence delay times. IEEE J Quantum Electron 22(11):2070–2074
22. Ludvigsen H, Tossavainen M, Kaivola M (1998) Laser linewidth measurements using self-homodyne detection with short delay. Optics Commun 155:180–186
23. Tsuchida H (2011) Laser frequency modulation noise measurement by recirculating delayed self-heterodyne method. Opt Lett 36(5):681–683
24. Rutman J, Walls FL (1991) Characterization of frequency stability in precision frequency sources. Proc IEEE 79(6):952–960
25. Kikuchi K (2012) Characterization of semiconductor-laser phase noise and estimation of bit-error rate performance with low-speed offline digital coherent receivers. Opt Express 20(5):5291–5302
26. Tourrenc JP, Signoret P, Myara M (2005) Low-frequency FM-noise-induced lineshape: a theoretical and experimental approach. IEEE J Quantum Electron 41(4):549–553
27. Zhou W, Chong KM, Guo H (2008) Linewidth measurement of Littrow structure semiconductor laser with improved methods. Phys Lett A 372:4327–4332
28. Huynh TN, Nguyen L, Barry LP (2013) Phase noise characterization of SGDBR lasers using phase modulation detection method with delayed self-heterodyne measurements. J Lightwave Technol 31(8):1300–1308
29. Ye Q, Pan Z, Wang Z et al (2015) Novel slow-light reflector composed of a fiber ring resonator and low-reflectivity fiber Bragg grating. J Lightwave Technol 33(14):3016–3022
30. Kikuchi K, T. Okoshi T (1985) Measurement of FM noise, AM noise, and field spectra of 1.3 pm InGaAsP DFB lasers and determination of the linewidth enhancement factor. IEEE J Quantum Electron 21(11):1814–1818
31. Pan Z, Ye Q, Cai H et al (2014) Fiber ring with long delay used as a cavity mirror for narrowing fiber laser linewidth. IEEE Photonics Technol Lett 26(16):1621–1624
32. Camatel S, Ferrero V (2008) Phase noise power spectral density measurement of narrow linewidth CW lasers using an optical phase-locked loop. IEEE Photonics Technol Lett 18(23):2529–2531
33. Huynh TN, Nguyen L, Barry LP (2012) Delayed self-heterodyne phase noise measurements with coherent phase modulation detection. IEEE Photonics Technol Lett 24(4):249–251
34. Takushima Y, Choi HY, Chung YC (2009) Measurement of differential phasor diagram of multilevel DPSK signals by using an adjustment-free delay interferometer composed of a 3 × 3 optical coupler. J Lightwave Technol 27(6):718–730
35. Dan X, Yang F, Chen D et al (2015) Laser phase and frequency noise measurement by Michelson interferometer composed of a 3 × 3 optical fiber coupler. Opt Express 23(17):22386–22393
36. Fang Z, Chin K, Qu R et al (2012) Fundamentals of Optical fiber Sensors. Wiley, Hoboken
37. Cliché JF, Allard M, Têtu M (2006) High-power and ultra-narrow DFB laser: the effect of linewidth reduction systems on coherence length and interferometer noise. Proc SPIE 6216: 62160C(1–11)
38. Butler T, Slepneva S, O'Shaughnessy B et al (2015) Single shot, time-resolved measurement of the coherence properties of OCT swept source lasers. Opt Lett 40(10):2277–2280
39. Domenico GD, Schilt S, Thomann P (2010) Simple approach to the relation between laser frequency noise and laser line shape. Appl Opt 49(25):4801–4807
40. Van der Ziel A (1986) Noise in solid state devices and circuits. Wiley, London
41. Baili G, Alouini M, Moronvalle C et al (2006) Broad-bandwidth shot-noise-limited class-A operation of a monomode semiconductor fiber-based ring laser. Opt Lett 31(1):62–64
42. Lauprêtre T, Proux C, Ghosh R et al (2011) Photon lifetime in a cavity containing a slow-light medium. Opt Lett 36(9):1551–1553
43. Amili AE, Alouini M (2015) Noise reduction in solid-state lasers using a SHG-based buffer reservoir. Opt Lett 40(7):1149–1152

44. Riley WJ (2003) Techniques for frequency stability analysis. In: IEEE international frequency control symposium, Tampa, Florid
45. Allan DW (1966) Statistics of atomic frequency standards. Proc IEEE 54(2):221–230
46. Vernotte F (2011) Variance measurements. In: Joint conference of the IEEE international frequency control symposium & European frequency & time forum. San Francisco
47. Wei F, Chen D, Xin G et al (2013) A compact and rugged tunable external cavity diode laser with Littman-Metcalf configuration. Chin J Lasers 40(11): 1102012(1–7) (in Chinese)
48. Dainty JC (1975) Laser speckle and related phenomena. Springer, Berlin
49. Dawkins ST, McFerran JJ, Luiten AN (2007) Considerations on the measurement of the stability of oscillators with frequency counters. IEEE Trans Ultrason Ferro-electronics Freq Control 54(5):918–925
50. Lesage P, Ayi T (1984) Characterization of frequency stability: analysis of the modified Allan variance and properties of its estimate. IEEE Trans Instrum Meas IM-33(4):332–336
51. Lorenz EN (1963) Deterministic nonperiodic flow. J Atmos Sci 20:130–148
52. Lorenz EN (1993) The essence of chaos. University of Washington Press
53. Sivaprakasam S, Shore KA (1999) Demonstration of optical synchronization of chaotic external-cavity laser diodes. Opt Lett 24(7):466–468
54. Heil T, Fischer I, Elsäer W et al (2003) Delay dynamics of semiconductor lasers with short external cavities: bifurcation scenarios and mechanisms. Phys Rev E 67:066214 (1–11)
55. Krauskopf B, Green K, H. Erzgräber H et al (2004) Dynamics and bifurcations of a semiconductor laser with short external cavity. Proc SPIE 5452:82–92
56. Hohl A, Gavrielides A (1999) Bifurcation cascade in a semiconductor laser subject to optical feedback. Phys Rev Lett 82(6):1149–1151
57. Chen D, Fang Z, Cai H et al (2008) Instabilities in a grating feedback external cavity semiconductor laser. Opt Express 16(21):17014–17020
58. Tkach RW, Chraplyvy AR (1986) Regimes of feedback effects in 1.5-μm distributed feedback lasers. J Lightwave Technol 4(11):1655–1661
59. Mørk J, Tromborg B, Mark J et al (1992) Instabilities in a laser diode with strong optical feedback. Proc SPIE 1837:90–104
60. Mørk J, Tromborg B, Mark J (1992) Chaos in semiconductor lasers with optical feedback: theory and experiment. IEEE J Quantum Electron 28(1):93–108
61. Ohtsu M, Teramachi Y (1989) Analyses of mode partition and mode hopping in semiconductor lasers. J Quantum Electron 25(1):31–38
62. Ahmed M, Yamada M (2002) Influence of instantaneous mode competition on the dynamics of semiconductor lasers. IEEE J Quantum Electron 38(6):682–693
63. Li J, Cai H, Geng J et al (2007) Specklegram in a multiple-mode fiber and its dependence on longitudinal modes of the laser source. Appl Opt 46(17):3572–3578
64. Wang Y, Cai H, Qu R et al (2008) Specklegram in a grapefruit fiber and its response to external mechanical disturbance in a single-multiple-single mode fiber structure. Appl Opt 47(20):3543–3548
65. Yin S, Ruffin PB, Yu FTS (2008) Fiber optic sensors, 2nd edn. CRC Press, Taylor & Francis Group
66. Fukuda M, Hirono T, Kurosaki T et al (1993) 1/*f* noise behavior in semiconductor laser degradation. IEEE Photonics Technol Lett 5(10):1165–1167
67. Pralgauskaite S, Matukas J, Palenskis V et al (2003) Low-frequency noise and quality prediction of MQW buried-heterostructure DFB lasers. Proc SPIE 5123:85–93
68. Hu G, Shi J, Shi Y et al (2003) An improved method and experimental results of noise used as reliability estimation for semiconductor lasers. Opt Laser Technol 35:481–483
69. Van Der Ziel JP (1981) Self-focusing effects in pulsating $Al_xGa_{1-x}As$ double-heterostructure lasers. IEEE J Quantum Electron 17(1):60–68
70. Ueno M, Lang R (1985) Conditions for self-sustained pulsation and bistability in semiconductor lasers. J Appl Phys 58(4):1689–1692
71. Yamada M (1993) A theoretical analysis of self-sustained pulsation phenomena in narrow-stripe semiconductor lasers. J Quantum Electron 29(5):1330–1336

Chapter 4
Monolithically Integrated Semiconductor Lasers

4.1 Distributed Feedback Semiconductor Laser

It is hard for ordinary F-P cavity LDs to work in a single longitudinal mode because its gain spectrum is broad and flat, and no mode selection is provided by the F-P cavity. Even if a certain longitudinal mode has the dominant superiority over others under a DC bias, possibly for the index guided LD, such superiority is not stable, and the side mode suppression ratio (SMSR) is not high enough. Moreover, the superiority will be diminished in high frequency modulations. In many practical applications, especially the optical communications, lasers must work in the single longitudinal mode dynamically under high frequency modulations.

Several schemes of LD structures with mode selections had been tried based on F-P cavity, but were not satisfactory for the performances, and for the possibility of mass production. As well known, the optical grating has strong wavelength selectivity. By combining the grating and the gain medium active region the idea of DFB was proposed earlier and demonstrated in dye laser first [1, 2]. Great efforts were paid to develop the DFB semiconductor laser, in which a grating is fabricated inside or closely near the active region, and parallel to the direction of laser beam propagation. After a series of difficulties in its fabrication was overcome, the DFB-LD was successfully developed, showing excellent performances and commercial availability [3]. Its cross section is shown in Fig. 4.1 schematically.

4.1.1 Coupled Mode Theory of DFB Laser

The theory and characteristics of DFB lasers have been expounded in numbers of papers and monographs [2, 4–8]. This section gives a brief introduction to its basic principle.

Z. Fang et al., *Single Frequency Semiconductor Lasers*, Optical and Fiber Communications Reports 9, DOI 10.1007/978-981-10-5257-6_4

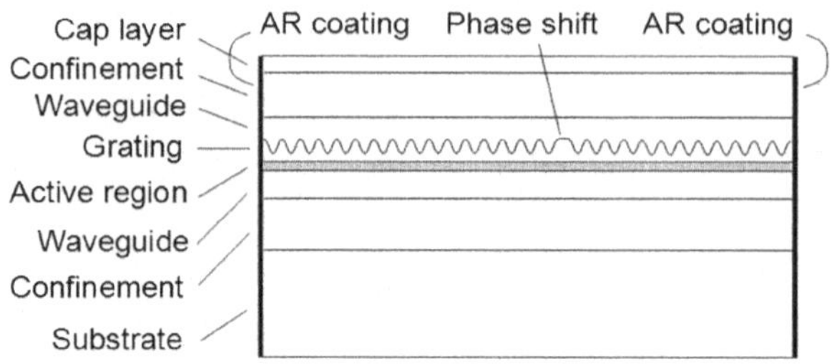

Fig. 4.1 Schematic diagram of DFB-LD structure

(1) **Threshold Condition of DFB-LD with Uniform Grating.**

The coupled mode theory (CMT) is the DFB laser [7]. The index of active region of DFB laser is expressed as

$$n(z) = n_0 + \sum_{m=1}^{\infty} n_m \cos(2\pi m z/\Lambda) \approx n_0 + n_1 \cos(2\beta_B z + \varphi), \tag{4.1}$$

where z is the position along the cavity, φ is the phase of origin point respect to the grating period. According to CMT, the second term of (4.1) is taken as a perturbation of the rate equation. The last expression of (4.1) is for the first order Fourier expansion, $\beta_B = \pi/\Lambda$ is Bragg wavevector of the grating. The index is composed of real and imaginary parts: $n_0 = n_{0r} + jn_{0i}$ with $n_{0i} = \alpha/2k_0$ for the loss, or $n_{0i} = -g/2k_0$ for the gain. The index modulation and gain modulation in the cavity are generally supposed with the same period, with $n_1 = n_{1r} + jn_{1i}$. Under the distributed feedback, the optical field is composed of forward and backward waves:

$$\boldsymbol{E}(\boldsymbol{r}) = \boldsymbol{F}(x, y)[A(z)\exp(j\beta_B z) + B(z)\exp(-j\beta_B z)]. \tag{4.2}$$

The transverse field $\boldsymbol{F}(x, y)$ obeys equation of $\nabla_{xy}^2 \boldsymbol{F} + (n_0^2 k_0^2 - \beta^2)\boldsymbol{F} = 0$, as eigen functions without perturbations, where $\beta = 2\pi n/\lambda$. Substituting (4.2) into Helmholtz equation $\nabla^2 \boldsymbol{E} + n^2(z) k_0^2 \boldsymbol{E} = 0$, with the squares of index perturbation and high order oscillations of $\exp(\pm j3\beta_B z)$ omitted, and with $\varphi = 0$ for simplicity, the coupled mode equations are reduced to:

$$\begin{aligned} A' - (g + j\delta)A - j\kappa B = 0 \\ B' + (g + j\delta)B + j\kappa A = 0, \end{aligned} \tag{4.3}$$

where the second order derivatives of fields are omitted as a model of slowly varying amplitude; $g = -n_0 n_{0i} k_0^2/\beta \approx -n_{0i} k_0$ is the net gain coefficient of field. For simplicity, g is defined as field gain here, being half of power gain coefficient; $\kappa = n_0 n_1 k_0^2/2\beta \approx n_1 k_0/2$ is the coupling coefficient; and $\delta = (\beta^2 - \beta_B^2)/2\beta_B \approx \beta - \beta_B$. Combining the two equations of (4.3), the solutions are expressed as:

$$A(z) = a_1 \exp(\gamma z) + a_2 \exp(-\gamma z)$$
$$B(z) = b_1 \exp(\gamma z) + b_2 \exp(-\gamma z), \tag{4.4}$$

where $\gamma = \sqrt{(g + j\delta)^2 + \kappa^2}$. The relations of coefficients are deduced from the equations: $j(g + j\delta - \gamma)a_1 = \kappa b_1$ and $j(g + j\delta + \gamma)a_2 = \kappa b_2$. We consider here a pure DFB laser without reflections at the two cleaved facets, i.e., no fields input at $z = \pm l/2$. The boundary conditions are written as $A(-l/2) = 0$ and $B(l/2) = 0$ with the origin of coordinate set at the middle point of grating; therefore $a_1 + a_2 e^{\gamma l} = 0$, $b_1 e^{\gamma l} + b_2 = 0$. The eigen value equation is then obtained, and the threshold condition, including the gain condition and the phase condition, is given as

$$\frac{g + j\delta - \gamma}{g + j\delta + \gamma} = e^{-2\gamma l}. \tag{4.5}$$

with $\gamma^2 = (g + j\delta)^2 + \kappa^2$, (4.5) can be rewritten as different forms of complex transcend equations: $g + j\delta \pm \gamma = j\kappa e^{\mp \gamma l}$; or $g + j\delta = j\kappa \cosh(\gamma l)$ and $\gamma = j\kappa \sinh(\gamma l)$.

(2) **Stop Band**

The grating in DFB semiconductor lasers is typically classified into two kinds: index grating and gain grating, corresponding to $n_{1r} \neq 0$, $n_{1i} \approx 0$ and $n_{1i} \neq 0$, $n_{1r} \approx 0$, respectively. Their characteristics are quite different. For index grating DFB, κ is real, and $\gamma = \sqrt{g^2 + \kappa^2}$ at $\delta = 0$. The threshold condition (4.5) is reduced to

$$\frac{g - \sqrt{g^2 + \kappa^2}}{g + \sqrt{g^2 + \kappa^2}} = \exp[-2\sqrt{g^2 + \kappa^2} l]. \tag{4.6}$$

This equation has no solution of real g. It means that the DFB laser with index coupled grating cannot work at Bragg wavelength $\lambda_B = n\Lambda/2$; lasing is stopped there.

The stop of lasing originates from the Bragg diffraction effect of one dimensional index grating. Reference [5] gives a qualitative explanation of the stop band, as shown in Fig. 4.2. The grating is divided into two sections arbitrary, where their starting phases respect to the grating are roughly half period different from each other, i.e. $\Delta\varphi = \pi/2$. If the reflection of right section to the left is r_1, the reflection of left section to the right should be $r_2 \exp(j\pi/2)$, corresponding to a negative product of the two intensity reflections, $R_1 R_2 = -r_1^2 r_2^2$, which goes against the lasing condition.

The lasing modes of index grating DFB laser have to be shifted from Bragg wavelength to its two sides. In the case of weak grating for qualitative discussions [5], parameter γ can be expanded as $\gamma \approx g + j\delta + \kappa^2/2(g + j\delta)$. Substitute it into (4.5), the lasing condition is approximated as

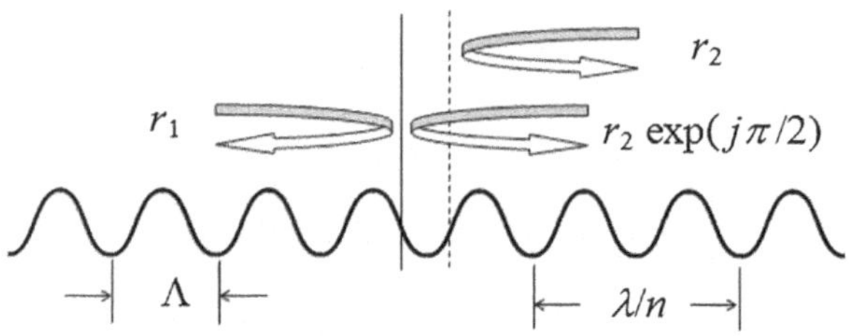

Fig. 4.2 Qualitative explanation of stop band

$$\frac{j\kappa}{2(g+j\delta)} \approx \exp[-(g+j\delta)l + \frac{\kappa^2 l/2}{g+j\delta}]. \tag{4.7}$$

The position of longitudinal modes and the gain condition is then obtained as:

$$\delta_m l \approx \tan^{-1}(\delta_m/g_m) + (m-1/2)\pi, \tag{4.8a}$$

$$\kappa^2 \exp(2g_m l) \approx 4(g_m^2 + \delta_m^2). \tag{4.8b}$$

It is seen from (4.8a) that δ_m cannot equal to zero, meaning the laser cannot work at Bragg wavelength. Two nearest modes is given by $m = 0$ and $m = 1$ with mode spacing of $\Delta\beta \approx \pi/l$; the band between them is called the stop band. It is seen also that the higher order modes have much higher thresholds and cannot lase usually, demonstrating the mode selection of grating.

For the case of gain coupled grating, $\kappa = jn_{1i}k_0/2$ is an imaginary. At Bragg wavelength $\gamma = \sqrt{g^2 - |\kappa|^2}$; the lasing condition is

$$\frac{g - \sqrt{g^2 - |\kappa|^2}}{g + \sqrt{g^2 - |\kappa|^2}} = \exp[-2\sqrt{g^2 - |\kappa|^2} l]. \tag{4.9}$$

This equation has real solutions, indicating DFB-LD with gain grating can work at Bragg wavelength with the lowest threshold. The fabrication technologies of index and gain gratings will be introduced in Sect. 4.1.2.

(3) **Phase-shifted DFB Laser**

The situation of working in two longitudinal modes at sides of stop band of the index coupled DFB LD is not welcome. The competition between the two modes will result in high mode partition noises. Even if one of them works dominantly over the other due to some asymmetric factors in fabrication or in working conditions, the production yield for a certain wavelength will be reduced to half averagely.

The qualitative explanation of stop band above implies that if a proper phase shift is made inside the grating, as depicted by the dashed line in Fig. 4.2, to get a positive product of intensity reflections, lasing at Bragg wavelength can be realized. It was just the trigger of developing the phase-shift DFB, i.e. making an index coupled grating with a $\pi/2$ phase-shift inside, as shown in Fig. 4.3 schematically, where the gray line is for the grating without phase shift for comparison.

The characteristics of phase-shifted index coupled grating are analyzed as follows. The left and right sections are uniform gratings with different starting phases. Similar to (4.1), their index distributions are written as

$$\begin{aligned} n(z) &= n_0 + n_1 \cos(2\beta_B z - \varphi) \cdots\cdots [0 \le z < l_2] \\ n(z) &= n_0 + n_1 \cos(2\beta_B z + \varphi) \cdots\cdots [-l_1 < z \le 0]. \end{aligned} \tag{4.10}$$

The optical fields are expressed as

$$\begin{aligned} \boldsymbol{E}_1(\boldsymbol{r}) &= \boldsymbol{F}(x,y)[A(z)e^{j\beta_B z} + B(z)e^{-j\beta_B z}] \\ \boldsymbol{E}_2(\boldsymbol{r}) &= \boldsymbol{F}(x,y)[C(z)e^{j\beta_B z} + D(z)e^{-j\beta_B z}]. \end{aligned} \tag{4.11}$$

Similar to the deduction of (4.3), two groups of coupled mode equations are obtained:

$$\begin{aligned} A' - (g + j\delta)A - j\kappa e^{-j\varphi} B &= 0 \\ B' + (g + j\delta)B + j\kappa e^{j\varphi} A &= 0, \end{aligned} \tag{4.12a}$$

$$\begin{aligned} C' - (g + j\delta)C - j\kappa e^{j\varphi} D &= 0 \\ D' + (g + j\delta)D + j\kappa e^{-j\varphi} C &= 0. \end{aligned} \tag{4.12b}$$

Their solutions are written in forms of

$$\begin{aligned} A(z) &= a_1 e^{\gamma z} + a_2 e^{-\gamma z} \quad & C(z) &= c_1 e^{\gamma z} + c_2 e^{-\gamma z} \\ B(z) &= b_1 e^{\gamma z} + b_2 e^{-\gamma z}, \quad & D(z) &= d_1 e^{\gamma z} + d_2 e^{-\gamma z}, \end{aligned}$$

with coefficient relations of

$$\begin{aligned} (\gamma - g)a_1 &= j\kappa e^{-j\varphi} b_1 \quad & (\gamma - g)c_1 &= j\kappa e^{j\varphi} d_1 \\ (\gamma + g)a_2 &= -j\kappa e^{-j\varphi} b_2, \quad & (\gamma + g)c_2 &= -j\kappa e^{j\varphi} d_2 \end{aligned}$$

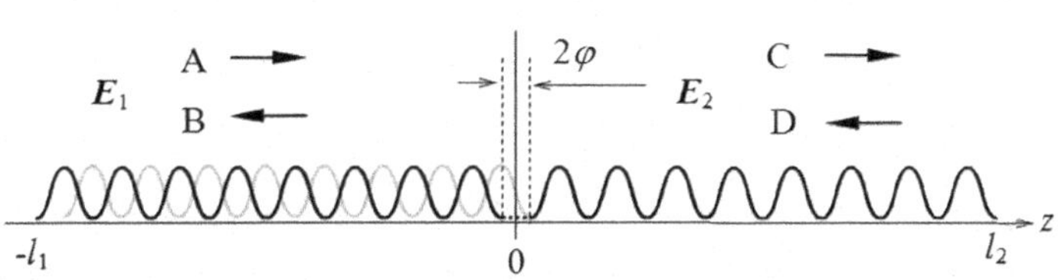

Fig. 4.3 Model of phase-shifted grating

For an ideal DFB laser with cleaved facets anti-reflection coated, the boundary conditions are $a_1 e^{-\gamma l_1} + a_2 e^{\gamma l_1} = 0$, $d_1 e^{\gamma l_2} + d_2 e^{-\gamma l_2} = 0$, $c_1 + c_2 = a_1 + a_2$, and $d_1 + d_2 = b_1 + b_2$, where the optical fields are continuous at the point of phase shift, z = 0. The eigen value equation is deduced to be [9, 10]

$$\rho^2 e^{2\gamma l} + \rho[e^{2\gamma l_1} + e^{2\gamma l_2} + (1 - e^{2\gamma l_1})(1 - e^{2\gamma l_2})e^{-j2\varphi}] + 1 = 0, \tag{4.13}$$

where $\rho = (\gamma - g - j\delta)/(\gamma + g + j\delta)$. The equation determines the threshold condition of DFB for different phase shift at different position in the grating. As qualitatively discussed above, the optimum value of phase shift is $2\varphi = \pi$, corresponding to groove's displacement of $\Lambda/2 = \Lambda\lambda_B/4$. For simplicity, $l_1 = l_2 = l/2$ is assumed; (4.13) is then reduced to

$$e^{\gamma l} = \frac{1 + \sqrt{\rho}}{\sqrt{\rho} - \rho} = \frac{g + j\delta + \kappa + \gamma}{g + j\delta + \kappa - \gamma}. \tag{4.14}$$

At Bragg wavelength, $\delta = 0$, $\gamma = \sqrt{g^2 + \kappa^2}$, the threshold condition is obtained:

$$\exp(\sqrt{g^2 + \kappa^2} l) = \frac{g + \kappa + \sqrt{g^2 + \kappa^2}}{g + \kappa - \sqrt{g^2 + \kappa^2}}. \tag{4.15}$$

The equation has real solutions, indicating that the DFB laser with phase shift index grating can work at Bragg wavelength. The threshold gain and mode positions are obtained as functions of κ and l. As a qualitative discussion, a weak grating is considered, $\kappa \ll g, \gamma \approx g + j\delta$, and $\exp[(g + j\delta)l] \approx 2(g + j\delta + \kappa)/\kappa$; the threshold conditions of gain and phase are obtained approximately as:

$$\delta_m l \approx \tan^{-1}(\delta_m/g_m) + 2m\pi, \tag{4.16a}$$

$$\kappa^2 \exp(2g_m l) \approx 4(g_m^2 + \delta_m^2). \tag{4.16b}$$

It is seen that $m = 0$ and $\delta = 0$ satisfy (4.16a); the threshold gain at Bragg wavelength is determined by equation $\kappa \exp(g_0 l) \approx 2g_0$; whereas the thresholds of $m \geq 1$ side modes are much higher than that of the 0th order mode.

In case the phase shift deviates from the ideal value with $\varphi = \pi/2 + x$, the lasing wavelength will move away from Bragg wavelength. For $l_1 = l_2 = l/2$, Eq. (4.13) is reduced to

$$e^{jx} = \sqrt{\rho}(e^{\gamma l} - 1)/(\rho e^{\gamma l} + 1). \tag{4.17}$$

It is deduced that for $x << \pi/2$ the lasing wavelength is moved to position of

$$\delta_L = \frac{\kappa^2(g^2+\kappa^2)}{g^2(g-\kappa+\kappa^2 l)}x \approx \frac{\kappa^2}{g}x. \tag{4.18}$$

The last approximation is for the case of $\kappa << g$. The threshold gain changes a little also from that for $\varphi = \pi/2$.

For the case with finite reflections of cleaved facets, the eigen equation is deduced as [10]:

$$\frac{\kappa^2\hat{\gamma}^2(1-e^{2\gamma l_1})+j\kappa\hat{\gamma}(\hat{\gamma}^2+\kappa^2 e^{2\gamma l_1})\hat{r}_1}{(\kappa^2+\hat{\gamma}^2 e^{2\gamma l_1})+j\kappa\hat{\gamma}(1-e^{2\gamma l_1})\hat{r}_1}\cdot\frac{\kappa^2\hat{\gamma}^2(1-e^{2\gamma l_2})+j\kappa\hat{\gamma}(\hat{\gamma}^2+\kappa^2 e^{2\gamma l_2})\hat{r}_2}{(\kappa^2+\hat{\gamma}^2 e^{2\gamma l_2})+j\kappa\hat{\gamma}(1-e^{2\gamma l_2})\hat{r}_2} = -\kappa^2\hat{\gamma}^2 e^{j2\varphi}, \tag{4.19}$$

where $\hat{\gamma} = g + j\delta - \gamma$, $\hat{r}_{1,2} = r_{1,2}\exp[j(2\beta_B l_{1,2} - \varphi)]$ with field reflection of $r_{1,2}$. It is noticed that the position of facets respect to grating period should also be taken into account. Simulation results of (4.19) can be read in Ref. [10].

4.1.2 Characteristics of DFB Lasers

(1) Field Distribution in Cavity and Rate Equations of DFB Laser

The threshold current and L-I characteristics depend mainly on gain coefficient, grating coupling coefficient, and spontaneous emission lifetime of carriers. The distribution of photon density in cavity of DFB laser is different from that of F-P LD, because they have different feedback ways. By using CME, optical field distribution for a uniform grating without phase shift is deduced as [5]:

$$\begin{aligned} A(z) &= \sqrt{I_0}\sinh[\gamma(z+l/2)]/\sinh(\gamma l) \\ B(z) &= -\sqrt{I_0}\sinh[\gamma(z-l/2)]/\sinh(\gamma l). \end{aligned} \tag{4.20}$$

The beam intensities in $\pm z$ directions have symmetric distributions; their sum is

$$I_+(z) + I_-(z) = I_0\frac{\cos(\gamma_i l)\cos(2\gamma_i z) - \cosh(\gamma_r l)\cosh(2\gamma_r z)}{2[\cos^2(\gamma_i l) - \cosh^2(\gamma_r l)]}, \tag{4.21}$$

where $\gamma_r + j\gamma_i = \gamma$. The total optical energy is obtained by integration:

$$\begin{aligned} P &= I_0\frac{\gamma_r^{-1}\sinh(\gamma_r l)\cosh(\gamma_r l) - \gamma_i^{-1}\sin(\gamma_i l)\cos(\gamma_i l)}{\sinh^2(\gamma_r l) + \sin^2(\gamma_i l)} \\ &= I_0/[\kappa\sinh(\gamma_r l)\sin(\gamma_i l)] = I_0/g_{th}. \end{aligned} \tag{4.22}$$

The last expression comes from relations of $\gamma = j\kappa \sinh(\gamma l)$ and $g + j\delta = j\kappa \cosh(\gamma l)$. The total output from the two facets is $I = I_+(l/2) + I_-(-l/2) = 2I_0$. The photon lifetime is defined as the ratio of cavity energy to lost power:

$$\tau_P = \frac{P}{v_g(I + \alpha_c P)} = \frac{1}{v_g(g_I + \alpha_c)}, \tag{4.23}$$

where $g_I = 2g_{\text{th}}$ denotes the intensity threshold gain, α_c is the loss coefficient of cavity.

The DFB laser with a phase shift grating can be analyzed similarly. For simplicity, we consider a grating with $\lambda/4$ phase shift at middle point, and the laser works at Bragg wavelength. Obviously, a symmetric distribution is expected. Referring to the above results, the field distribution in region $(-l/2, 0)$ is deduced as

$$\begin{aligned} A &= a_1[\mathrm{e}^{\gamma z} - \mathrm{e}^{-\gamma(z+l)}] \\ B &= a_1[\mathrm{e}^{\gamma z} + \frac{\gamma + g}{\gamma - g}\mathrm{e}^{-\gamma(z+l)}]\frac{\gamma + \kappa - g}{g + \kappa + \gamma}, \end{aligned} \tag{4.24}$$

where $\gamma = \sqrt{g^2 + \kappa^2}$ is a real. The optical intensity distributions of $\pm z$ direction propagations are obtained as

$$\begin{aligned} I_{1+} &= I_0(\kappa/8g\gamma^2)[(g + \kappa + \gamma)^2 \mathrm{e}^{2\gamma z} + (g + \kappa - \gamma)^2 \mathrm{e}^{-2\gamma z} - 4g\kappa] \\ I_{1-} &= I_0(\kappa/8g\gamma^2)[(\gamma + \kappa - g)^2 \mathrm{e}^{2\gamma z} + (\gamma + g - \kappa)^2 \mathrm{e}^{-2\gamma z} + 4g\kappa], \end{aligned} \tag{4.25}$$

where I_0 is the output from end of $z = -l/2$. Their sum is

$$I_1(z) = I_0(\kappa/g\gamma)[\gamma \cosh(2\gamma z) + \kappa \sinh(2\gamma z)]. \tag{4.26}$$

The optical energy (proportional to the total number of photons) in the region is obtained by integration:

$$P_1 = \frac{\kappa I_0}{4g\gamma^2}[(\gamma - \kappa)\mathrm{e}^{\gamma l} - (\gamma + \kappa)\mathrm{e}^{-\gamma l} + 2\kappa] = \frac{I_0}{2g_{th}}. \tag{4.27}$$

The last expression comes from the threshold condition. The photon lifetime is expressed by the same formula as (4.23), which is coincident with that given in Chap. 2. The external quantum efficient is also the same as that for F-P LD: $\eta_d = g_I/(\alpha_c + g_I)$. The difference is that the threshold gain is given by a transcend equation, Eq. (4.15), not an analytic expression for F-P LD. The asymmetric DFB laser with $l_1 \neq l_2$ can be analyzed by the same method. The characteristics are similar to that of F-P LD with different reflectivity on its two facets.

The rate equation written with the deduced photon lifetime has the same forms as that of F-P cavity LD:

$$\frac{dN}{dt} = \frac{J}{ed} - \frac{N}{\tau_{\text{sp}}} - v_g \Gamma g_N (N - N_{\text{tr}}) p, \tag{4.28a}$$

$$\frac{dp}{dt} = v_g \Gamma g_N (N - N_{\text{tr}}) p - \frac{p}{\tau_p} + \frac{\gamma_{\text{sp}} N}{\tau_{\text{sp}}}. \tag{4.28b}$$

The output power and slope efficient are then obtained by solving the rate equation. The threshold current, L-I characteristics, transverse mode, polarization performance, and temperature dependence of threshold of DFB-LD are with behaviors similar to those of F-P cavity LD, so long as the reflection of cavity mirror is replaced by the coupling coefficient of grating κl. The DFB laser is now commercially available; the detailed specifications can be found in their data sheets.

(2) **High Stability of Longitudinal Mode**

As described by (4.8b) and (4.16b), the threshold of higher order mode increases with mode number m^2. Therefore the side mode suppression ratio (SMSR) of DFB laser is much higher than the ordinary F-P cavity laser. Even working at high frequency modulation, the side modes will be suppressed to low levels. It is the most important advantage of DFB-LD for applications of high speed optical fiber communications, because multiple longitudinal modes cause optical pulse broadening due to fiber dispersion, which is the main limitation to the speed and distance of communication systems. The market of optical communications is the primary push to DFB LD development and its industry.

The high SMSR brings about other advantages, e.g. a better thermal stability and a wider continuous tuning range than the ordinary F-P LD. The temperature dependence of longitudinal modes is determined by temperature coefficient of index and thermal expansion coefficient of material. The modes of F-P LD and DFB LD are given by $\lambda_{FP} = 2nl/m$ and $\lambda_B = 2n\Lambda$, respectively. They have similar temperature dependences in a certain linear range:

$$\frac{\partial \lambda_{\text{FP}}}{\partial T} = \left(\frac{1}{n} \frac{\partial n}{\partial T} + \frac{1}{l} \frac{\partial l}{\partial T} \right) \lambda_{\text{FP}}; \tag{4.29a}$$

$$\frac{\partial \lambda_B}{\partial T} = \left(\frac{1}{n} \frac{\partial n}{\partial T} + \frac{1}{\Lambda} \frac{\partial \Lambda}{\partial T} \right) \lambda_B. \tag{4.29b}$$

The lasing wavelengths both of DFB and F-P LD are red-shifted with temperature increase. However, the DFB mode has continuous tuning range with temperature much wider than F-P LD made of the same material system. The red-shift tuning is attributed to the temperature dependence of refractive index. On the other hand, the peak of gain spectrum is also red-shifted due to the band-gap shrinking. For the F-P LD, the threshold of a certain longitudinal mode is increasing with temperature; and then the next adjacent mode tends to lase. Such a mode hopping is suppressed for DFB LD, giving much wider continuous tuning range, as shown in Fig. 4.4.

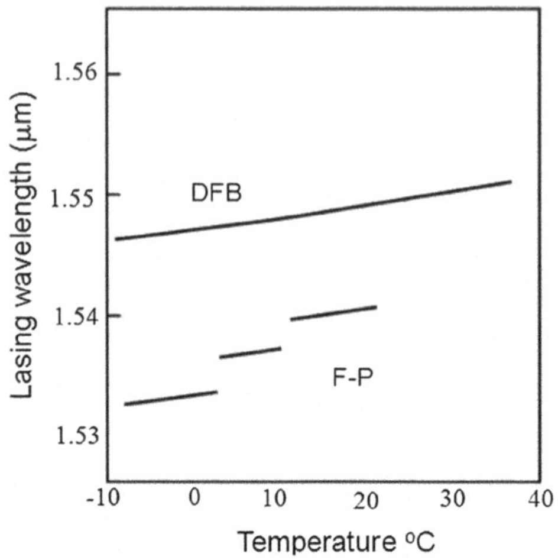

Fig. 4.4 Temperature dependence of lasing wavelength

Similar to F-P LD, the mode of DFB LD varies also with the injected current based on two factors: one is junction temperature variation due to Joule heating; the other is variation of index with carrier concentration. It is noticed that the grating for index coupled DFB is located in the waveguide layer instead of in active layer; the influence of carrier concentration on the waveguide index will be weaker than the case of F-P LD. The current tuning rate is mainly determined by its thermal effect rather than the dependence of index on carrier concentration. However the measured data of current tuning rate were somewhat divergent [11], that is probably due to other factors, such as the lateral structure and difference of fabrication processing.

The wavelength of laser emitter should coincide with channel standards in applications of DWDM communications, and in some other applications, such as optical sensors. The DFB laser may be fabricated by incorporated with a micro heating resistor to be used for precisely tuning.

(3) **Linewidth and Noises**

In principle, mechanisms of DFB laser phase noise are the same as F-P LD. Its linewidth is expressed as the product of Schawlow-Townse linewidth and linewidth enhancement factor of semiconductor laser, as given in Sect. 3.2:

$$\Delta v = \frac{v_g^2 h v \gamma_{sp} g \alpha_m}{8\pi P_0}(1+\alpha^2), \tag{4.30}$$

where $g = \alpha_c + 2g_{th} = \alpha_c + \alpha_m$, g_{th} is the net threshold gain of optical field, given by (4.16b). Experimental measurements show that DFB LD linewidth is usually in the order of MHz, much narrower than that of F-P LD [11]. It is because the lasing wavelength is locked at the Bragg wavelength with much higher stability. The noise due to fluctuation of injected current is much weak for DFB LD because the fluctuation of carrier concentration is quite weak in waveguide region where the grating is located.

The experimental measurements demonstrated that the characteristics of intensity noise and frequency noise of DFB are also superior to F-P LD, based on the mechanism of lasing frequency locking [12]. Besides, frequency chirping with current modulation, which is one of the concerned specifications in optical communications, is weaker in DFB LD than in F-P LD. However, at code rates higher than 2.5 Gbit/s, the chirping of DFB LD may be usually beyond the requirement of communications. External modulation schemes should be adopted then, such as by using $LiNbO_3$ optical waveguide modulator, or by monolithically integrating DFB LD with an electro-absorption modulator (EAM).

(4) **Effects of Photon Density Distribution and Spatial Hole-Burning**

The phase shift grating makes DFB laser with perfect wavelength locking. However a negative aspect has to be noticed, that is, the spatial hole-burning of gain coefficient in cavity due to high accumulation of photons near the phase shift point [13–15]. Formulas (4.20) and (4.26) describe the photon density distributions inside the cavity for gratings without and with phase shift, respectively. At Bragg wavelength, formula (4.26) is reduced to

$$I_1(z) = (I_0\kappa/g)\exp(-2\kappa|z|). \tag{4.31}$$

Figure 4.5 gives the photon density distributions for gratings with and without phase shift schematically.

Such a feature can be explained qualitatively as follows. In the DFB LD with phase shifted grating, the right region can be regarded as a high reflective mirror to the left region, whereas its left facet is with no mirror or with very low reflection,

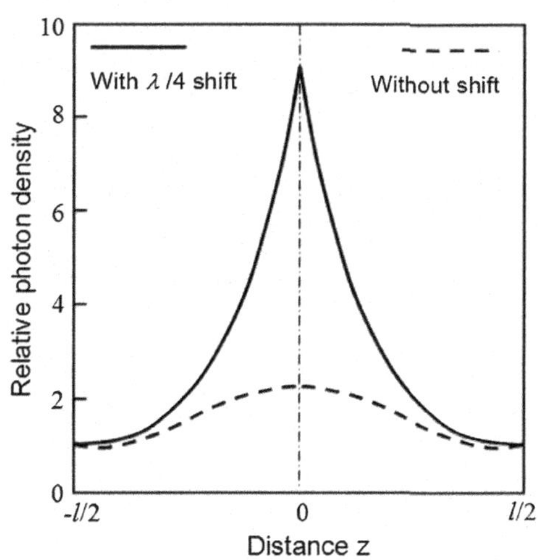

Fig. 4.5 Photon distributions inside cavity

resulting in a highly non-uniform distribution of photon density, as discussed in Chap. 2. The same distribution with opposite direction is formed in the right region. It is meant that the photon density near the phase shift point is much higher than the regions near the cleaved facets. Thus the stimulated recombination rate and the depletion rate of inversed population are much higher there, leading to a non-uniform spatial distribution of gain coefficient, i.e. the spatial hole-burning. It is seen from (4.26) that the ratio of intensities at the middle point to that at the end points is $I_1(0)/I_1(-l/2) = \kappa/g_{\mathrm{th}}$. Therefore, the hole-burning is stronger in lasers with higher grating coupling coefficient κ and lower threshold gain g_{th}.

The highly non-uniform distribution of photon density and carrier concentration causes also a non-uniform temperature distribution. The coupled mode equations and rate equations have to be modified for the non-uniformities. The carrier concentration, photon density and gain coefficient are now functions of position z. Formulas of photon lifetime and linewidth should be modified also. The waveguide index is with a spatial distribution, and the uniform grating becomes a chirped grating. The index spreading will cause line broadening. The phase shift may deviate from $\pi/2$ since it is related to the index there. The lasing wavelength will thus shift from the Bragg wavelength, as shown in formula (4.18). Moreover, these parameters will vary with laser's working point; lasing wavelength will vary with the pump current in a certain degree. Detailed analysis and discussions can be found in literature.

A longer photon lifetime is the necessary condition for narrow linewidth. It is seen from (4.23) that the photon lifetime of DFB-LD increases with decreasing of threshold gain g_{th}; whereas higher grating coupling coefficient and larger grating length are the main factors for a lower threshold. However, a larger κl will lead to a stronger hole burning, and also to lower output efficiency. In case of strong coupling grating with $\kappa \gg g$, the threshold condition for DFB without phase shift is approximately expressed as $g_{\mathrm{th}} \approx \pi^2/\kappa l^3$ [2, 10]; whereas it is $g_{\mathrm{th}} \approx 2\kappa e^{-\kappa l}$ for DFB with π phase shift. A comprehensive survey and careful trade-off between linewidth and output power is therefore necessary in laser design.

In order to avoid or mitigate the effect of hole-burning, several technical schemes have been proposed and developed, such as with multiple phase shift point, with special distributions of index modulation depth, with special changes of stripe width, and with chirped grating [16, 17]. This book will not go further in details. Another important performance is the transient characteristics of DFB laser, especially the modulation frequency response, the dynamic range for high power output. Quite many papers were published for those subjects [11, 12].

4.1.3 Structures and Fabrication of DFB Laser

Fabrication of DFB lasers is based on the conventional two-step epitaxial technologies of the material growth. The most widely used and most mature methods include the metal organic chemical vapor deposition (MOCVD) and the molecular beam epitaxy (MBE). In the first step, N-type confinement layer, N-type waveguide

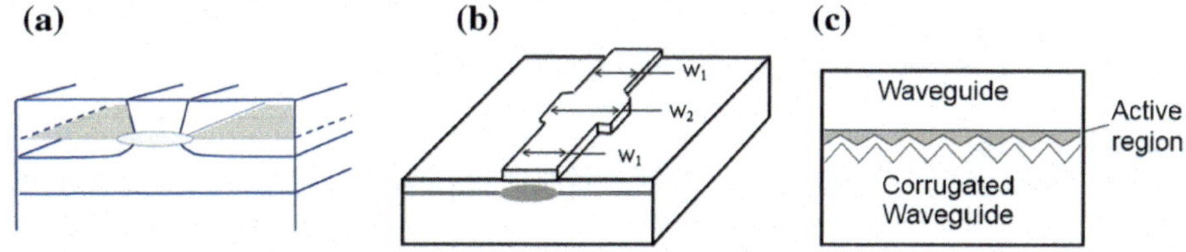

Fig. 4.6 **a** DFB-LD with lateral gratings. **b** Phase shift induced by change of stripe width. **c** Gain coupled grating

layer, quantum well active layer and part of P-type waveguide are grown successively on an N-type substrate. Then the surface is corrugated periodically as a grating. In the second epitaxial growth, P-type waveguide layer with composition a little different from the top layer, P-type confinement layer and the cap layer for Ohmic contact are grown successively on the corrugated surface, as shown in Fig. 4.1. The corrugated interface between different composition materials induces an effective index perturbation, playing a role of index grating. An alternative structure is with grating made inside N-type waveguide layer. Apart from the above typical structure, the grating can be made on the two side of laser stripe, as shown in Fig. 4.6a [18]. Two-step epitaxial growth is avoided in this configuration, so that beneficial to the quality of active layer, but with weak coupling strength of grating. Quite many technical progresses are developed for different material systems, such as InGaAsP/InP and AlGaAs/GaAs for different lasing wavelengths, and for different growth machines. Detailed technology and processing can be found in journals and monographs.

The fabrication of grating is one of the most important technologies for DFB lasers. Two approaches are usually adopted. One is the holographic photolithography, in which a high quality laser beam is divided into two beams with a proper angle to form an interference pattern, which is then transferred to the photoresist coated on the surface of material. The other uses a phase mask with fixed patterns. The $\lambda/4$ phase shift can be designed and drawn into the mask. The phase mask method has advantages of higher reproducibility and higher yield; but requires a high precision printer. After the exposure and development, the pattern should be transferred to material surface by wet etching or by dry etching. Reactive ion etching (RIE) is a widely used dry etching method. The coupling coefficient of grating is affected by the channel profile, such as the ratio of its depth to width; careful design and experiments are needed for processing optimization.

The second order diffraction of grating may be utilized for shorter wavelength DFB lasers to lower the fabrication difficulty of narrow period grating, that is, the required Bragg wavelength corresponds to the second order Fourier expansion of the gating with $\lambda_B = n\Lambda$. The detailed theory and experimental research can be found in literature [19].

Bragg wavelength is determined by the grating period and the index of material. This property may be used to induce a phase shift in a grating with uniform period. Figure 4.6b shows such a design, where the stripe width of middle section is made wider than two other sections, $w_2 > w_1$, resulting in a difference of effective index, as analyzed in Sect. 2.3 for the ridge waveguide stripe LD. When the length and width of middle section satisfy $(n_2–n_1)\delta l = \lambda/4$, the section will play a role of $\lambda/4$ phase shift [16, 20].

The fabrication technology of gain coupled grating is somewhat more complicated than that of the index grating. A typical structure is shown in Fig. 4.6c [21], in which the active layer is grown on a corrugated surface of waveguide layer, making the thickness of active layer changed periodically. The injected carrier concentration will thus be changed periodically in cavity length, resulting in a gain modulation. The DFB-LD with gain coupled grating has excellent mode characteristics, but with higher fabrication difficulty and production cost. The mass produced and widely used laser is still the DFB-LD with phase shifted index grating now.

Other procedures of device processing of DFB-LD are basically the same as that of F-P LD, including fabrication of stripes, typically the ridge waveguide, and Ohmic contact, etc. The cleaved facets of DFB LD are coated anti-reflective to avoid disturbance of facet reflection to the action of grating. In practice a certain residual reflection is inevitable. Its influence on the longitudinal mode has been studied in detail [10, 17].

4.2 Distributed Bragg Reflector Semiconductor Laser

4.2.1 Structures and Principle of DBR Laser

In the structure of DFB laser the grating is fabricated parallel to the active layer, the distributed feedback and optical gain are tightly and directly integrated with each other. Such a configuration lacks flexibility. It is often needed to control the wavelength and output power separately, or to optimize and modify one of the performances, whereas leave the other unchanged. The distributed Bragg reflector (DBR) laser is then developed [22–24], in which the waveguide gratings are fabricated at the ends of active region. Two typical structures are depicted in Fig. 4.7. Figure (a) is with two gratings fabricated at both ends of active region; Fig. (b) is with grating at one of the ends and with a cleave facet at the other end.

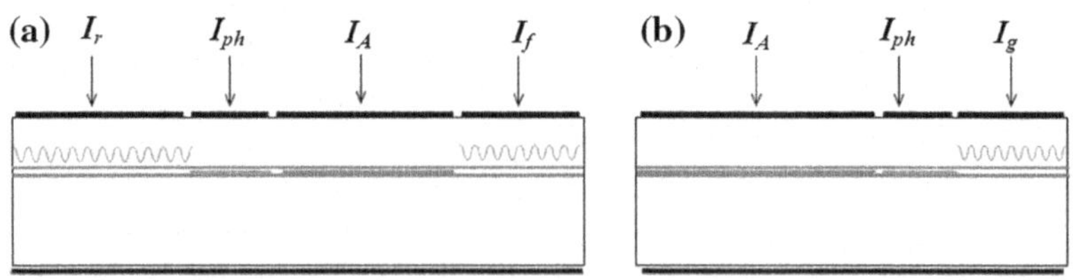

Fig. 4.7 Structure of DBR laser, **a** with gratings at both ends; **b** with grating at one end

Four currents of I_f, I_r, I_A and I_{ph} can be applied into the four regions with electric isolated electrodes. Since the Bragg wavelength is dependent on waveguide index, it can be tuned by the injected current, while the gain of active region is adjusted by a separate current, making the two parameters controlled separately. A region without grating is usually inserted for phase-controlling, which provides an adjustable phase by the injected current, or by a reversed bias to deplete carrier concentration, playing a role of tuning the wavelength precisely, or/and a role of expanding tunable range without mode hopping [25, 26].

Topologically, DBR LD is the same as F-P LD. The difference is that narrow linewidth reflectors are used as cavity mirror instead of broad band reflection mirrors, so that much higher SMSR is ensured. The technologies of DBR laser fabrication, including the material growth and device structure processing, are basically similar to that of DFB laser. Detailed techniques can be read in monographs and journals.

4.2.2 *Reflection Spectrum of Passive Waveguide Grating*

The grating used in DBR lasers is a passive waveguide grating. Its characteristics obey Bragg diffraction of a one-dimensional lattice, and can be analyzed by the coupled mode equations (CME), as used in fiber Bragg gratings. The index distribution is written by the same expression as (4.1), but both of indexes, n_0 and n_1, are real now. The coupled mode equations and the solutions are written as

$$\begin{aligned} A' - j\delta A - j\kappa B = 0 \\ B' + j\delta B + j\kappa A = 0 \end{aligned}. \tag{4.32}$$

$$\begin{aligned} A = a_1 e^{\sigma z} + a_2 e^{-\sigma z} \\ B = b_1 e^{\sigma z} + b_2 e^{-\sigma z} \end{aligned}. \tag{4.33}$$

where $\sigma = \sqrt{\kappa^2 - \delta^2}$ is denoted. Relations between the coefficients are obtained from the equations as $j\kappa b_1 = (\sigma - j\delta)a_1$ and $-j\kappa b_2 = (\sigma + j\delta)a_2$. For an incident wave propagating in $+z$ direction, the boundary conditions are $a_1 + a_2 = 1$ at $z = 0$ and $b_1 e^{\sigma l} + b_2 e^{-\sigma l} = 0$ at the end of $z = l$. The field reflectance is $r = b_1 + b_2$, and its transmission is $t = (a_1 e^{\sigma l} + a_2 e^{-\sigma l})e^{j\beta_B l}$. The axial distributions of forward and backward propagating fields are reduced to [27]:

$$\begin{aligned} A &= A(0)\frac{\sigma\cosh\sigma(l-z) - j\delta\sinh\sigma(l-z)}{\sigma\cosh\sigma l - j\delta\sinh\sigma l}e^{-j\delta z} \\ B &= A(0)\frac{-j\kappa\sinh\sigma(l-z)}{\sigma\cosh\sigma l - j\delta\sinh\sigma l}e^{j\delta z} \end{aligned} \quad \begin{aligned} &(\delta^2 \le \kappa^2, \\ &\sigma = \sqrt{\kappa^2 - \delta^2}) \end{aligned}, \tag{4.34a}$$

$$\begin{aligned} A &= A(0)\frac{s\cos s(l-z)-j\delta\sin s(l-z)}{s\cos sl-j\delta\sin sl}e^{-j\delta z} \\ B &= A(0)\frac{-j\kappa\sin s(l-z)}{s\cos sl-j\delta\sin sl}e^{j\delta z} \end{aligned} \quad \left(\delta^2\ge\kappa^2,\ s=\sqrt{\delta^2-\kappa^2}\right). \tag{4.34b}$$

The reflectance and transmission of field and power are obtained as:

$$\begin{aligned} r=\frac{E_-(0)}{E_+(0)} &= \frac{-j\kappa\sinh\sigma l}{\sigma\cosh\sigma l-j\delta\sinh\sigma l} & (\delta^2\le\kappa^2) \\ &= \frac{-j\kappa\sin sl}{s\cos sl-j\delta\sin sl} & (\delta^2\ge\kappa^2) \end{aligned}, \tag{4.35a}$$

$$\begin{aligned} t=\frac{E_+(l)}{E_+(0)} &= \frac{\sigma e^{j\beta_B l}}{\sigma\cosh\sigma l-j\delta\sinh\sigma l} & (\delta^2\le\kappa^2) \\ &= \frac{s e^{j\beta_B l}}{s\cos sl-j\delta\sin sl} & (\delta^2\ge\kappa^2) \end{aligned}. \tag{4.35b}$$

$$\begin{aligned} R=|r|^2 &= \frac{\kappa^2\sinh^2\sigma l}{\kappa^2\cosh^2\sigma l-\delta^2} & (\delta^2\ge\kappa^2) \\ &= \frac{\kappa^2\sin^2 sl}{\delta^2-\kappa^2\cos^2 sl} & (\delta^2\le\kappa^2) \end{aligned}, \tag{4.36a}$$

$$\begin{aligned} T=|t|^2 &= \frac{\kappa^2-\delta^2}{\kappa^2\cosh^2\sigma L-\delta^2} & (\delta^2\le\kappa^2) \\ &= \frac{\delta^2-\kappa^2}{\delta^2-\kappa^2\cos^2 sl} & (\delta^2\ge\kappa^2) \end{aligned}. \tag{4.36b}$$

It is seen that $T = 1–R$ for the lossless grating, the reflection peak is at $\delta = 0$, i.e. at Bragg wavelength of $\lambda_B = 2n\Lambda$, and the peak power reflectance is

$$R_p = \tanh^2 \kappa l. \tag{4.37}$$

A series side bands exist centered at $\tan sl = sl$. Their peak reflectance declines as Lorentzian function $R_m = \kappa^2 l^2/(1+\delta_m^2 l^2)$. The left and right zero points of main lobe are located at $\delta_{\pm1} = \pm\sqrt{\kappa^2+\pi^2/l^2}$. The bandwidth at half maximum of the main lobe is determined by equation of $\kappa l\ \mathrm{sinc}(s_{1/2}l) = \sqrt{R_p/(2-R_p)}$, decreasing with increase of grating length. The properly designed waveguide grating is a narrow bandwidth filter. Figure 4.8 a shows a calculated reflection spectrum.

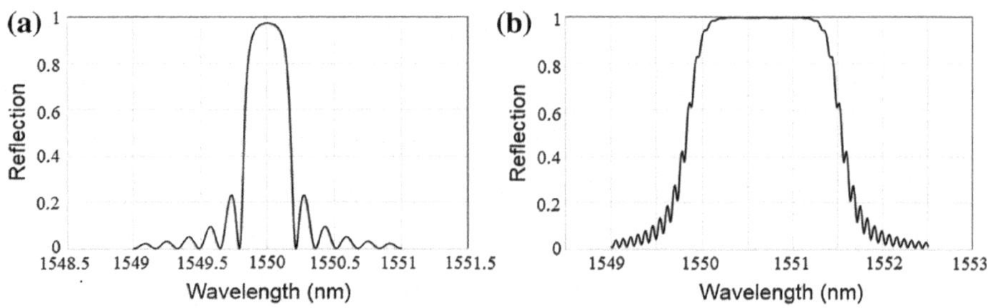

Fig. 4.8 **a** A typical reflection spectrum of uniform grating. **b** Reflection spectrum of a chirped grating

The field reflectance (4.35a) can be written as $r = |r|\exp(j\phi_g)$, with a phase factor of

$$\begin{aligned}\phi_g &= \tan^{-1}[\tfrac{-\sigma}{\delta}\coth(\sigma l)] \quad (\sigma = \sqrt{\kappa^2 - \delta^2}) \\ &= \tan^{-1}[\tfrac{-s}{\delta}\cot(sl)] \quad (s = \sqrt{\delta^2 - \kappa^2}).\end{aligned} \tag{4.38}$$

At the peak, $\delta = 0$, it is $-\pi/2$, coincident with the analysis of stop band in DFB laser. When the waveguide grating is used in lasers, it is necessary to take its phase factor into account.

Based on the plain Bragg grating, gratings with sub-structures are developed, such as chirped gratings and sampled gratings. The period of chirped grating is not uniform, but varied with the axial position; mostly in a linear relation: $\Lambda = \Lambda_0 + \rho z$. The spectral range of Bragg wavelength is thus broadened to $\Delta\lambda_B = 2n\rho L$. Figure 4.8 b shows a calculated reflection spectrum for a chirped grating with length of 10 mm and chirping rate ρ = 0.0001 μm/mm, as an example. Since the reflection occurs at different positions, the chirped grating has a group delay designable for particular applications: $\tau_{GD} = (\lambda - \lambda_0)/\rho c$ with $\lambda_0 = 2n\Lambda_0$. The chirped grating is fabricated usually by using a designed mask in process of photolithography and with other precise technologies incorporated.

4.2.3 Characteristics of DBR Lasers

(1) Threshold Conditions

The threshold condition of DBR laser takes the same form as that of F-P cavity laser:

$$|r_1 r_2|\exp[j2n_r k l_g + g_{net} l_g + j2\phi + j(\phi_{g1} + \phi_{g2})] = 1, \tag{4.39}$$

where $g_{net} = g - \alpha_c$, l_g is the length of gain section, ϕ is the phase factor induced by phase controlling section; r_1 and r_2 are the reflectance of the two end gratings, and the cleaved facets are assumed fully anti-reflective coated. The peak wavelengths of r_1 and r_2 may be coincident with each other, or with a certain deviation. In the latter case the lasing wavelength is determined by their composite spectrum. For structure of Fig. 4.7 b, one of them is a broadband reflector. The phase condition of threshold is related not only to the propagation phase in the cavity length, but also to the composite phase of gratings; the phase controlling section is used to adjust the working point of DBR laser for optimization of performance.

Similar to the F-P cavity laser, the distribution of photon density in cavity can be described by a hyperbolic function: $P \propto \cosh(g_{net}z)$. Its uniformity is much better than that of DFB laser, and the spatial hole-burning effect is much weaker, which is beneficial to higher power output, higher side mode suppression ratio (SMSR) and narrower linewidth than DFB lasers.

(2) **Superstructure Waveguide Grating**

One of the attractive advantages of DBR laser is its agile and broadband tunability. By adjusting the current injected in the grating section, the Bragg wavelength $\lambda_B = 2n\Lambda$ can be changed due to the effect of index on the carrier concentration and temperature. A long uniform grating is helpful for a narrow linewidth laser with high SMSR, but its tuning range is often not large enough. The tuning range of InGaAsP laser working at 1500 nm band is roughly within 15 nm, limited by the thermo-optic coefficient of semiconductor.

In order to expand the tuning range, DBR lasers with sampled gratings were developed [28–31]. The sampled grating is composed of multiple sections of uniform grating, as shown in Fig. 4.9(a). Because of the interferences between waves reflected from multiple grating sections, the reflection of sampled grating displays a comb-like spectrum with multiple peaks. Figure 4.9(b) shows a calculated spectrum.

The sampled grating is a rectangular function chopped uniform grating, with repetition period of p and width of p_g. Its index distribution can be described by a Fourier series:

$$\Delta n(z) = n_1\eta[\cos\frac{2\pi z}{\Lambda} + \sum_m \frac{\sin(\pi m\eta)}{\pi m\eta}(\cos\frac{2\pi z}{\Lambda_{m-}} + \cos\frac{2\pi z}{\Lambda_{m+}})], \tag{4.40}$$

where $\eta = p_g/p$ is termed the sampling rate, and $\Lambda_{m\pm} = \Lambda/(1 \mp m\Lambda/p) \approx \Lambda(1 \pm m\Lambda/p)$. The sampled grating can be regarded as a composite grating composed of a series of gratings with different periods of $\Lambda_{m\pm}$, which generates a series of peak wavelengths with interval of:

$$\Delta\lambda = 2n\Lambda^2/p = \lambda_0^2/(2np), \tag{4.41}$$

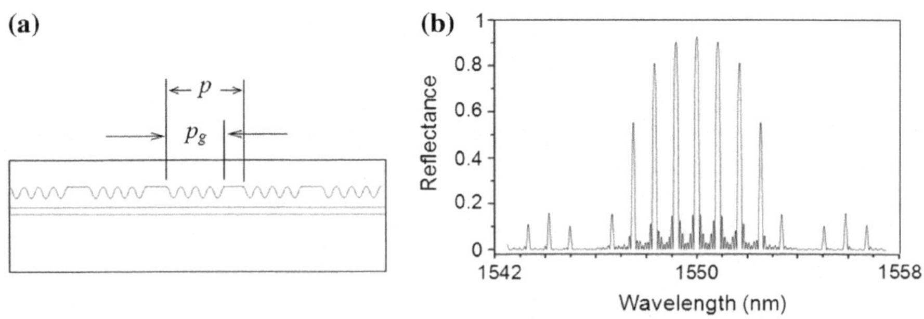

Fig. 4.9 **a** Structure of sampled grating. **b** Its calculated reflection spectrum

which is inversely proportional to period p. The index modulation amplitudes of the multiple peaks in (4.40) has an envelope of sinc function: $\sin(\pi m\eta)/(\pi m\eta)$. The spectral width of its main lobe ($|m\eta| \leq 1$) is $\Delta\lambda_{\text{env}} = \lambda_0^2/(np_g)$; the number of peaks within the main lobe is $M = 2p/p_g$. A lower sampling rate is thus needed to get more reflection peaks within the main lobe; and a larger whole grating length is also preferred to get a narrower linewidth of peaks. Similarly, the multiple reflection peaks can be tuned by injected current and temperature.

The uniformity of reflection peak amplitudes in the envelope is one of the requirements in practical applications. For the purpose a special sampled grating has been developed, which is composed of multiple chirper gratings, as shown in Fig. 4.10 [32]. The multiple reflection peaks distributes in the range between λ_a and λ_b with uniform amplitudes and interval of $\lambda^2/(2np)$. Such a grating is termed superstructure grating (SSG). Reference [32] reported a SSG DBR laser with chirping rate adjusted by applied current through the multiple electrodes.

For better performances, such as lower side lobes, careful designs are needed. For example, the index modulation depth n_1 may be modified at the grating ends with a smoothed variation to replace interrupt changes between grating region and blank region to reduce Fresnel reflection. Different sampling functions may also be adopted. Such a technique is named the apodization. Phase shift can also be introduced in sampled gratings [31]. Simulations have to be carried in designs by using approximation method, such as the transfer matrix method (TMM) [27].

(3) **Quasi-continuous Tuning**

DBR LD with SSG has an advantage of wide tuning range, based on the principle of vernier effect, i.e. two sampled gratings with different peak intervals are used at the two ends of DBR laser, respectively. When two peaks from the two sampled gratings are coincident with each other in spectrum, the composite peak will be much higher than other peaks, and the laser will work at the coincident peak, as shown in Fig. 4.11 [33, 34]. When one of the two gratings is tuned, the adjacent peaks are coincident with each other, and the laser will work at that wavelength. Such a tuning is not a continuous tuning of single longitudinal mode, but in a step way with basically equal intervals. It is usually called a quasi-continuous tuning.

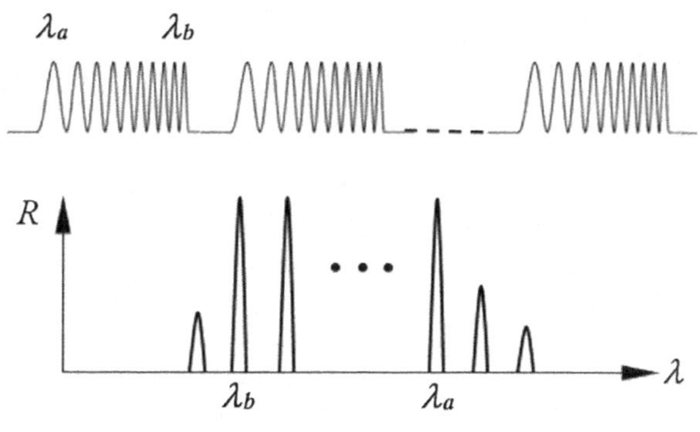

Fig. 4.10 Illustration of superstructure grating

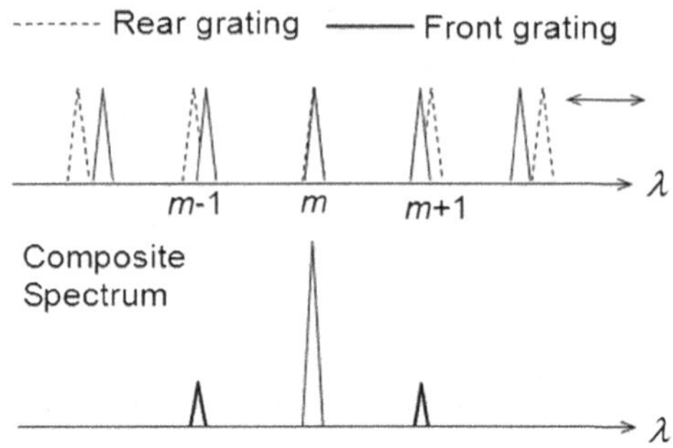

Fig. 4.11 Large tuning range by a pair of sampled gratings

According to (4.40) the spacing between peaks of sampled grating can be designed by the sampling period p, and adjusted by the bias current. According to the principle of vernier caliper, the number of tuned wavelength is $M \approx \Delta\lambda_f/\delta\lambda \sim \Delta\lambda_r/\delta\lambda$, where $\Delta\lambda_f$ and $\Delta\lambda_r$ are the peak intervals of two sampled grating, and $\delta\lambda = |\Delta\lambda_f - \Delta\lambda_r|$. Therefore, the tunable wavelength number increases with decreasing of $\delta\lambda$ [33]. The limitation is the linewidth of each reflection peak, because $\delta\lambda$ as small as the linewidth is not applicable. It was reported in journals that the whole tuning range can reach 100 nm. Such wide tuning range is useful in multi-channel DWDM communication system and fiber sensor systems.

Other components with different functions can be integrated with DBR LD, such as electro-absorption modulator (EAM) and semiconductor optical amplifier (SOA). By using a pair of waveguide grating, high speed wavelength switching can be realized based on their tunability [35]. The techniques of waveguide grating fabrication are developed continuously, such as surface Bragg grating and twin waveguide [36]. As the development of material researches and fabrication techniques, DBR lasers are now commercially available, and used widely in practical applications.

4.3 Vertical Cavity Surface Emitting Laser

4.3.1 Principles of VCSEL

In the ordinary F-P cavity LD, DFB-LD and DBR-LD, the laser oscillation is parallel to the P-N junction; so is the output beam direction. Such a configuration makes full use of the gain in active region, and has advantages of high output power and high energy efficiency. However, the shortcomings are obvious. First, the output beam is highly divergent, and with an elliptical shape. Careful beam reshaping has to be carried out for most applications, such as for fiber coupling, and for beam collimating or focusing. Second, the cleaving of semiconductor crystal is necessary in laser's fabrication, which is a time-consuming process, hardly to be replaced by automatic processing, especially in mass production. Therefore, to develop a laser with output emitting from the surface of semiconductor material had been an attractive subject. Several schemes were tried. One of them is to make a 45° mirror inside the material by dry-etching to deflect the laser beam by 90°. Another method is to make a grating on the surface of material, by the second order diffraction of grating the laser beam is deflected from horizontal direction to near vertical. These structures are too complicated for production. Inspired by the structure of light emitting diodes (LED), with lots of technical difficulties overcome, the vertical cavity surface emitting laser (VCSEL) has been successfully developed based on the principle of multi-layer dielectric thin film filters (TFF) and technology of hetero-structure growth [37–39].

The success of VCSEL relies on maturations of high efficiency quantum well active region and multi-layer hetero-structure reflector. Figure 4.12a, b depict two basic structures of VCSEL with different output ways. The active region is usually a high gain multiple quantum well (MQW), which is sandwiched by distributed Bragg reflectors (DBR) as the laser cavity. The direction of laser oscillation is perpendicular to the P-N junction and to the surface of material. Since the cleaved facet is not necessary as cavity mirrors, the laser chips can be diced by conventional sawing machine in semiconductor device fabrication, with much higher efficiency of mass production. What's more, the performance of every laser chip can be tested before dicing, that is beneficial to enhance the quality and yield. The aperture of laser chip is formed by mask design of photolithography, which is usually in circular or square shape, making the output a circular beam instead of long elliptical beam in edge emitting LD. The structure is also beneficial to make two dimensional VCSEL array with high power output, and to integrate with other elements, such as two dimensional (2D) high speed optical interconnection sources, and addressable 2D sources.

Pushed by various applications, research and development of VCSEL are speed up quickly [40–42]. Its theory and technologies have been expounded in details in many monographs and review articles [43–48].

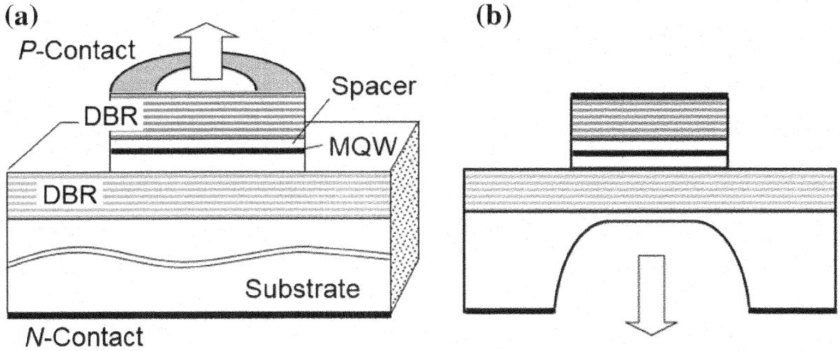

Fig. 4.12 Structures of VCSEL with output from epitaxial layer (**a**) and from *bottom* of substrate (**b**)

4.3.2 Design and Fabrication of VCSEL

(1) Design and Fabrication of DBR

The laser oscillation is perpendicular to P-N junction in VCSEL, the length of gain region is very short; therefore it is a critical task to meet the basic requirements for lasing: the gain coefficient must be high enough, and the mirror reflectance must be as high as near unity. Multiple layer dielectric film, i.e. the thin film filter (TFF), is the widely used cavity mirror in various lasers. It is composed of layers with lower index and higher index materials grown one over the other. When their index and thickness satisfy condition of $n_H d_H = n_L d_L = \lambda/4$, the reflectance of TFF is expressed as [49]:

$$R = \left[\frac{n_H^2 (n_H/n_L)^{2N} - n_0 n_{\text{sub}}}{n_H^2 (n_H/n_L)^{2N} + n_0 n_{\text{sub}}}\right]^2, \tag{4.42}$$

where subscripts H and L stand for the higher and lower index; n_0 and n_{sub} are the indexes of external medium and substrate, respectively, N is the number of layer pairs. It is indicated that the reflectance will be towards unity when the index ratio and N are high enough. The quarter wavelength condition of the layer thickness is also critical for a reflectance towards unity.

The ordinary TFF is fabricated by vacuum evaporation machines, and is made of poly-crystals or amorphous materials. The method is not suitable for VCSEL, since the active layer must be grown on one of the reflectors, which must be a mono-crystal matching with the laser semiconductor. Therefore at least one of the DBR must be fabricated by epitaxial growing machine, such as the metal organic chemical vapor deposition (MOCVD) or the molecular beam epitaxy (MBE) . The crystal constants of lower and higher index materials must be close enough to compose heterostructures. For example, for VCSEL of 800 nm wavelength $Al_xGa_{1-x}As/GaAs$ heterostructure is used; the index of $Al_xGa_{1-x}As$ decreases with increasing of x in the range between $n_{GaAs} = 3.5 \sim 3.6$ and $n_{AlAs} = 2.9 \sim 3.0$ [39, 40]. When the layer number reaches $N = 20 \sim 30$, a TFF with reflectance over 99% can be obtained. The stack of AlGaAs/GaAs heterostructures was just used as cavity mirrors in the earliest developed VCSEL.

For lasers in 1300 and 1550 nm wavelength range, the active material is InGaAsP-based quaternary compound; ideal quantum well can be grown with the material system. However the index ratio of InGaAsP with different compositions is too small to make a high reflective TFF; in other words, the layer number needed is too many to be a reality. In order to overcome this difficulty a different technology, called wafer bonding or wafer fusion, has been developed, which relies on attractive forces between molecules, i.e. van de Waals force. When two materials with different crystal constants are piled up with perfectly flat surfaces, and put in certain conditions of temperature and pressure for a certain period, they will be combined into a whole. By such a way, one material with quantum well grown can be fused with another material with TFF grown, though they do not have the same or near crystal constant. Next, one of the two original substrates can be removed by physical approaches and wet or dry chemical etching; and a planar laser diode is fabricated at last [50, 51]. This method allows the TFF made of material totally different from the quaternary crystal, and to optimize laser material and mirrors separately. The most often used TFF, or called DBR, include Si/SiO_2 films, MgO/ZnSe films, etc. and even with highly reflective metal coatings. Based on such advanced technologies VCSELs in various wavelength bands have been developed, not only in 850 nm, 1300–1550 nm, and 980 nm bands, but also in visible bands, blue and violet bands (GaN-based compound), and in mid-infrared bands [52–55].

(2) **Design of Active Region Structure**

As discussed in Chap. 2, the injected carriers can be confined in quantum wells (QW) by the heterostructure barriers; the QW thickness is in the order of de Broglie wavelength, so that the carrier concentration and the gain coefficient are greatly increased. The QW and strained layer QW are the core of semiconductor lasers, especially of the VCSEL. The design and growth of QW involve a series of theories and special technologies, beyond the scope of this book. Interested readers can refer related textbooks and monographs [40–42].

The threshold condition of VCSEL is the same as that of DBR LD, expressed as $r_{DBR1}r_{DBR2}\exp(j2\beta L_z) = 1$, where r_{DBR1} and r_{DBR2} are complex with a phase factor as shown in (4.38). The threshold condition of gain and phase should be written as [5]:

$$2n_r k L_z + \phi_{DBR1} + \phi_{DBR2} = 2m\pi, \tag{4.43}$$

$$\Gamma g = \alpha_c + L_z^{-1}\ln(1/|r_{DBR1}r_{DBR2}|), \tag{4.44}$$

where α_c is the loss coefficient inside the cavity, including material absorption, scattering at interfaces, and so on. α_c is actually the loss averaged over the cavity length. Similar to edge emitting LD, the contribution of active region to cavity gain has to be integrated over cavity length with a weighting factor of photon density distribution, that is, to multiply the confinement factor Γ. The difference from the edge emitting LD is that it is necessary to take the overlap of gain and transverse photon distribution in P-N junction plane into account. The confinement factor is now the product of two factors: $\Gamma = \Gamma_{\parallel} \cdot \Gamma_{\perp}$ [41]. The transverse confinement factor $\Gamma_{\perp}$ is related to current distribution in the P-N plane and to the transverse mode of VCSEL.

The longitudinal factor $\Gamma_{\parallel}$ depends on the position of active region respect to the standing wave of optical field. Since the thickness of active layer cavity is extremely short, usually only one period of standing wave exists in the cavity, i.e. the

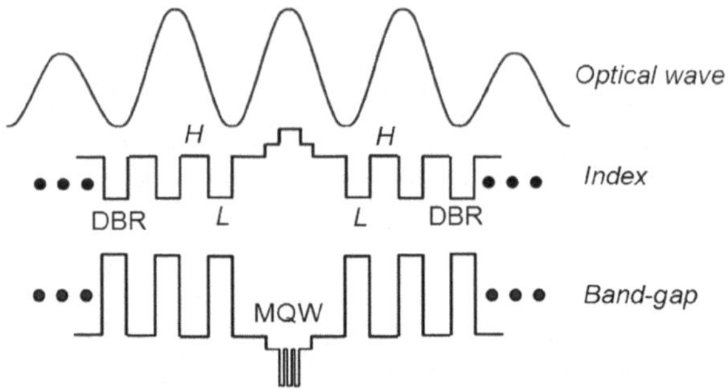

Fig. 4.13 Illustration of material structure of VCSEL

cavity length is in the order of half wavelength. Obviously, it is better to control the peak of standing wave coincident with the active layer. Therefore two spacer layers are designed at the two sides of MQW to adjust its position related to the standing wave. Figure 4.13 is a schematic diagram of VCSEL material [38, 39, 56], showing energy band structure and index distribution. These geometrical sizes are dependent on the wavelength, index and their temperature relations; it is necessary to design all the parameters carefully and to ensure them realized in fabrication processing.

(3) **Arrangement of Current Injection and Aperture Design**

Since the laser beam of VCSEL emits from the top surface in direction perpendicular to the P-N junction, the electrodes for current injection must be arranged as a ring, leaving the central area as the aperture of laser beam output. On the other hand, the TFFs (DBR) should be made of high resistive material to reduce their optical losses. The pump current has to go through a curved path from all around to the center and to bypass regions of TFF. Ingenious designs are needed for the purpose, including the arrangement of conductive layers and isolative layers, and the process of their fabrication.

Figure 4.14 shows typical designs of transverse structures. The structure of Fig. 4.14a adopts ion implantation process to build an isolation region to confine the area of current injection [51]. In the processing, high energy ions are implanted into the material deeply, to generate impurities as recombination centers of carriers, so that the resistance of material is greatly increased. The isolation region, such as oxide, can also be formed by chemical vapor deposition (CVD), as shown in Fig. 4.14b. Polymer is often used to bury the mesa with emitting element inside, and to flatten the device surface, making electrode fabrication and wire bonding easier. The performances of VCSEL are affected greatly by the transverse structure designs and the adopted materials. Detailed analyses and technologies can be found in literature, e.g. in Ref. [47].

The structure of Fig. 4.14c adopts highly doped materials ($p^+$and n^+regions) as current channels, and oxide isolation layer to restrict the current injection area. Its notable feature is that both of the positive and negative electrodes are arranged on

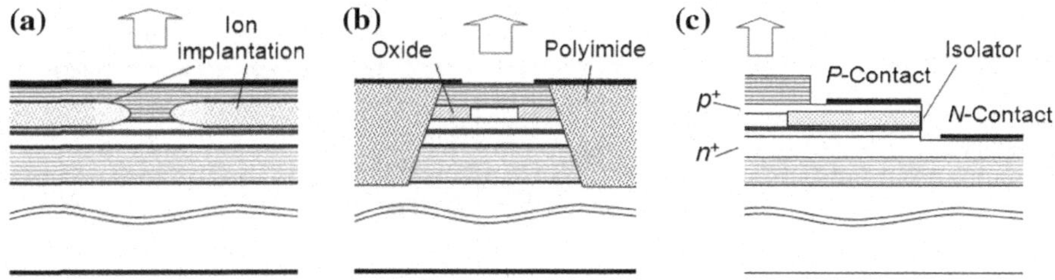

Fig. 4.14 Transverse structure of VCSEL: **a** Confinement by ion implantation; **b** Mesa buried by polymer; **c** Structure of in-face electrodes

the same top surface. Such a structure has important advantages. It is a structure suitable for addressable two dimensional integrated sources; and beneficial to enhance modulation frequency response. The high resistive DBR, especially the bottom one, is put out of current path in the structure, beneficial to reduce the series resistance. It is important for VCSEL made of different materials, such as InGaAsP QW with AlAs/GaAs DBR, GaN-based blue and violet VCSEL on silicon substrate, and so on.

Lots of processing techniques of semiconductor device technology are used in VCSEL fabrications, such as proton and ion implantation, selective oxidation, two-step epitaxy, wafer bonding. Some physical effects are utilized, e.g., the tunneling effect is used to reduce the series resistance [57, 58].

The emitting area of VCSEL is usually larger than that of edge emitting LD. However, the aperture size is limited by some reasons. One is the uniformity of injected current density in the gain area. Since the current goes the way from all around to the center, a concave distribution is inevitable. An annular beam may be yielded if the aperture is too big [45]. Generally a smaller gain area is beneficial to get an ideal Gaussian beam. A trade-off should be set between beam shape and output power. Most VCSELs are used in fiber optical communications, an aperture near the size of fiber core is preferable, which is about 9 μm in diameter.

(4) **Polarization Control**

Most applications require lasers with stable polarization and high degree of polarization (DOP). The beam of edge emitting LD is strongly polarized by the asymmetry of structure, whereas the VCSEL has no such a polarization-selectivity, due to its usual round shape of emitting area. It is necessary to design a special mechanism to ensure the output beam with high DOP, and to avoid random fluctuations of polarization. One of the schemes is to use a rectangular shape emitting area [59]. More widely used scheme is to make a surface planar grating on the top [40, 60–63], as shown in Fig. 4.15a.

The period of grating is designed less than laser's wavelength to eliminate the unwanted diffraction, and to induce birefringence; that is, the normally incident TE

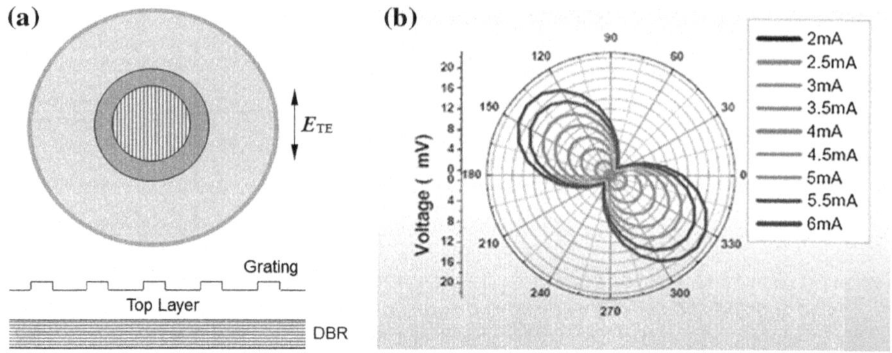

Fig. 4.15 **a** Surface grating on top of polarized VCSEL. **b** Measured polarization extinction ratio of a 850 nmVCSEL

mode beam with field parallel to the grating grooves experiences different reflection from that of TM modes. The effective dielectric constants of the grating is expressed as [60–63]:

$$\varepsilon_{TE} = f\varepsilon_1 + (1 - f)\varepsilon_2, \tag{4.45a}$$

$$\varepsilon_{TM} = \varepsilon_1\varepsilon_2/[f\varepsilon_2 + (1 - f)\varepsilon_1], \tag{4.45b}$$

where ε_1 and ε_2 are electric constants of the top layer and the surround medium, respectively; f is the ratio of grating width to its period.

Different effective dielectric constants generate different reflections and thus different thresholds between TE and TM modes, giving good polarization selectivity. It is indicated that the TE mode will be superior to TM mode for a careful designed VCSEL. Figure 4.15b shows a measured polarization of a commercial VCSEL in polar coordinator, where the detected signals of output intensity vs. rotated angles are depicted for different working currents. The polarization extinction ratio reaches 200:1, not worse than that of ordinary edge emitting LD.

(5) **Structure of Tunable VCSEL**

The tunability is one of the important specifications concerned widely in laser applications. VCSELs can be tuned by variations of temperature and working current, similar to the other semiconductor lasers. Besides, the technology of micro-electro-mechanical system (MEMS) has been successfully used to build a wide range tunable VCSEL [64, 65].

Figure 4.16 gives one of the typical structures schematically, where the upper DBR is built on the center of a cross cantilever, separated from the active material; and the latter is anti-reflective coated. The cantilever can be moved up-and-down by static electric force with the applied voltage between the cantilever and the substrate, and the cavity length is varied correspondingly. It is indicated by elasticity

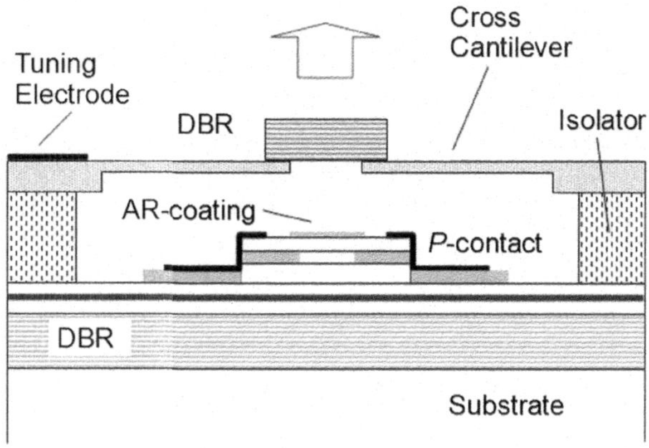

Fig. 4.16 Tunable VCSEL with a cross cantilever

analysis that the displacement Δz of cantilever is dependent on the sizes and Young's modulus expressed as [64]:

$$\Delta z \propto \frac{l^3}{Ywt^3}\frac{V^2}{(z_0 - z)^2}, \tag{4.46}$$

where l, w, and t are the length, width and thickness of cantilever, V is the applied voltage, z_0 is the original distance between the upper DBR and the active region.

4.3.3 Characteristics of VCSEL

VCSEL has many features different from the edge emitting LD. The following characteristics are mostly concerned in its applications.

(1) **Low Threshold and Low Energy Dissipation**

One of the important advantages of VCSEL is its extremely low threshold. It is because the active area of a fundamental transverse mode VCSEL is much smaller than that of edge emitting LD; it is usually less than 100 μm^2, whereas the latter is in the order of 1000 μm^2 for a narrow stripe geometry structure. On the other hand, relying on the improvements of process, the loss is minimized and the DBR reflectance near unity is realized. The threshold current density has been decreased to as low as sub-milliampere. Quite many papers reported their researches on ultra-low threshold VCSEL [46, 67–70]. Figure 4.17 gives measured L-I curve and V-A curve of a typical 850 nm band AlGaAs VCSEL [67], showing its threshold current as low as 0.14 mA.

The output power of VCSEL is usually quite low, compared with ordinary edge emitting lasers. The power at 10 times threshold is less than 1 mW in Fig. 4.17. The

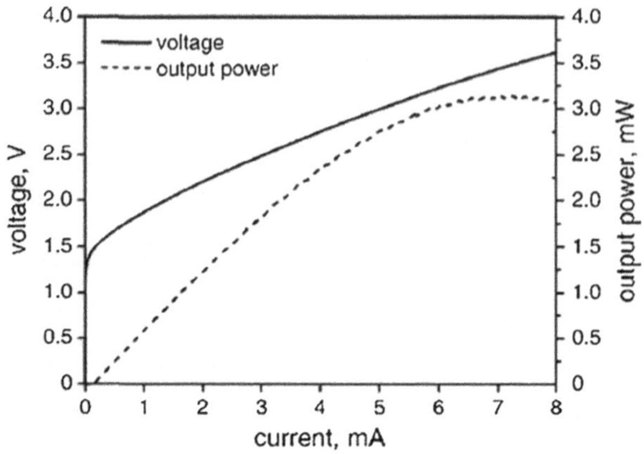

Fig. 4.17 L-I curve and V-A curve of a typical VCSEL. Reproduced from [67] by permission of the Institution of Engineering & Technology

reason is that the reflectance of mirror is very close to unity, resulting in a low external differential efficiency [56]:

$$\eta_d = \eta_i \frac{\ln(1/R_{DBR})}{\alpha_c L_z + \ln(1/R_{DBR})} \approx \eta_i \frac{1 - R_{DBR}}{\alpha_c L_z + 1 - R_{DBR}}. \tag{4.47}$$

Obviously the output power $P = \eta_d V_J (I - I_{th})$ is low since R_{DBR} is towards unity. However, such a power can already meet the requirement of many applications, especially the optical interconnection between computers, between data processors, and in short distance access networks.

The thermal performance of VCSEL is one of the concerned characteristics. The VCSEL with output beam emitting from substrate, as shown in Fig. 4.12b, can be bonded on heatsink by the epitaxial surface, and therefore has better heat dissipation condition. This configuration is adopted widely for high power VCSEL and arrays. The other structure with output beam emitting from the epitaxial surface may have higher thermal resistance. Fortunately, the thermal effect is mitigated by its very low threshold. Besides, DBR itself has thermal behaviors different from the cleaved facet in ordinary LD. Detailed investigations of VCSEL thermal characteristics are given in papers and monographs [47, 71].

The low threshold of VCSEL is extremely important to applications of large volume optical communication and interconnection. As the amount of data being processed has been increasing explosively, the demand on equipment volumes in data centers increases continuously. The power dissipation of active devices in the equipment becomes one of the bottlenecks. The low threshold, low power dissipation, the capability of large-scale two-dimensional integration, and possibility of being integrated with CMOS (Complementary metal oxide semiconductor) drivers are remarkable superiority of the VCSEL, attracting wide attentions of the information technology (IT) industry [68–70]. To evaluate the performance of energy dissipation of lasers, indexes of energy-to-data ratio (EDR) and heat-to-bit rate ratio (HBR) are defined as

$$EDR = IV/BR, \tag{4.48}$$

$$HBR = (IV - P)/BR; \tag{4.49}$$

where I and V are the working current and voltage, BR stands for the bit rate, P is the output power; the unit of EDR and HBR is Joule per bit (*J/bit*). EDR and HBR are required to be minimized. As examples, 850 nm VCSELs with threshold current as low as 0.12 mA, output power over 0.5mW were reported in Ref. [69, 70]. The energy-to-data ratio was lowered to less than 100 *fJ/bit*, where $fJ = 10^{-15}$ J. The laser was used to transport data with BR of 17 Gb/s over multimode fiber of 1 km in experiments.

(2) **High Side Mode Suppression Ratio**

Another prominent advantage of VCSEL is its single longitudinal mode operation with extremely high SMSR due to its very short cavity length. Its mode spacing $\Delta\lambda = \lambda^2/(2n_{\text{eff}}L_z)$ can be estimated to reach hundreds nanometers, larger than the gain spectrum width of active semiconductor, so that no side modes can work, thus no mode partition and mode hopping occur. Especially, the single mode operation maintains under high frequency modulations.

Although requirements of linewidth may not be high in many applications of VCSEL, such as data transportations, but narrow line VCSEL are demanded in some cases, such as for frequency swept optical coherent tomography (OCT) and for physical researches of cooled atom [42, 61]. The linewidth is inversely proportional to the square of photon lifetime, as indicated by formula (4.30). Although the cavity length is very short, the reflectance of DBR is very high, resulting in a great number of roundtrips, beneficial to enlarge the photon lifetime. Reference [42] used a 1300 nm VCSEL in OCT technology; its coherent length was measured larger than 10 cm, which corresponds to linewidth of 0.8 pm, superior to ordinary edge emitting LD.

High power VCSELs with large aperture often work with multiple transverse modes; and the broadened linewidth may be allowed in related applications.

(3) **Tunability**

Wide range tunable VCSELs are developed based on structures of Fig. 4.16. Reference [42, 66] showed experimental results of 1300nm and 1550nm VCSELs, with tunable range larger than 100 nm. The lasing wavelength of a 900 nm VCSEL varied with the applied voltage was given by Ref. [64], showing a tunable range of 20 nm. Parabolic curves of tuning versus applied voltage were observed, coincident with formula (4.44). It is worthy of noticing that the laser maintains the single mode operation all the way; but as an external cavity laser the mode hopping may occur during the tuning. Such a wide tunable range is large beyond that of edge emitting LD. It is necessary to note that the threshold and output efficiency are changed with the applied voltage; one of the reasons is that the peak of standing wave in the cavity may move with the voltage, making the effective gain changed.

(4) **High Quality Beam**

The output beam of VCSEL has ideal near field and far field patterns because of its round aperture. The beam should obey Helmholtz equation in a cylindrical coordinate. Generally the near field is described by Laguerre polynomials, expressed as

$$L_{l,v}(r) = \frac{e^r}{l!r^v}\frac{d^l}{dr^l}(e^{-r}r^{v+l}).$$

When the aperture of VCSEL is small enough, the fundamental mode can be obtained, with $L_{0,v}(r) = 1$. The beam will be an ideal Gaussian beam, expressed as

$$E = E_0\frac{w_0}{w(z)}\exp\frac{-r^2}{w^2(z)}\exp j[kz + \frac{kr^2}{2R(z)} + \Gamma(z)], \tag{4.50}$$

where $w(z) = (w_0^2 + 4z^2/k^2w_0^2)^{1/2}$ is the beam width with waist of w_0 at its near field, $R(z) = z(1 + z_0^2/z)^{1/2}$ is the curvature radius of wave front in far field, with Rayleigh range of $z_0 = kw_0^2/2$; and $\Gamma(z) = \tan^{-1} z/z_0$ is called Gouy phase shift. As a comparison, the field of edge emitting laser has an elliptical beam, proportional to $\exp[-(x^2/w_x^2 + y^2/w_y^2)]$ with much different beam widths in x and y directions.

Reference [61] showed a measured far field pattern with an excellent fundamental transverse mode; the full width at half maximum (FWHM) divergent angle of about 12° is read. No asymmetry is observed, that is very attractive to laser applications, such as fiber coupling, collimating and focusing with lenses. Higher order transverse modes may appear in high power VCSELs with large apertures. The composed beam of multiple modes is usually affected by gain distribution in the active region. Reference [72] reported a TEM_{00} mode VCSEL with beam quality parameter $M^2 \sim 1.2$. It worked at 980 nm with output power more than 1 W, very suitable to be used as a pump of erbium doped fiber amplifier (EDFA) and laser (EDFL).

(5) **Broadband Modulation Response**

The modulation frequency response of VCSEL is higher than the edge emitting LD because its structure, including electrode configuration, ensures very low electrical capacity. On the other hand, its low working current is beneficial to high frequency modulation. As stated in Chap. 2, the modulation response is characterized by the 3 dB bandwidth $f_{3dB} \approx 1.55 f_{osc}$, where f_{osc} is the resonance frequency of relaxation oscillation, which is basically proportional to $\sqrt{I - I_{th}}$. An index, termed the modulation current efficiency factor (MCEF), is defined to measure the modulation performance of lasers, expressed as [61, 73, 74]:

$$MCEF = \frac{f_{3dB}}{\sqrt{I - I_{\text{th}}}} \propto \sqrt{\frac{\eta_{\text{inj}} \Gamma v_g g_N}{\mathrm{e} V_M}}. \tag{4.51}$$

The last term of (4.51) is written by referring the result of rate equation analysis, where η_{inj} is the current injection efficiency; Γ is the confinement factor describing the overlap degree between optical field and gain region; v_g is the group velocity; $g_N = \partial g / \partial N$ is the differential gain coefficient; and V_M is the mode volume.

A lot of papers reported their researches on the modulation response of VCSEL for high speed optical interconnection applications [68–70, 75, 76]. Figure 4.18 displays the measured frequency spectrum of modulation response of a 850 nm VCSEL, its 3 dB bandwidth is higher than 20 GHz, *MCEF* reaches 7.3 GHz/mA$^{1/2}$ [73].

(6) **Arrays and High Power Output**

In applications of high power laser, VCSEL has its unique attractions. The planar VCSEL arrays are suitable to side-pump the solid-state laser material, whereas the edge emitting LD bars have to be packaged in stacks. Its round output beam is suitable to match with usual optical components; and suitable to couple with fibers used as pumps for fiber amplifiers and fiber lasers. Technological progresses were reported in papers [41, 77, 78]. As an example, Fig. 4.19 shows a measured L-I curve, V-A curve, and electro-to-optic conversion efficiency of a VCSEL array with aperture of $2.8 \times 2.8\ \text{mm}^2$ and with the output power up-to 58 W [41].

By using two dimensional addressable VCSEL arrays, optical interconnection with ultra-high data speed up to 100 Gbps was realized [78], showing a good prospect. VCSEL has been used in variety of applications, including high speed optical communications, optical interconnections, optical fiber sensors, and optical storage technology. It is also used as sources in opto-electronic integrated chips, as

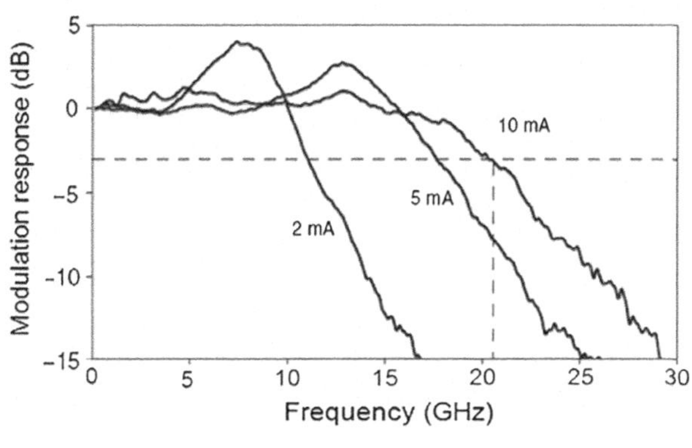

Fig. 4.18 Measured modulation frequency spectrum of VCSEL. Reproduced from [72] by permission of the Institution of Engineering & Technology

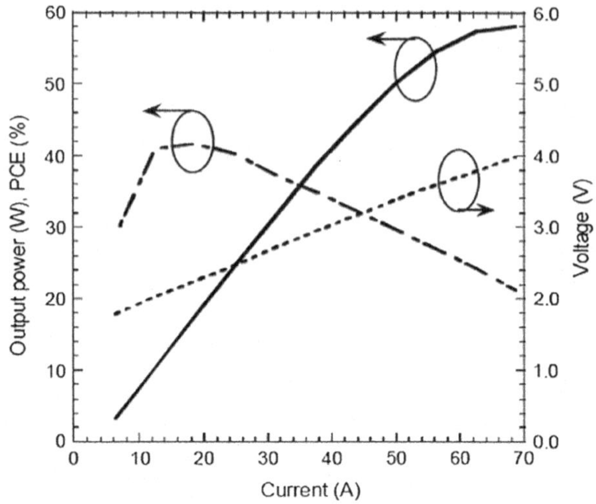

Fig. 4.19 L-I, V-A, and E-O conversion efficiency curves of a high power VCSEL array. Reprinted from [41] by courtesy of Princeton Optronics, Inc

pumps of fiber amplifiers, fiber lasers and solid state lasers. Modified VCSEL itself can play a role of optical amplifier [79], similar to SOA. VCSEL has its unique features and prospect also in the technology of cold atom frequency standard [80].

References

1. Kogelnik H, Shank CV (1971) Stimulated emission in a periodic structure. Appl Phys Lett 18:152–154
2. Kogelnik H, Shank CV (1972) Coupled-wave theory of distributed feedback lasers. J Appl Phys 43(5):2327–2335
3. Scifres DR, Burnham RD, Streifer W (1974) Distributed-feedback single heterojunction GaAs diode laser. Appl Phys Lett 25:203–206
4. Ghafouri-Shiraz H, Lo BSK (1996) Distributed feedback laser diodes. Wiley
5. Yariv A (1997) Optical electronics in modern communications. Chapter 16. 5th edn, Oxford University Press, Inc
6. Carroll JE, Plumb D, Whiteaway J (1998) Distributed feedback semiconductor lasers. Institution of Electrical Engineers
7. Yariv A (1973) Coupled-mode theory for guided-wave optics. IEEE J Quantum Electron 9 (9):919–933
8. Haus HA, Shank CV (1976) Antisymmetric taper of distributed feedback lasers. IEEE J Quantum Electron 12(9):532–539
9. Utaka K, Akiba S, Sakai K et al (1984) Analysis of quarter-wave-shifted DFB laser. Electron Lett 20(8):326–327
10. Kojima K, Kyuma K, Nakayama T (1985) Analysis of the spectral linewidth of distributed feedback laser diodes. J Lightwave Technol 3(5):1048–1055
11. Vankwikelberge P, Buytaert F, Franchois A et al (1989) Analysis of the carrier-induced FM response of DFB lasers: theoretical and experimental case studies. IEEE J Quantum Electron 25(11):2239–2254

12. Hirayama Y, Morinaga M, Onomura M et al (1992) High-speed 1.5 μm compressively strained multi-quantum well self-aligned constricted mesa DFB lasers. J Lightwave Technol 10(9):1272–1280
13. Wu M, Lou Y, Wang S (1988) Linewidth broadening due to longitudinal spatial hole burning in a long distributed feedback laser. Appl Phys Lett 52(14):1119–1121
14. Kimura T, Sugimura A (1987) Linewidth reduction by coupled phase-shift distributed-feedback lasers. Electron Lett 23(19):1014–1015
15. Rabinovich WS, Feldman BJ (1989) Spatial hole burning effects in distributed feedback lasers. IEEE J Quantum Electron 25(1):20–30
16. Nakano Y, Tada K (1998) Analysis, design, and fabrication of GaAlAs/GaAs DFB lasers with modulated stripe width structure for complete single longitudinal mode oscillation. IEEE J Quantum Electron 24(10):2017–2033
17. Agrawal GP, Bobeck AH (1988) Modeling of distributed feedback semiconductor lasers with axially-varying parameters. IEEE J Quantum Electron 24(12):2407–2414
18. Rennon S, Bach L, Reithmaier JP et al (2001) Complex coupled distributed-feedback and Bragg-reflector lasers for monolithic device integration based on focused-ion-beam technology. IEEE J Sel Top Quantum Electron 7(2):306–311
19. Zhang LM, Yu SF, Nowell MC et al (1994) Dynamic analysis of radiation and side-mode suppression in a second-order DFB laser using time-domain large-signal traveling wave model. IEEE J Quantum Electron 30(6):1389–1395
20. Yu SF (1996) A quasi-three-dimensional large-signal dynamic model of distributed feedback lasers. IEEE J Quantum Electron 32(3):424–432
21. Luo Y, Nakano Y, Tada K et al (1991) Fabrication and characteristics of gain-coupled distributed semiconductor lasers with a corrugated active layer. IEEE J Quantum Electron 27 (6):1724–1731
22. Broberg B, Nilsson S (1988) Widely tunable active Bragg reflector integrated lasers in InGaAsP/InP. Appl Phys Lett 52(16):1285–1287
23. Coldren LA, Fish GA, Akulova Y et al (2004) Tunable semiconductor lasers: a tutorial. J Lightwave Technol 22(1):193–202
24. Todt R, Jacke T, Meyer R et al (2004) Wide wavelength tuning of sampled grating tunable twin-guide laser diodes. Electron Lett 40(23):1491–1492
25. Numai T (1992) 1.5 μm phase-controlled distributed feedback wavelength tunable optical filter. IEEE J Quantum Electron 28(6):1508–1512
26. Numai T (1992) 1.5 μm phase-shift-controlled distributed feedback wavelength tunable optical filter. IEEE J Quantum Electron 28(6):1513–1519
27. Fang Z, Chin K, Qu R et al (2012) Fundamentals of optical fiber sensors. Wiley
28. Tohmori Y, Yoshikuni Y, Ishii H et al (1993) Broad-range wavelength-tunable superstructure grating (SSG) DBR lasers. IEEE J Quantum Electron 29(6):1817–1823
29. Kano F, Ishii H, Tohmori Y et al (1993) Characteristics of super structure grating (SSG) DBR lasers under broad range wavelength tuning. IEEE Photonics Technol Lett 5(6):611–613
30. Tohmori Y, Yoshikuni Y, Ishii H (1993) Over 100 nm wavelength tuning in superstructure grating (SSG) DBR lasers. Electron Lett 29(4):352–354
31. Morthier G, Moeyersoon B, Baets R (2001) A λ/4-shifted sampled or superstructure grating widely tunable twin-guide laser. IEEE Photonics Technol Lett 13(10):1052–1054
32. Ward AJ, Robbins DJ, Busico G et al (2005) Widely tunable DS-DBR laser with monolithically integrated SOA: design and performance. IEEE J Sel Top in Quantum Electron 11(1):149–152
33. Ishii H, Tanobe H, Kano F et al (1996) Quasicontinuous wavelength tuning in super-structure-grating (SSG) DBR lasers. IEEE J Quantum Electron 32(3):433–441
34. Akulova YA, Fish GA, Koh P et al (2002) Widely tunable electroabsorption-modulated sampled-grating DBR laser transmitter. IEEE J Sel Top Quantum Electron 8(6):1349–1357
35. Phelan R, Guo W, Lu Q et al (2008) A novel two-section tunable discrete mode Fabry-Pérot laser exhibiting nanosecond wavelength switching. IEEE J Quantum Electron 44(4):331–337

36. Fricke J, Bugge F, Ginolas A et al (2010) High-power 980-nm broad-area lasers spectrally stabilized by surface Bragg gratings. IEEE Photonics Technol Lett 22(5):284–286
37. Jewell JL, Harbison JP, Scherer A et al (1991) Vertical-cavity surface-emitting lasers: design, growth, fabrication, characterization. IEEE J Quantum Electron 27(6):1332–1345
38. Geels RS, Corzine SW, Coldren LA (1991) InGaAs vertical-cavity surface-emitting lasers. IEEE J Quantum Electron 27(6):1359–1367
39. Hasnain G, Tai K, Yang L et al (1991) Performance of gain-guided surface emitting lasers with semiconductor distributed Bragg reflectors. IEEE J Quantum Electron 27(6):1377–1385
40. Grabherr M, King R, Jäger R et.al. (2008) Volume production of polarization controlled single-mode VCSELs. Proceeding SPIE 6908:690803(1–9)
41. Seurin J, Xu G, Khalfin V et.al. (2009) Progress in high-power high-efficiency VCSEL arrays. Proceeding SPIE 7229:722903(1–11)
42. Jayaraman J, Jiang J, Potsaid B et al (2012) Design and performance of broadly tunable, narrow line-width, high repetition rate 1310 nm VCSELs for swept source optical coherence tomography. Proc SPIE 8276:82760D(1–11)
43. Lee TP (ed) (1995) Current trends in vertical cavity surface emitting lasers. World Scientific Publishing Co., Singapore
44. Li HE, Iga K (ed) (2003) Vertical-cavity Surface-emitting laser devices. Springer
45. Yu SF (2003) Analysis and design of vertical-cavity surface-emitting lasers. Wiley
46. Morgan RA (1997) Vertical-cavity surface-emitting lasers: present and future. Proc SPIE 3003:14–26
47. Chow WW, Choquette KD, Crawford MH et al (1997) Design, fabrication, and performance of infrared and visible vertical-cavity surface-emitting lasers. IEEE J Quantum Electron 33 (10):1810–1824
48. Michalzik R, Grabherr M, Jäger R et al (1998) Progress in high power VCSELs and arrays. Proc SPIE 3419:187–195
49. Born M, Wolf E (1999) Principles of optics. Seventh edn. Cambridge University Press
50. Yeh HJ, Smith JS (1994) Integration of GaAs vertical cavity surface emitting laser on Si by substrate removal. Appl Phys Lett 64(12):1466–1468
51. Babi DI, Dudley JJ, Streubel K et al (1995) Double fused 1.52 m vertical cavity lasers. Appl Phys Lett 66(9):1030–1032
52. Iga K (2000) Surface-emitting laser—its birth and generation of new optoelectronics field. IEEE J Sel Top Quantum Electron 6(6):1201–1205
53. Lu TC, Kao CC, Kuo HC et al (2008) CW lasing of current injection blue GaN-based vertical cavity surface emitting laser. Appl Phys Lett 92(14):141102(1–3)
54. Alford WJ, Raymond TD, Allerman AA (2002) High power and good beam quality at 980 nm from a vertical external-cavity surface-emitting laser. J Opt Soc of Am B 19(4):663–666
55. Mereuta A, Iakovlev V, Caliman A et al (2008) In(Al)GaAs-AlGaAs wafer fused VCSELs emitting at 2 μm wavelength. IEEE Photonics Technol Lett 20(1):24–26
56. Michalzik R, Ebeling KJ (1993) Modeling and design of proton-implanted ultralow-threshold vertical-cavity laser diodes. IEEE J Quantum Electron 29(6):1963–1973
57. Yang GM, MacDougal MH, Dapkus PD (1995) Ultralow threshold current vertical-cavity surface-emitting lasers obtained with selective oxidation. Electron Lett 31(11):886–888
58. Huffaker DL, Deppe DG (1997) Improved performance of oxide-confined vertical-cavity surface-emitting lasers using a tunnel injection active region. Appl Phys Lett 71(11):1449–1451
59. Yoshikawa H, Kosaka H, Kurihara K et al Complete polarization control of 8 × 8 vertical cavity surface emitting laser matrix arrays. Appl Phys Lett 66(8):908–910
60. Verschuuren MA, Gerlach P, van Sprang HA et al (2011) Improved performance of polarization-stable VCSELs by monolithic sub-wavelength gratings produced by soft nano-imprint lithography. Nanotechnol 22:505201(1–9)

61. Miah MJ, Al-Samaneh A, Kern A et al (2013) Fabrication and characterization of low-threshold polarization-stable VCSELs for Cs-based miniaturized atomic clocks. IEEE J Sel Top in Quantum Electron 19(4):1701410(1–10)
62. Chou SY, Schablitsky S, Zhuang L (1997) Subwavelength transmission gratings and their applications in VCSELs. Proc SPIE 3290:73–81
63. Debernardi P, Ostermann JM, Feneberg M et al (2005) Reliable polarization control of VCSELs through monolithically integrated surface gratings: a comparative theoretical and experimental study. IEEE J Sel Top Quantum Electron 11(1):107–116
64. Chang-Hasnain CJ (2000) Tunable VCSEL. IEEE J Sel Top Quantum Electron 6(6):978–987
65. Huang MCY, Zhou Y, Chang-Hasnain CJ (2008) A nanoelectromechanical tunable laser. Nat Photonics 2:180–184
66. Gierl C, Gruendl T, Debernardi P et al (2011) Surface micromachined tunable 1.55 μm-VCSEL with 102 nm continuous single-mode tuning. Opt Express 19(18):17336–17343
67. Chang YC, Wang CS, Coldren LA (2007) High-efficiency, high-speed VCSELs with 35 Gbit/s error-free operation. Electron Lett 43(19):1022–1023
68. Imai S, Takaki K, Kamiya S et al (2011) Recorded low power dissipation in highly reliable 1060-nm VCSELs for “green” optical interconnection. IEEE J Sel Top Quantum Electron 17 (6):1614–1620
69. Moser P, Hofmann W, Wolf P et al (2011) 81 fJ/bit energy-to-data ratio of 850 nm vertical-cavity surface-emitting lasers for optical interconnects. Appl Phys Lett 98:231106(1–3)
70. Moser P, Lott JA, Wolf P et al (2012) 99 fJ/(bit km) energy to data-distance ratio at 17 Gb/s across 1 km of multimode optical fiber with 850-nm single-mode VCSELs. IEEE Photonics Technol Lett 24(1):19–21
71. Osinski M, Nakwaski W (1995) Thermal effects in vertical-cavity surface-emitting lasers. Selected paper of “Current Trends in Vertical Cavity Surface Emitting Lasers”. In: Lee TP (ed) World Scientific Publishing Co., Singapore
72. Mooradian A. (2001). High brightness cavity-controlled surface emitting GaInAs lasers operating at 980 nm. Proc OFC PD17–3, Anahaim, USA
73. Westbergh P, Gustavsson JS, Haglund A et al (2008) Large aperture 850 nm VCSELs operating at bit rates up to 25 Gbit/s. Electron Lett 44(15):907–908
74. Al-Samaneh A, Renz S, Strodl A et al (2010) Polarization-stable single-mode VCSELs for Cs-based MEMS atomic clock applications. Proc SPIE 7702:770206(1–14)
75. Gustavsson J, Westbergh P, Szczerba K et al (2010) High-speed 850-nm VCSELs for 40 Gb/s transmission. Proc SPIE 7720:772002(1–11)
76. Mutig A, Lott JA, Blokhin SA et al (2011) Modulation characteristics of high-speed and high-temperature stable 980 nm range VCSELs operating error free at 25 Gbit/s up to 85°C. IEEE J Sel Top Quantum Electron 17(6):1568–1575
77. Miller M, Grabherr M, Jäger R et al (2001) High-power VCSEL arrays for emission in the watt regime at room temperature. IEEE Photonics Technol Lett 13(3):173–175
78. Tatum JA, Johnson RH, Guenter JK et al (2010) High data throughput VCSELs. Proc SPIE 7720:772004(1–6)
79. Karim A, Björlin S, Pipre J et al (2011) Long-wavelength vertical-cavity lasers and amplifiers. IEEE J Sel Top Quantum Electron 6(6):1244–1253
80. Keeler GA, Geib KM, Serkland DK et al (2007) VCSEL polarization control for chip-scale atomic clocks. Sandia Report

Chapter 5
External Cavity Semiconductor Lasers

5.1 General Characteristics and Theory of External Cavity Diode Lasers

It is known that the ordinary F-P cavity semiconductor laser works often in multiple longitudinal modes, because its gain spectrum is as wide as several nanometers and the cavity mirror lacks wavelength selectivity. The spectrum of such lasers changes with variation of temperature and injection current, and mode hopping occurs frequently. Although the electro-optic conversion efficiency of semiconductor laser is much higher than other lasers because of its higher gain coefficient and lower reflectance of cavity mirror, this advantage brings about shortcomings of larger linewidth due to a lower Q of cavity, and higher susceptibility to external optical feedback. Such characteristics are not welcome in practical applications of single frequency lasers. It is found that a proper stable feedback of laser output is helpful for mode selectivity and linewidth reduction. Such a laser is called the external cavity diode laser or the extended cavity diode laser (ECDL). Quite many types of ECDL have been developed, including the cavities composed of a simple planar mirror and of different mode selective reflectors. Among them, the planar diffractive grating, the Bragg grating, and Fabry-Perot cavity are mostly often used as frequency selective elements.

The theory and structures of ECDL and the effect of external optical feedback on LD performances have been studied and expounded in many monographs and papers [1–4]. This section introduces the general characteristics and theory of ECDL, and explains its basic properties and physical mechanisms.

5.1.1 Basic Model of ECDL

A simplified model of ECDL is given in Fig. 5.1, where r_i (i = 1, 2, ex) is the field reflectance, L and l are the lengths of LD chip and extended section, respectively.

Z. Fang et al., *Single Frequency Semiconductor Lasers*, Optical and Fiber Communications Reports 9, DOI 10.1007/978-981-10-5257-6_5

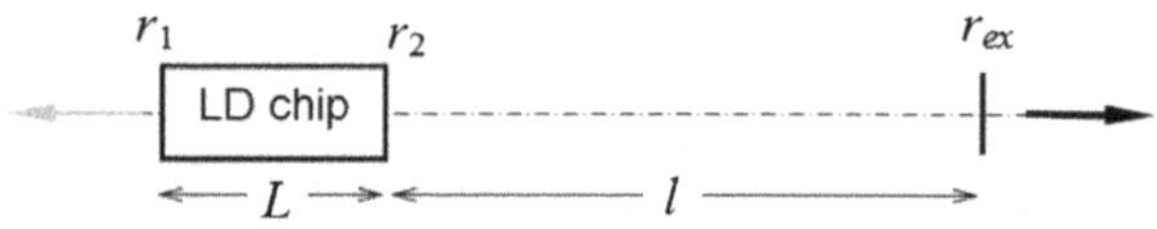

Fig. 5.1 Basic structure of ECDL

The composite role of the external mirror (r_{ex}) and the right facet (r_2) can be described as an effective reflectance [5]:

$$r_{eff} = \frac{r_2 + r_{ex} e^{j\omega\tau}}{1 + r_2 r_{ex} e^{j\omega\tau}} = |r_{eff}| e^{j\phi_{eff}}, \tag{5.1a}$$

$$|r_{eff}|^2 = \frac{r_2^2 + r_{ex}^2 + 2r_2 r_{ex}\cos\omega\tau}{1 + r_2^2 r_{ex}^2 + 2r_2 r_{ex}\cos\omega\tau}, \tag{5.1b}$$

$$\tan\phi_{eff} = \frac{r_{ex}(1 - r_2^2)\sin\omega\tau}{r_2(1 + r_{ex}^2) + r_{ex}(1 + r_2^2)\cos\omega\tau}. \tag{5.1c}$$

where $\tau = 2\,l/c$ is the roundtrip time delay in the extended section, which is assumed with medium of the air. The amplitude of effective reflectance is spectrally modulated for $r_2 \neq 0$. The effective phase shift is just equal to $\omega\tau$ in case of $r_2 = 0$.

The operation of ECDL must meet gain and phase conditions, the same as any other lasers:

$$\Gamma g = \alpha_c + L^{-1}\ln(1/r_1 r_{eff}), \tag{5.2}$$

$$2n_{LD}kL + \phi_{eff} = 2m\pi, \tag{5.3}$$

where $k = 2\pi/\lambda$ is the wavevector in vacuum, n_{LD} is the index of LD chip. Usually the extended cavity length l is much longer than LD chip length L, and the mode spacing of ECDL is much narrower than that of the solitary LD. If the external mirror is a broadband reflector without wavelength selectivity, the ECDL will operate in multiple longitudinal modes, unless the internal longitudinal mode of LD chip coincides with one of the external cavity modes, making the main mode dominating over the other side modes.

Even if in such a multimode operation case, the external cavity will benefit linewidth reduction because the photon lifetime in the cavity is extended. The laser cavity is usually modeled as an F-P cavity; the linewidth of resonant peak is written as $\delta\lambda = \Delta\lambda/F$, where $\Delta\lambda = \lambda^2/(2nd)$ is the free spectral range with cavity length of d, $F = \pi\sqrt{R}/(1 - R)$ is the finesse of F-P cavity. The frequency width of passive cavity is inversely proportional to the cavity length, $\delta\nu_c \propto \delta\lambda \propto 1/d$. From Schawlow–Townes formula $\delta\nu = 2\pi h\nu(\delta\nu_c)^2/P$ [7], the laser's intrinsic linewidth will decrease with the cavity length.

The contribution of external cavity can be specified with the relation of photon lifetime with the propagation phase shift. If the external mirror r_{ex} has no frequency selectivity, the photon lifetime is given as

$$\tau_p = \frac{\partial}{\partial\omega}[\phi_{\text{LD}} + \phi_{\text{eff}}] = \frac{2n_g L}{c} + \frac{u(v\cos\omega\tau + w)}{u^2\sin^2\omega\tau + (v + w\cos\omega\tau)^2}\tau, \tag{5.4}$$

where $u = r_{\text{ex}}(1 - r_2^2)$, $v = r_2(1 + r_{\text{ex}}^2)$, $w = r_{\text{ex}}(1 + r_2^2)$. For $r_2 = 0$, $\tau_p = 2(n_g L + l)/c$. Therefore, the photon lifetime is basically proportional to the cavity's optical path, but modulated by a function related to parameters r_2, r_{ex}, and $\omega\tau$.

Another important feature of ECDL is the kinks of its output versus current (*L–I*) curve, as shown in Fig. 5.2a. The phenomenon was reported experimentally and studied theoretically earlier in Ref. [6]. The behavior can be understood by investigating the lasing conditions (5.2) and (5.3). The index of LD chip is a function of carrier concentration, temperature, and wavelength; the position of longitudinal mode of ECDL is dependent on the three factors. If one of the LD longitudinal modes is located between two external cavity modes, the laser may work at the *m*th mode with a higher threshold, or at the (*m* + 1)th mode with a lower threshold, as shown in Fig. 5.2b. The Joule heating of injected current induces an additional effect on the behavior, making bistability possible. Mode hopping will occur in such cases, and output power will jump up and down, showing a bistability behavior in a small region. To avoid kinks, laser's working point has to be moved to between the kinks by precisely adjusting the extended length and pumping bias, to get a good single mode performance and continuous tunability by current variation.

The structure of Fig. 5.1 is usually called linear external cavity laser; differently, ring cavity lasers are also developed, which uses a LD chip with both facets fully anti-reflective coated, as the gain element of a semiconductor optical

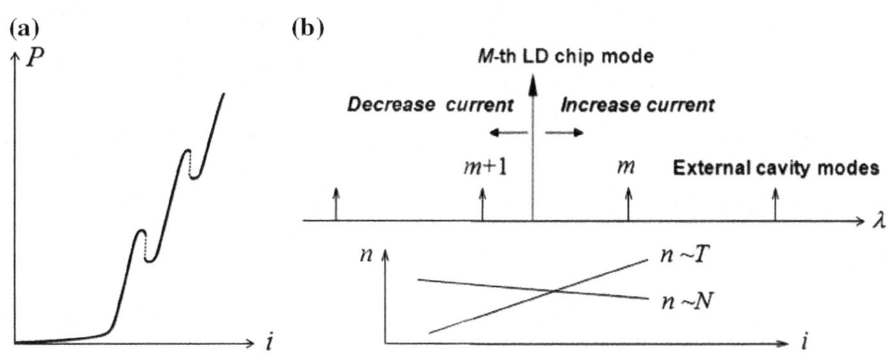

Fig. 5.2 **a** *L–I* curve kinks of ECDL. **b** Explanation of mode competition

amplifier (SOA). Components with different functions can be inserted in the ring, such as wavelength selector, phase modulator, amplitude modulator, and output coupler. The configuration and characteristics will be introduced in related sections.

5.1.2 Linewidth Reduction of ECDL with Weak Feedback

For weak feedback ECDLs with a broad spectrum external mirror, Ref. [6, 8–10] presented detailed investigations. A brief introduction to the theories is given here. The rate equations of optical field and phase should be taken into account separately and the external feedback should be added into the equations. The equation is now written as [9]:

$$\frac{\mathrm{d}}{\mathrm{d}t}\left[E(t)\mathrm{e}^{-j\omega t}\right] = \left(-j\omega_{\mathrm{LD}} + \frac{1}{2}G_{\mathrm{net}}\right)E(t)\mathrm{e}^{-j\omega t} + \kappa E(t-\tau)\mathrm{e}^{-j\omega(t-\tau)}, \tag{5.5}$$

where $\omega_{\mathrm{LD}} = \pi Mc/(nL)$ is the resonant frequency of LD chip, $G_{\mathrm{net}} = v_g(\Gamma g - \alpha_c)$ is the net gain in time domain, and Γ is the confinement factor. The loss coefficient α_c here includes the loss of output from the left facet, $\alpha_1 = L^{-1}\ln(1/r_1^2)$, to make the influence of right facet clearer. The last term of (5.5) describes the feedback contribution, denoted by coefficient κ, which is obtained from the difference between with and without the external mirror. For the case of weak feedback, $r_2 r_{\mathrm{ex}} \ll 1$, only one roundtrip propagation between chip's right facet and the external mirror is needed to be taken into account; thus $\kappa \approx r_{\mathrm{ex}}(1-r_2^2)/r_2\tau_{\mathrm{LD}}$ is given, where $\tau_{\mathrm{LD}} = 2n_g L/c$ is the roundtrip time of LD chip. The beam in the extended path is assumed without divergence here for simplicity. Denoting the optical intensity with $I = |E|^2$ and $E = \sqrt{I}\mathrm{e}^{-j\phi}$, the rate equations of intensity and phase are obtained as [9, 10]:

$$\dot{I} = G_{\mathrm{net}}I + 2\kappa If\cos(\omega\tau - \Delta\phi) + R_1 + F_I(t), \tag{5.6a}$$

$$\dot{\phi} = \omega_{\mathrm{LD}} - \omega - \kappa f\sin(\omega\tau - \Delta\phi) + F_\phi(t). \tag{5.6b}$$

where $f(t,\tau) = \sqrt{I(t-\tau)/I(t)}$ and $\Delta\phi = \phi(t-\tau) - \phi(t)$; $R_1 = \gamma_{\mathrm{sp}}R_{\mathrm{sp}} = \gamma_{\mathrm{sp}}N/\tau_{\mathrm{sp}}$ is the contribution of spontaneous emission to the laser mode. $F_I(t)$ and $F_\phi(t)$ are Langevin forces of intensity and phase noises. The rate equation of carrier concentration is written as:

$$\dot{N} = C - R - GI + F_N(t), \tag{5.6c}$$

where $C = J/ed$, $G = v_g g$; $R = R_{\mathrm{sp}} + R_{\mathrm{nr}} = N/\tau_N$ is the recombination rate, including nonradiative recombination; τ_N is the lifetime of carriers. $F_N(t)$ is the Langevin force describing fluctuations of injected carrier concentration.

Under the stationary state, $f = 1$, $\Delta\phi = 0$, and all Langevin forces are averaged to zero, the stationary parameters are given as

$$\bar{I} = \frac{R_1}{-G_{\text{net}} - 2\kappa \cos \overline{\omega}\tau}, \tag{5.7a}$$

$$\overline{\omega} = \omega_{\text{LD}} - \kappa \sin \overline{\omega}\tau, \tag{5.7b}$$

$$\overline{R} + \overline{G} \cdot \bar{I} = C. \tag{5.7c}$$

It is noted that the net gain under stationary state is a negative; the sum of net gain and the contribution of spontaneous emission equals to zero to maintain the laser with a stationary output. The spontaneous emission coefficient is very small; under the usual bias, $\bar{I} \gg \tau_{\text{LD}} R_1$, therefore, $G_{\text{net}} \approx -2\kappa \cos \overline{\omega}\tau$ from (5.7a). The longitudinal modes are determined by (5.7b), in which both of ω_{LD} and κ are functions of chip index, resulting in mode hopping at certain points in current variation, where the mode hopping causes kink of *L–I* curve, from (5.7a).

It is shown by (5.7a), (5.7b) and (5.7c) that the effect of external feedback on longitudinal modes and on threshold gain depends on phase shift $\omega\tau$. The ECDL can thus be tuned by precisely adjusting the extended length l by mechanical methods. Piezoelectric transducers (PZT) are often used for the tuning, and also for laser frequency stabilization.

The intensity noise and linewidth of ECDL are determined by the dynamic rate Eqs. (5.6a) and (5.6b). In case the extended cavity length is less than laser coherence length, and the variation of LD output is small during the delay $\tau = 2l/c$, approximations are taken as $f(t,\tau) \approx 1 - \tau\dot{p}/2\bar{I}$ with $p = I - \bar{I}$ and $\Delta\phi = \phi(t-\tau) - \phi(t) \approx -\tau\dot{\phi}$. The gain coefficient is expanded over carrier concentration as $G_{\text{net}} = \overline{G}_{\text{net}} + G_N \delta N$ with $\delta N = N - \overline{N}$ and $G_N = \partial G/\partial N$. The LD frequency and feedback coefficient are expanded as $\omega_{\text{LD}} = \overline{\omega}_{\text{LD}}[1 - (h_N \delta N + h_\omega \dot{\phi})]$ and $\kappa = \overline{\kappa}[1 - (h_N \delta N + h_\omega \dot{\phi})]$ with $h_N = n^{-1}\partial n/\partial N$ and $h_\omega = n^{-1}\partial n/\partial \omega$. The rate equations are then rewritten as

$$(1+a)\dot{I} + (2b - h_\omega G_{\text{net}}) I \dot{\phi} = (G_N - h_N G_{\text{net}}) I \delta N + F_I, \tag{5.8a}$$

$$(n_g/n + a)\dot{\phi} - (b/2I)\dot{I} = (n_g/2n)\alpha G_N \delta N + F_\phi, \tag{5.8b}$$

$$\dot{N} = -G_N I \delta N - G \delta I + F_N. \tag{5.8c}$$

where $a = \overline{\kappa}\tau \cos \omega\tau$, $b = \overline{\kappa}\tau \sin \omega\tau$. $\alpha = 2\omega_0(\partial n/\partial N)/(n_g G_N) = \Delta n_r/\Delta n_i$ is just the linewidth enhancement factor [9]. Langevin forces have properties of [7, 10]:

$$\begin{gathered} \langle F_i(t) \rangle = 0 \\ \langle F_i(t) F_j(u) \rangle = 2D_{ij}\delta(t-u), \end{gathered} \tag{5.9}$$

where $D_{II} = \gamma_{sp} I$, $D_{\phi\phi} = \gamma_{sp}/4I$, $D_{NN} = (\gamma_{sp} I + R)$, $D_{IN} = -\gamma_{sp} I$, $D_{\phi I} = 0$, and $D_{\phi N} = 0$. Equations (5.8a), (5.8b) and (5.8c) can be solved by Fourier transform, and the power spectral densities (PSD) of intensity noise, phase noise, and carrier concentration noises are then obtained. The ECDL linewidth is deduced in Ref. [9, 10], expressed as

$$\delta v = \frac{\delta v_{\mathrm{LD}}}{\left[1 + \kappa\tau(\cos\omega\tau - \alpha\sin\omega\tau)\right]^2}, \tag{5.10}$$

where $\delta v_{\mathrm{LD}} = v_g^2 h v \gamma_{sp} g \alpha_m (1 + \alpha^2)/(8\pi P)$ is the linewidth of solitary LD.

It is indicated from (5.10) that the linewidth reduction effect of external feedback is enhanced by a larger external cavity length, and depends on the phase shift. The best state is obtained when $\omega\tau$ is in the fourth quadrant [9], where the positive $\cos\omega\tau$ gives a lower threshold gain from (5.7a), and negative $\sin\omega\tau$ is beneficial to linewidth reduction. It is shown also from (5.7b) and (5.10) that the linewidth reduction requires $\kappa \geq \sqrt{1+\alpha^2}|(\omega - \omega_{\mathrm{LD}})|$; i.e., the feedback coefficient needed for the ECDL is proportional to the difference between the lasing frequency and the LD chip longitudinal mode.

Another effect of the extended cavity is the lowered relaxation oscillation frequency due to the increased photon lifetime, $\omega_R \approx \sqrt{v_g \Gamma g_N p_0/\tau_p}$. Based on the coupling between photons and carriers, described by the dynamic rate equation, the power spectral densities of intensity noise and frequency noise are affected by the cavity extension. Ideally, the noises are suppressed in a certain degree, as discussed in Sect. 3.3.3. However, the larger cavity length may cause a higher susceptivity to the external disturbances and the intensity noise have a tend to increase [11]. Obviously, the mechanical stability of external cavity is important, and intentional active stabilization measures are necessary for low noise and narrow line lasers.

5.1.3 Characteristics of ECDL with Frequency Selective Feedback

Although the extension of cavity length benefits to line narrowing, multimode operation due to reduced mode spacing is not welcome for a single mode laser. An external mirror with narrow bandwidth frequency selectivity is thus necessary. Since the tunability of ECDL is one of the concerned functions, the frequency selective element is required to provide wide tuning ranges. On the other hand, the intracavity mode of LD chip will bring about unwanted influences on the frequency selectivity of external mirror; the chip facet facing the external mirror should be coated with anti-reflective film to weaken the oscillation between the two cleaved facets. It is meant that the external mirror cannot be regarded as a weak feedback, and the analysis above is no longer valid for such a strong feedback case.

(a) **Simplified Model**

The frequency selective elements used widely include diffractive planar gratings, Bragg gratings, Fabry-Perot cavity, multilayer dielectric films, and so on. They have different spectral characteristics. For simplicity, their reflectivity near the top of reflection peak can be described approximately as Lorentzian shape, expressed as [12–14]:

$$r_{\mathrm{ex}} = \frac{r_0}{1 + j(\omega - \omega_0)/\Delta\omega}. \tag{5.11}$$

The intensity reflectivity is $R_{\mathrm{ex}} = |r_{\mathrm{ex}}|^2 = r_0^2/[1 + (\omega - \omega_0)^2/\Delta\omega^2]$, where $\Delta\omega$ is the half width at half maximum of R_{ex}. The effective reflectance of external cavity is now reduced to

$$r_{\mathrm{eff}} = \frac{r_2[1 + j(\omega - \omega_0)/\Delta\omega] + r_0 \mathrm{e}^{j\omega\tau}}{1 + j(\omega - \omega_0)/\Delta\omega + r_2 r_0 \mathrm{e}^{j\omega\tau}} \approx \frac{r_0 \mathrm{e}^{j\omega\tau}}{1 + j(\omega - \omega_0)/\Delta\omega}. \tag{5.12}$$

The last form is the approximation for $r_2 \ll r_0 < 1$. The lasing condition is then written as

$$r_1 r_{\mathrm{ex}} \exp[g_{\mathrm{net}} L + j2k(nL + l)] = 1 \tag{5.13}$$

where g_{net} is the net gain of LD chip. Its real part and imaginary part are written as:

$$r_1 r_0 \mathrm{e}^{g_{\mathrm{net}} L} \cos(2nkL + 2kl) = 1, \tag{5.14a}$$

$$r_1 r_0 \mathrm{e}^{g_{\mathrm{net}} L} \sin(2nkL + 2kl) - (\omega - \omega_0)/\Delta\omega = 0. \tag{5.14b}$$

The gain and phase lasing conditions are obtained as

$$g_{\mathrm{net}} = \frac{1}{L} \ln \frac{1}{r_1 r_0 \cos(\omega\tau_T)}, \tag{5.15a}$$

$$\Delta\omega \tan(\omega\tau_T) = \omega - \omega_0, \tag{5.15b}$$

where $\tau_T = 2(nL + l)/c$ is the roundtrip time of propagation in the composite cavity. It is noticed that the narrow bandwidth external mirror affects not only the gain condition, but also the phase condition. The mode positions are given by Eq. (5.15b), whose solution can be obtained with graphic method, as shown in Fig. 5.3. The mode spacing is now a spectral function; and the mode number decreases one in the band of external reflection, compared with a broadband external mirror, as shown in Fig. 5.4.

The LD chip with residue reflectance r_2 can be regarded as a low finesse active F-P cavity with the gain spectrum modulated by roundtrip propagations. The active reflection to the extended cavity can be expressed as

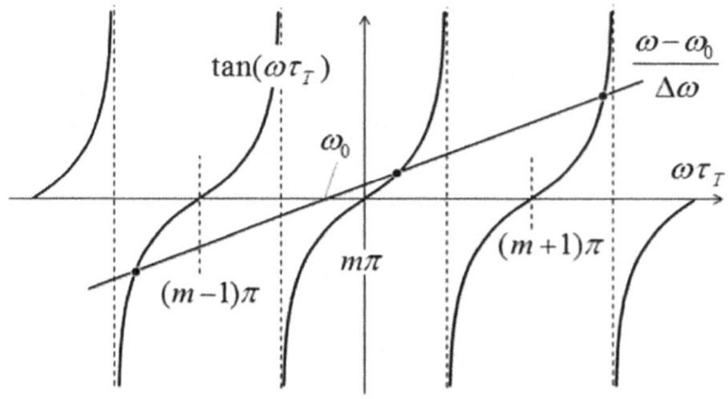

Fig. 5.3 Schematic diagram of longitudinal mode positions

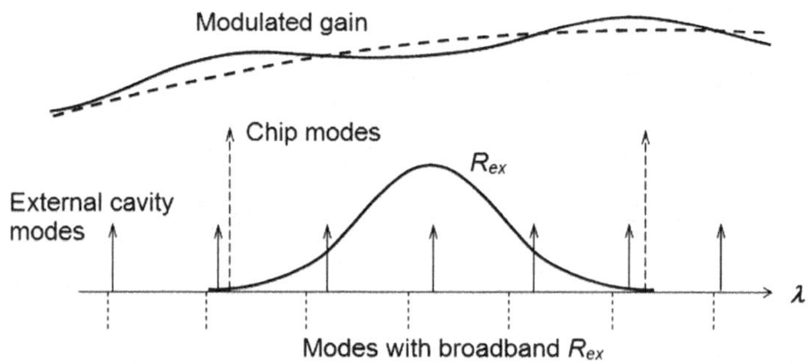

Fig. 5.4 Schematic diagram of ECDL modes, chip modes, external reflection spectrum and modulated gain spectrum

$$\begin{aligned} R_{\mathrm{LD}} &= \left| \frac{r_2 + r_1 \mathrm{e}^{g_{\mathrm{net}}L/2 + jnkL}}{1 + r_2 r_1 \mathrm{e}^{g_{\mathrm{net}}L/2 + jnkL}} \right|^2 = \frac{R_2 + R_1 \mathrm{e}^{g_{\mathrm{net}}L} + 2 r_2 r_1 \mathrm{e}^{g_{\mathrm{net}}L/2} \cos nkL}{1 + R_2 R_1 \mathrm{e}^{g_{\mathrm{net}}L} + 2 r_2 r_1 \mathrm{e}^{g_{\mathrm{net}}L/2} \cos nkL}. \\ &\approx R_1 \mathrm{e}^{g_{\mathrm{net}}L} + R_2 + 2 r_2 r_1 \mathrm{e}^{g_{\mathrm{net}}L/2} \left(1 - R_1 \mathrm{e}^{g_{\mathrm{net}}L}\right) \cos nkL \end{aligned} \tag{5.16}$$

The last expression is the approximation with only a linear term of r_2. The lasing condition can now be written as $R_{\mathrm{LD}} R_{\mathrm{ex}} = 1$. The modulated gain will affect the ECDL mode competition, as shown schematically in Fig. 5.4.

(b) **Linewidth Reduction:**

The above qualitative discussion is mainly with respect to the lasing conditions. The effect of external cavity on laser noises is more concerned in single frequency lasers. The linewidth of ECDL with a narrowband cavity mirror will be reduced not only by the extended cavity length, but also by other mechanisms related to the spectrum of feedback. First, the phase factor of external reflectance will induce an increasing of group delay and the photon lifetime. This is so-called "slow-light" effect [15, 16]. Second, the spectrally varying reflectance of external mirror changes the threshold gain and related carrier concentration; and the induced index variation causes the lasing frequency chirping, as analyzed by Ref. [17–20]. A brief introduction is given as follows.

Differentiating the threshold gain condition (5.2) gives relation of

$$\Delta g_{\text{net}} = -\frac{\Delta\omega}{L}\frac{\partial}{\partial\omega}\ln r_{\text{eff}}. \tag{5.17}$$

To take into account the intrinsic stochastic phase fluctuation, such as those induced by spontaneous emission and other mechanisms [18], denoted by $\Delta\phi_{\text{in}}$, the phase condition (5.3) is rewritten as

$$2n_{\text{LD}}kL + \phi_{\text{eff}} + \Delta\phi_{\text{in}} = 2m\pi. \tag{5.18}$$

With the relations between g_{net}, n_i and linewidth enhancement factor α, the chip index is expanded as

$$n_{\text{LD}} = n_0 + \frac{\partial n}{\partial\omega}\Delta\omega + \frac{\partial n}{\partial N}\Delta N = n_0 + \frac{\partial n}{\partial\omega}\Delta\omega - 2k\alpha\Delta g_{\text{net}}. \tag{5.19}$$

By differentiating (5.18) a relation is generated as

$$\tau_{\text{rt}}\Delta\omega - \alpha L\Delta g_{\text{net}} + \frac{\partial\phi_{\text{eff}}}{\partial\omega}\Delta\omega + \tau_{\text{rt}}\Delta\omega_0 = 0, \tag{5.20}$$

where $\tau_{\text{rt}} = 2n_g L/c$ is the roundtrip time of LD cavity, $n_g = n_0 + \omega(\partial n/\partial\omega)$ is the group index. The intrinsic phase fluctuation of solitary LD is expressed as $\Delta\phi_{\text{in}} = \tau_{\text{rt}}\Delta\omega_0$ by the same consideration. Combining (5.17) and (5.20), the reduction of frequency chirping by external cavity is obtained:

$$\Delta\omega = \frac{\Delta\omega_0}{1 + A + B}, \tag{5.21}$$

where $A = \tau_{\text{rt}}^{-1}(\partial\phi_{\text{eff}}/\partial\omega)$ and $B = \alpha\tau_{\text{rt}}^{-1}(\partial\ln|r_{\text{eff}}|/\partial\omega)$. The two parameters reflect the abovementioned two factors of frequency selectivity of external mirror.

Expression (5.21) gives a basic relation for the effect of external cavity on frequency fluctuation, but it cannot be regarded as frequency noise quantitatively, which should be obtained by statistically averaging the second moment of stochastic variable, i.e., its variance. To understand the noises as a whole, including the intensity noise, the deduction should start from the rate equations. The ECDL with frequency selective mirror can be modeled by a Fabry-Perot oscillator, whose transfer function is written as $E(\omega) = F(\omega)/(1 - r_1 r_{\text{eff}} e^{j2nkL})$ [18], where $F(\omega)$ is the Fourier component of Langevin force. Under the DC bias high enough above threshold, approximation of $(r_1 r_{\text{eff}} e^{j2nkL})_0 \approx 1$ is valid. The denominator of $E(\omega)$ is expanded in spectral domain, written as

$$1 - r_1 |r_{eff}| e^{j(2nkL + \phi_{eff})}$$
$$= \left(\frac{1}{|r_{eff}|} \frac{\partial |r_{eff}|}{\partial \omega} + j \frac{\partial \phi_{eff}}{\partial \omega} \right) \Delta\omega - j \frac{2n_g L}{c} \Delta\omega - j2kL \left(\frac{\partial n_r}{\partial N} + j \frac{\partial n_i}{\partial N} \right) \Delta N .$$
$$= -j\tau_{rt}(A - jB/\alpha)\Delta\omega - j\tau_{rt}\Delta\omega - jL(\alpha + j)\Delta g$$

The transfer function $E(\omega) = F(\omega)/(1 - r_1 r_{\mathrm{eff}} e^{j2nkL})$ is then reduced to

$$\left(1 + A - j\frac{B}{\alpha}\right)[-j\Delta\omega E(\omega)] = \frac{1}{2}(1 - j\alpha)E(\omega)\Delta G + F(\omega)/\tau_{\mathrm{rt}}.$$

By inverse Fourier transform, the rate equation in time domain is obtained:

$$\left(1 + A - j\frac{B}{\alpha}\right)\dot{E}(t) = \frac{1}{2}(1 - j\alpha)\Delta G E(t) + F(t). \tag{5.22}$$

With $E = \sqrt{I} e^{-j\phi}$, the rate equations of photon density and phase are obtained as

$$(1 + A)\dot{I} - \frac{2B}{\alpha} I\dot{\phi} = I\Delta G + R_{\mathrm{sp}} + F_I, \tag{5.23a}$$

$$(1 + A)\dot{\phi} + \frac{B}{\alpha}\frac{\dot{I}}{2I} = \frac{\alpha}{2}\Delta G + F_\phi, \tag{5.23b}$$

$$\dot{N} = -G_N I \Delta N - G\Delta I + F_N. \tag{5.23c}$$

The contribution of spontaneous emission R_{sp} is added in (5.23a) for a complete expression. The rate equation of carrier concentration (5.23c) is the same as that in Sect. 5.2. Compared with the intrinsic frequency noise induced by the spontaneous emission, the intensity fluctuation is much slower, $\dot{I} \approx 0$ can be taken approximately. The phase equation is then reduced to an expression similar to (5.21), but with noise source included:

$$(1 + A + B)\dot{\phi} = F_\phi - \alpha F_I/2I. \tag{5.24}$$

The frequency width, determined by the variance of phase fluctuation, is thus derived to be

$$\delta\nu = \frac{\delta\nu_{\mathrm{LD}}}{(1 + A + B)^2} = \frac{1 + \alpha^2}{(1 + A + B)^2}\delta\nu_{\mathrm{ST}}, \tag{5.25}$$

where $\delta\nu_{\mathrm{LD}}$ is the frequency width of solitary LD, as given in Chap. 3.

It is noticed if the external reflectance is not a real factor, but a complex as $r_{\mathrm{ex}} = |r_{\mathrm{ex}}| e^{j\phi_{\mathrm{ex}}}$, formula (5.1a), (5.1b) and (5.1c) of effective reflectance has to be modified, expressed as:

$$|r_{\text{eff}}|^2 = \frac{r_2^2 + |r_{\text{ex}}|^2 + 2r_2|r_{\text{ex}}|\cos(\omega\tau + \phi_{\text{ex}})}{1 + r_2^2|r_{\text{ex}}|^2 + 2r_2|r_{\text{ex}}|\cos(\omega\tau + \phi_{\text{ex}})}, \tag{5.26a}$$

$$\tan\phi_{\text{eff}} = \frac{|r_{\text{ex}}|(1 - r_2^2)\sin(\omega\tau + \phi_{\text{ex}})}{r_2(1 + |r_{\text{ex}}|^2) + |r_{\text{ex}}|(1 + r_2^2)\cos(\omega\tau + \phi_{\text{ex}})}. \tag{5.26b}$$

It means that factor A and B are now coupled with each other, and the coupling declines as the reflection of LD chip facet r_2 decreases. In the limit case of strong feedback, $r_2 \ll |r_{\text{ex}}(\omega_0)|$, approximations of $|r_{\text{eff}}|^2 \approx |r_{\text{ex}}|^2$ and $\tan\phi_{\text{eff}} \approx \tan(\omega\tau + \phi_{\text{ex}})$ are valid. Parameters A and B in (5.21) are then expressed with relations to the reflectance of external mirror:

$$\begin{aligned} A &= \tau_{\text{rt}}^{-1}(\tau + \partial\phi_{\text{ex}}/\partial\omega) \\ B &= \alpha\tau_{\text{rt}}^{-1}\partial\ln|r_{\text{ex}}|/\partial\omega. \end{aligned} \tag{5.27}$$

For weak feedbacks, $|r_{\text{ex}}(\omega_0)| \ll r_2$, we have $|r_{\text{eff}}|^2 \approx r_2^2[1 + 2\rho\cos(\omega\tau + \phi_{\text{ex}})]$ and $\tan\phi_{\text{eff}} \approx \rho\sin(\omega\tau + \phi_{\text{ex}})$ with $\rho = |r_{\text{ex}}|(r_2^{-1} - r_2)$. Parameters A and B are reduced to the following approximate forms:

$$A = \frac{\rho}{\tau_{\text{rt}}}\left[\left(\tau + \frac{\partial\phi_{\text{ex}}}{\partial\omega}\right)\cos(\omega\tau + \phi_{\text{ex}}) + \frac{\partial\ln|r_{\text{ex}}|}{\partial\omega}\sin(\omega\tau + \phi_{\text{ex}})\right], \tag{5.28a}$$

$$B = \frac{\alpha\rho}{\tau_{\text{rt}}}\left[\frac{\partial\ln|r_{\text{ex}}|}{\partial\omega}\cos(\omega\tau + \phi_{\text{ex}}) - \left(\tau + \frac{\partial\phi_{\text{ex}}}{\partial\omega}\right)\sin(\omega\tau + \phi_{\text{ex}})\right]. \tag{5.28b}$$

The linewidth (5.25) is now expressed in a form somewhat similar to (5.10) for a plain external reflector; but more complicated with its spectral property taken into account.

Detailed analyses of the rate equations indicate that the intensity noise and relaxation oscillation will be suppressed by the extended cavity [20], consistent with the discussion in Sect. 3.3. Based on the rate equations the noises and stabilities of ECDL induced by variety of factors can be analyzed, such as temperature variation, mechanical vibrations, and bias current, and its fluctuations. Parameters A and B of different external feedback components have different characteristics, as discussed in related sections below.

5.1.4 Injection Locking of Semiconductor Lasers

The structure of injection-locked semiconductor lasers is similar to that of external cavity laser, as shown in Fig. 5.5. The mechanisms inside the LD chip are also similar to that of ECDL, though their functions are different. The master laser is

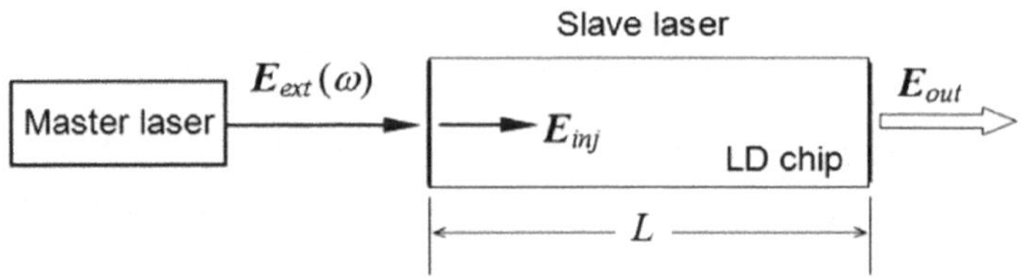

Fig. 5.5 Structure of injection locked LD

usually a continuous wave (cw) laser with excellent spectral characteristics, working in a stable single longitudinal mode with narrow linewidth and low noise. The slave laser has features to meet requirements of applications, such as capability of high speed modulation, high power output, etc., however its spectral performances are not good enough. Injection locking is to make the slave laser with both advantages.

The mechanism of injection locking can be analyzed by the same rate equations as those in ECDL with some modifications. The system can be modeled as a forced oscillation; the optical frequency ω of slave laser will be the same as that of master laser, which may be different from the frequency of slave laser without injection ω_{LD}. The rate equations are written as [21–23]

$$\frac{\mathrm{d}}{\mathrm{d}t}\left[E(t)\mathrm{e}^{-j\omega t}\right] = \left(-j\omega_{\mathrm{LD}} + \frac{1}{2}G_{\mathrm{net}}\right)E(t)\mathrm{e}^{-j\omega t} + \frac{1}{\tau_{\mathrm{LD}}}E_{\mathrm{inj}}(t)\mathrm{e}^{-j\omega t}, \tag{5.29a}$$

$$\frac{\mathrm{d}}{\mathrm{d}t}N(t) = C - \frac{N(t)}{\tau_{\mathrm{sp}}} - G(N)\left[|E(t)|^2 + \left|E_{\mathrm{inj}}(t)\right|^2\right], \tag{5.29b}$$

where $E_{\mathrm{inj}}(t)\mathrm{e}^{-j\omega t}$ is the injected field from master laser. To take field fluctuations into account, the fields of slave laser and injecting laser are written as $E(t) = [E_s + a(t)]\mathrm{e}^{-j\phi(t)}$ and $E_{\mathrm{inj}}(t) = [E_m + b(t)]\mathrm{e}^{-j\phi_m(t)}$; the rate equations are then reduced to

$$\dot{a} = \left(\frac{1}{2}G_{\mathrm{net}} + \kappa\cos\Delta\phi\right)(E_s + a), \tag{5.30a}$$

$$\dot{\phi} = \omega_{\mathrm{LD}} - \omega - \kappa\sin\Delta\phi, \tag{5.30b}$$

where $\Delta\phi = \phi - \phi_m$, $\kappa = (E_m + b)/[\tau_{\mathrm{LD}}(E_s + a)] \approx E_m/(\tau_{\mathrm{LD}}E_s)$. In stationary cases,

$$G_{\mathrm{net}} = \Delta G = -2\kappa\cos\Delta\phi, \tag{5.31a}$$

$$\omega_{\mathrm{LD}} + \kappa\sin\Delta\phi = \omega, \tag{5.31b}$$

where $\Delta G = G - G_0$ is the gain coefficient change caused by injection. The phase of injection-locked laser is locked on the phase of master laser as $\phi = \phi_m - \tan^{-1}[2(\omega_{LD} - \omega)/\Delta G]$; and the condition of frequency locking is $|\omega_{\mathrm{LD}} - \omega| \leq \kappa$. It is meant that the higher the injected master laser power, the larger the frequency locked range can be obtained. If the frequency difference and the injected power do not meet the locking conditions, the slave laser will work out of lock and unstably. The gain change from free running to locked operation is $\Delta G = -2\sqrt{\kappa^2 - (\omega - \omega_{\mathrm{LD}})^2}$. Therefore, the gain coefficient of slave laser will decrease when locked, and its threshold decreases.

The above analysis gives basic features of injection locking, but it is not exact, because the factor of carrier concentration is omitted. From (5.29b), the injection causes carrier concentration reduction; in case of slave laser working under a constant current, gain change of $\Delta G = G_N \Delta N$ occurs correspondingly. It makes also change of oscillation frequency of slave laser, due to dependences of index on carrier concentration. The frequency of Mth mode of free-running slave laser ω_M will be changed to $\omega_{\mathrm{LD}} = \omega_M(1 - h_N \Delta N - h_\omega \Delta\omega)$ by injection. It is indicated by detailed analysis that the locking condition should take different forms for shorter and longer wavelength sides, expressed, respectively, as [24]:

$$\omega - \omega_M \leq \kappa + \frac{1}{2}\alpha G_N \Delta N, \quad \text{for} \quad \omega \geq \omega_M, \tag{5.32a}$$

$$\omega_M - \omega \leq \kappa - \frac{1}{2}\alpha G_N \Delta N, \quad \text{for} \quad \omega \leq \omega_M, \tag{5.32b}$$

where α is the linewidth enhancement factor. That is, the locking ranges are not symmetric to the free-running mode. Since the injection causes carrier concentration decrease under the constant pump current, $\Delta N < 0$, the locking range at longer wavelength side is larger than that at shorter wavelength side. Even more, injection locking may not be realized at shorter wavelength side for a fixed operation current, due to the large α of semiconductor lasers. Such an asymmetric behavior is a unique feature of semiconductor lasers, different from other kind lasers.

For good performances of injection-locked laser, it is necessary to design and control the parameters of device structure carefully. One of them is the optimization of input facet reflectance of slave laser. A lower reflectance is helpful for higher injection of master laser, but increase the slave laser threshold due to its larger mirror loss. Details of the technical issues can be read in references.

It is indicated that the injection-locked laser has a series advantages over the free-running laser [24–26]. The following are the most noticeable features and concerns in practical applications.

(a) As stated above, the nearer the mode closes to the frequency of master laser, the lager the threshold gain reduction is. Therefore, the side mode suppression ratio (SMSR) is enhanced by the injection. It is an important feature, especially for high frequency modulated lasers, since the modulation deteriorates

SMSR of ordinary LD greatly, leading to pulse broadening and bit error rate (BER) increasing in pulse code modulation (PCM) communications. Injection locking is one of the schemes for laser operation in a dynamically single longitudinal mode.

(b) Injection locking by using a narrow linewidth and stable master laser will effectively reduce noises of slave laser. Enhanced stimulated amplification suppresses greatly the phase noise of slave laser. The effect is used for optical phase locking loop (OPLL), as one of the frequency stabilization schemes [27], which will be discussed further in Chap. 8.

(c) Injection locking can be realized not only for cases of the mode with frequency near the master laser, but also for side mode locking under certain conditions. When one of the side modes is locked by injection of master laser, the other modes, including the main mode will be stabilized also [28]. The side mode injection is often used in wavelength translation and conversion.

(d) Injection locking can improve not only the longitudinal mode characteristics, but also the transverse mode of the slave laser; especially for the broad area high power diode laser [29]. The latter often works in multiple transverse modes, with poor output beam. Careful designed injection of master laser with good beam quality can effectively suppress the high-order transverse modes of slave laser, while maintaining the high power output.

5.2 Planar Grating External Cavity Diode Laser

The planar diffractive grating is one of the most widely used external mirrors. It has good performances, especially the feasibility of wide tuning range, high spectral resolution, and flexible and precise tunability. Great progresses in the external cavity semiconductor laser with planar grating mirror have been achieved, and the results of research and development in the field are continuously transferred to industrializations.

5.2.1 Basic Characteristics of Planar Grating

The diffraction of planar grating is attributed to the interference of multiple optical beams scattered from its numerous grooves. By denoting $b = (\sin\theta - \sin\theta_i)a$ with grating period a, incident angle θ_i, and observation angle θ, the composed field is expressed as [5]:

$$E(b) = E_0 \sum_{n=0}^{N-1} e^{jnkb} = E_0 \frac{1 - e^{jNkb}}{1 - e^{jkb}} = E_0 \frac{\sin(Nkb/2)}{\sin(kb/2)} e^{j(N-1)kb/2}, \tag{5.33}$$

where $N \gg 1$ is the total number of grooves which the incident beam covers. If light scattering strength at grating grooves varies with its positions and with the spatial orientation, the amplitude E_0 should be expressed as an integral over grating width W: $E_0(b) \propto \int_W F(x,b)e^{-jkbx/a}dx$. For a constant F, $E_0(b) \propto W\sin(kbW/2a)/(kbW/2a)$, it is just the diffraction pattern of grating aperture. The normalized intensity distribution of grating diffraction is obtained from (5.33):

$$I \propto \frac{\sin^2(Nkb/2)}{N^2\sin^2(kb/2)}, \tag{5.34}$$

which gives a series of maxima at $kb/2 = m\pi$:

$$(\sin\theta_d - \sin\theta_i)a = m\lambda_m \quad m = 0, \pm1, \pm2, \ldots, \tag{5.35}$$

where θ_d is the diffraction angle, and m is the diffraction order. Formula (5.35) is called the grating equation. For fixed incident and diffraction angles, the formula can be approximately expressed as Lorenzian function of frequency near one of the diffraction orders, $b = m\lambda_m = 2m\pi c/\omega_m$:

$$I \propto \frac{1}{1+(\omega-\omega_m)^2/\omega_{1/2}^2} = \frac{1}{1+(\lambda-\lambda_m)^2/\lambda_{1/2}^2}, \tag{5.36}$$

where $\omega_{1/2} = 2\sqrt{3}c/(Nb)$, $\lambda_{1/2} = \sqrt{3}b/(m^2N\pi) = \sqrt{3}\lambda_m/(|m|N\pi)$. The line shape may also be approximated by a Gaussian function for a simplified and convenient analysis of its characteristics.

The angular dispersion is obtained from the grating equation:

$$\frac{d\theta_d}{d\lambda} = \frac{m}{a\cos\theta_d}. \tag{5.37}$$

The resolution of grating is given by $\lambda_{1/2}$, which is proportional to the grating period $a = W/N$. Conventionally, the groove number per millimeter, $1/a$, is called the grating constant, which is usually in range of 1000–300 mm^{-1} for visible and near infrared bands.

The phase factor of diffraction is $\phi = Nkb/2 = N\omega b/2c = N|m|\pi\omega/\omega_m$, given by (5.33). Its contribution to linewidth suppression factor A is thus obtained:

$$A \propto Nb/2c = N|m|\pi/\omega_m, \tag{5.38}$$

indicating that the larger the groove's number N, the more the linewidth is reduced by grating external cavity. Linewidth suppression factor B is obtained from (5.36):

$$B \propto \frac{\partial}{\partial \omega} \frac{\omega_{1/2}}{\left[(\omega - \omega_m)^2 + \omega_{1/2}^2\right]^{1/2}} = \frac{-(\omega - \omega_m)\omega_{1/2}}{\left[(\omega - \omega_m)^2 + \omega_{1/2}^2\right]^{3/2}}. \tag{5.39}$$

Its maxima $\pm 2/(3\sqrt{3}\omega_{1/2})$ are obtained at $\omega - \omega_m = \pm\omega_{1/2}/\sqrt{2}$, deviated from the peak wavelength. Such a behavior was observed in a diffractive grating ECDL experimentally [30].

The diffraction efficiency, its spatial distribution and its variation with wavelength are the main concerns in applications. These characteristics are dependent on grating's fabrication process. Two techniques are usually adopted. One is that the grooves are carved on a polished plane of substrate by a precise machine. Triangle grooves can thus be formed. Their diffraction efficiency is a function of incident angles, determined by the triangle shape. Blaze gratings with strong diffraction at a certain angle are usually made by this technique. The other method is by using holography and chemical etching. The groove shape is near a sinusoidal function, with diffraction efficiency roughly uniform in orientations and with a weaker spectral dependence, beneficial for a large tuning range.

5.2.2 Littrow and Littman ECDL

Wavelength selectivity and tunability of grating ECDL can be realized by adjusting the incident angle to grating plane. Different applications put forward different tuning requirements; some require large tuning range with hopping allowed, whereas others require fine continuous tuning without hopping. This book focuses mainly on the latter. The continuous tuning range and precision are dependent on design of optics and related mechanics. Two configurations are well developed and widely used: Littrow structure and Littman–Metcalf structure, introduced as follows.

(1) **ECDL with Littrow Configuration**

The optical path of Littrow configuration ECDL is shown in Fig. 5.6a [31, 32]. The grating here is used as the mirror of external cavity, and as the output mirror also. The first-order diffractive beam returns back to the LD chip, i.e., $\theta_i = \theta = -\theta_d$; while the zero order is used as the output beam in the reflection direction of grating plane. The grating obeys diffraction equation of

$$2a \sin\theta = \lambda. \tag{5.40}$$

The grating is required with a strong diffraction on backward direction so that the blaze grating is preferable.

The laser frequency can be tuned by rotating the grating, so long as the pivot is perpendicular to the incident plane OO_1G shown in Fig. 5.6, where O is the

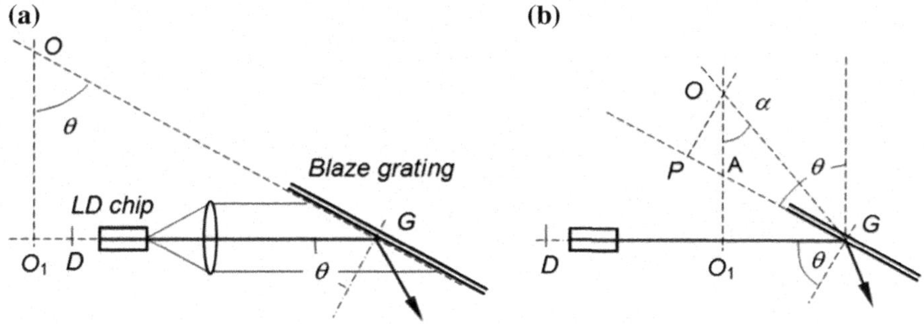

Fig. 5.6 **a** Optical path of Littrow structure ECDL; **b** In case of pivot not in grating plane

position of pivot. It can be either in grating plane as in Fig. 5.6a, or out of grating plane as in Fig. 5.6b. The incident angle θ can be adjusted by rotating the grating; however, the cavity length is changed at the same time. For continuous tuning without mode hopping, the position of pivot has to meet a certain condition, as discussed in Ref. [3, 31–33]. It is deduced as follows.

In Fig. 5.6a, line OO_1 is perpendicular to O_1G. Denoting $h = OO_1$, the length of composite cavity is $DG = L = h\tan\theta + O_1D$, where D is the virtual mirror position when the indexes of LD chip and other optical components are taken into account. The cavity length varies with the incident angle as $\mathrm{d}L/\mathrm{d}\theta = h/\cos^2\theta$. From $\mathrm{d}\lambda_L/\mathrm{d}L = \lambda_L/L$, the mode variation is obtained as $\mathrm{d}\lambda_L/\mathrm{d}\theta = \lambda_L h/(L\cos^2\theta)$. On the other hand, the angular dispersion of grating is $\mathrm{d}\lambda_g/\mathrm{d}\theta = \lambda_g/\tan\theta$. Continuous tuning requires $\mathrm{d}\lambda_g/\mathrm{d}\theta = \mathrm{d}\lambda_L/\mathrm{d}\theta$. Since the laser works usually at the peak of grating reflection spectrum, $\lambda_{g0} \approx \lambda_{L0}$, the continuous tuning condition is then deduced to

$$h/L = \cos^3\theta/\sin\theta. \tag{5.41}$$

A typical design uses a blaze grating with θ = 45°; h = L/2 is then needed.

In case of pivot with distance of OP to the grating plane, as shown in Fig. 5.6b, by denoting p = OP, h = OO_1, l = GO_1, and h_1 = AO_1, the following geometrical relations are obtained: $\tan\alpha = l/h$, $\tan\theta = l/h_1$, $l/p = h_1/(h - h_1)$, and $\tan\theta = \tan\alpha + p/h$. Angle α and θ change with grating rotation around axis O; sections l and h_1 change as well. The cavity length varies as $\mathrm{d}L/\mathrm{d}\alpha = h(1 + \tan^2\alpha)$. $\mathrm{d}\lambda_L/\mathrm{d}\alpha = (\lambda_L h/L)[1 + (\tan\theta - p/h)^2]$ is obtained. The peak wavelength of grating reflection varies as $\mathrm{d}\lambda_g/\mathrm{d}\alpha = \lambda_g/\tan\theta$. If the start point of tuning of $\lambda_{g0} \approx \lambda_{L0}$ is assumed, the continuous tuning condition is deduced to

$$h\tan\theta[1 + (\tan\theta - p/h)^2] = L. \tag{5.42}$$

The formula determines the geometrical parameters of ECDL structure. For a blaze grating with $\theta = 45°$, it is reduced to $h[1 + (1 - p/h)^2] = L$.

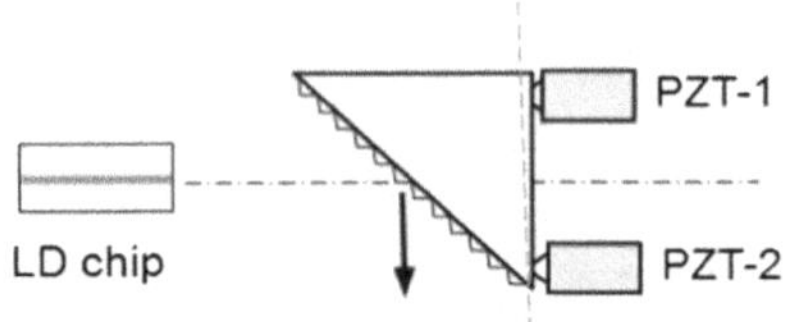

Fig. 5.7 Littrow ECDL with a rotation and translation separately controlled grating

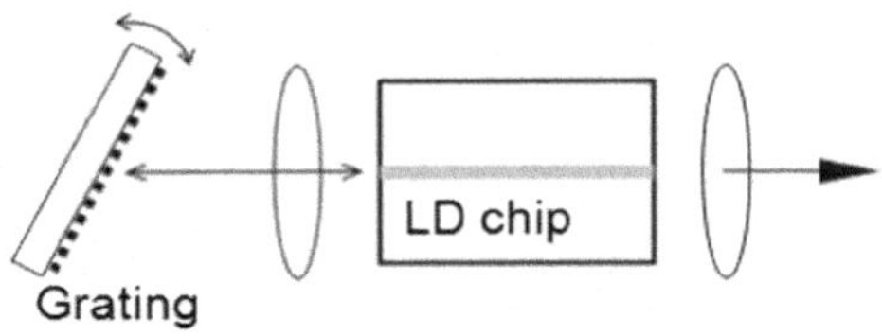

Fig. 5.8 Littrow ECDL with a stable output direction

For higher freedom of continuous tuning, the rotation and translation can be controlled separately, instead of only one pivot. An example is shown in Fig. 5.7 [34, 35]. The grating is moved by two PZT, whose driving voltages are controlled by a computer with programs written in advance for continuous tuning, quasi-continuous tuning, or step tuning.

Littrow ECDL is widely used in many applications, due to its advantages of simple structure and convenient operation. However, the direction of output beam will rotate in tuning. This shortcoming must be overcome for many applications. Reference [36] reported a design for compensating the output deflection. A simpler design is with both of two ends of LD chip utilized, as shown in Fig. 5.8 [37], where the laser beam outputs from the chip directly, so that a stable beam is obtained in tuning. The grating and the cavity parameters of such a configuration should be designed differently from the structure of Fig. 5.6.

(2) **ECDL with Littman–Metcalf Configuration**

Compared with Littrow structure, Littman–Metcalf structure ECDL (shortened as Littman ECDL) gives output beam with a stable direction. Figure 5.9a shows its schematic diagram [3, 38–41]. The feedback beam is from the second order diffraction of grating; the incident angle α of LD beam to the grating and the

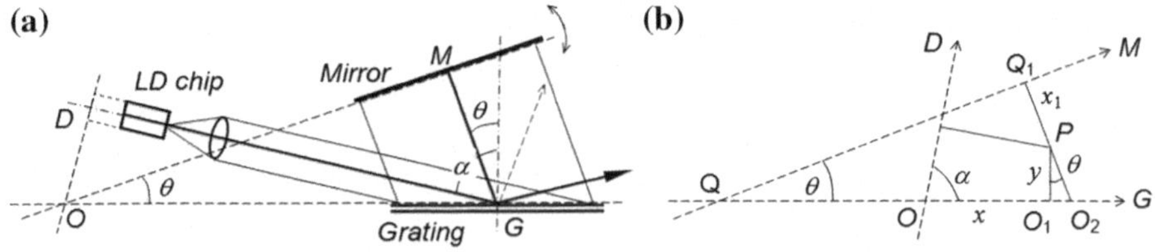

Fig. 5.9 **a** Schematic diagram of Littman ECDL; **b** Pivot dislocated from its optimum

diffraction angle θ meet relation of $\sin\theta = \lambda/a - \sin\alpha$. The zero-order diffraction is used as the output, shown in the figure by the arrow. Mirror M reflects the diffraction beam back to the grating and then to the LD chip, building up an external cavity. The zeroth order diffraction of the beam from the mirror, marked by a dashed line, goes out of the cavity as a loss, which should be minimized in the design.

Tuning of Littman ECDL is realized by rotating mirror M. Since it does not change the incident angle α to the grating, the direction of output beam is stable. Reference [33, 38] analyzed the influence of pivot position on tuning. An explicit explanation is given here. The pivot is marked by O in Fig. 5.9a; the lengths of lines are denoted by $OG = l_1$, $DG = l_0$, $MG = h$, and $OM = l_2$; angle OMG is a right angle. Point D is a virtual position of cavity mirror, taking the indexes of LD chip and optical components into account. The cavity length of ECDL is now $L = l_0 + h = l_0 + l_1 \sin\theta$. From relation of $\partial\lambda_L/\partial L = \lambda/2L = (\partial\lambda_L/\partial h)/2 = (\partial\lambda_L/\partial\theta)/(2l_1\cos\theta)$, the tuning rate is written as

$$\frac{\partial\lambda_L}{\partial\theta} = \frac{\lambda\cos\theta}{\sin\theta + l_0/l_1}. \tag{5.43}$$

On the other hand, the angular dispersion of grating is

$$\frac{\mathrm{d}\lambda_g}{\mathrm{d}\theta} = a\cos\theta = \frac{\lambda\cos\theta}{\sin\theta + \sin\alpha}. \tag{5.44}$$

Continuous tuning requires equality of the two rates, giving condition of $l_0/l_1 = \sin\alpha$, indicating that angle ODG should be a right angle. The pivot should be located at the corner of right triangle OMG and ODG, and unchanged during grating rotating.

It is needed to understand what will happen if the pivot deviates from its optimal position. In Fig. 5.9b, pivot P is at $OO_1 = x$ and $PO_1 = y$ from the ideal position O. The mirror's rotation changes angle MQG (θ), while angle DOG (α) and PQ_1 (x_1) are kept unchanged. With geometrical relations of $O_1O_2 = y\tan\theta$, $OO_2 = x + y\tan\theta$, $Q_1Q_2 = x_1 + y/\cos\theta$, $QO_2 = Q_1O_2/\sin\theta$, $h = QG\sin\theta$, and $QG = l_1 + QO = l_1 + QO_2 - OO_2$, the composite cavity length is expressed as

$$L = l_1(\sin\theta + \sin\alpha) - x\sin\theta + x_1 + y\cos\theta. \tag{5.45}$$

The tuning rate is then deduced as

$$\frac{\mathrm{d}\lambda_L}{\mathrm{d}\theta} = \frac{\lambda_L}{L}(l_1\cos\theta - x\cos\theta - y\sin\theta). \tag{5.46}$$

By denoting $f(\theta) = \mathrm{d}\lambda_L/\mathrm{d}\theta - \mathrm{d}\lambda_g/\mathrm{d}\theta$ as the difference between the tuning rate and angular dispersion of grating, and with $\lambda_{L0} = \lambda_{g0} = \lambda_0$ and $PO \ll OG$ assumed, we arrive at

$$f(\theta) \approx \frac{-a^2}{\lambda_0 l_1}[y(1 + \sin\alpha\sin\theta) + x\sin\alpha\cos\theta + x_1\cos\theta]. \tag{5.47}$$

It shows the influence of pivot deviation from its optimum. The continuous tuning without mode hopping requires the difference between mode and diffraction peak less than half of mode spacing: $f(\theta)\Delta\theta \leq \Delta\lambda_L/2$, which puts forward the requirement of mechanical design.

The above analyses of Littrow and Littman ECDL have not taken the chromatic dispersions of LD chip and optical components into account, which will affect the tuning characteristics and other performances. If a large tuning range is required, the higher order tuning rate $\mathrm{d}^2\lambda/\mathrm{d}\theta^2$ has to be considered. Detailed analysis can be read in references [3, 41–43].

(3) **Typical Characteristics of Planar Grating ECDL**

Planar grating ECDLs show good performances, especially with single frequency and narrow linewidth output. Figure 5.10 gives measured spectra of a Littman ECDL [44], as an example. Figure 5.10a is its static spectrum measured with an optical spectrum analyzer (OSA), showing peak wavelength of 780.184 nm, SMSR of 53.05 dB, and linewidth of 0.027 nm ($\sim$3.4 GHz). The linewidth is limited by the resolution of OSA. Figure 5.10b shows the results measured by a scanning F-P etalon, whose finesse is 300, free spectral range (FSR) is 1 GHz. The exact linewidth (full width at half maximum, FWHM) is less than 3 MHz.

Linear tuning is realized by applying a saw wave voltage on PZT. The tuning continuity is checked by the F-P etalon with its cavity length fixed, as shown in Fig. 5.11, where the bold line is the voltage on PZT; thin lines are the signal of F-P. The measured tuning range is 22 GHz, calculated from FSR of F-P cavity. No mode hopping occurs in the range.

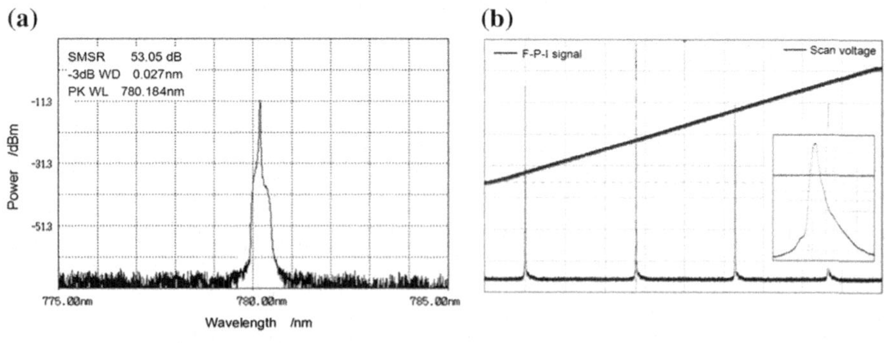

Fig. 5.10 **a** Spectrum measured by OSA. **b** Waveform measured by F-P-I; *Insert* Expended line shape

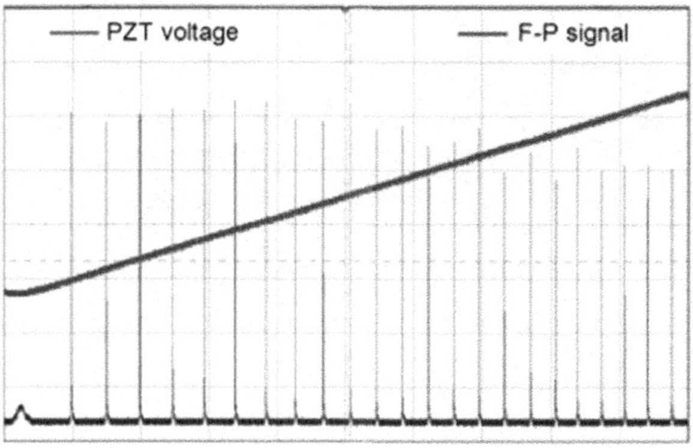

Fig. 5.11 Continuous tuning spectrum of ECDL

5.2.3 *Technical Issues of Grating ECDL*

Besides the optimized designs, a series of technical problems in optics and mechanics have to be noticed and solved for building a practical Littrow or Littman ECDL.

(1) **Design of Mechanical Structure and Tuning Driver**

One of the keys to planar grating ECDL is the pivot structure for precise tuning, as discussed above. The structure of axis with bearing bush, driven usually by a step motor, is used usually in some ECDL designed for wide range quasi-continuous tuning. Monolithic structures are mostly used for precise and stable tuning, where elastic deformation of material is utilized to rotate the angle of grating or mirror. Figure 5.12a shows a Littman ECDL with a cantilever turner with mirror fixed on the beam driven by a piezoelectric transducer (PZT). The beam material at its fixed end is thinned for reducing the pushing force and confining the pivot position.

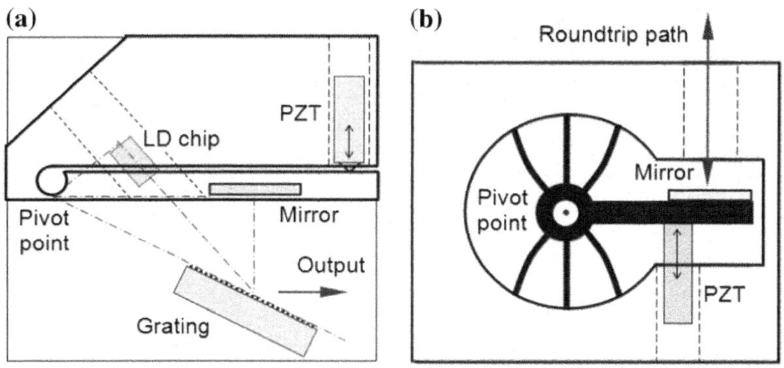

Fig. 5.12 **a** Littman ECDL with cantilever turner; **b** Flexible hinge turner

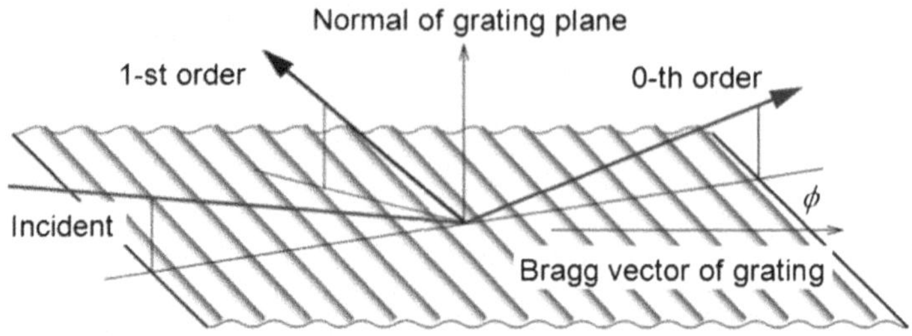

Fig. 5.13 Diffractions in three-dimensional space

Figure 5.12b depicts a flexible hinge structure, which composed of several symmetrically arranged thin beams for a larger turned angle with a smaller force, and maintaining the pivot at the center stably.

Apart from the static deformation range and the force needed, the resonance frequency of dynamic deformation has to be taken into account in designs of the turner. It determines the response speed of tuning, especially for applications of frequency stabilization. Reference [44] reported the simulated and measured data of dynamic response.

(2) **Optical Beam Collimation**

The divergent output beam of LD chip must be carefully collimated, usually by using a large aperture aspheric lens. A broad collimated beam is preferred if permitted by the requirement of whole device volume, to get not only a better beam quality, but also a larger grating area illuminated, which is important for higher spectral resolution. For the same reason, a larger incident angle is also preferred to enhance spectral resolution and tuning precision.

The normal of reflective mirror in Littman structure must be in the incident plane of grating to get higher coupling efficiency. Instead of a planar mirror, a right angle prism may be used to ensure the reflected beam exactly backward to the grating, so that the number of adjustable components is reduced and the mechanical stability is improved.

Attention should also be paid to the orientation of Bragg vector of grating respect to the incident plane. The grating Eq. (5.35) holds in case of incident plane perpendicular to grating grooves, i.e., parallel to Bragg vector of grating. If the angle between them is not zero, the diffraction beam will not be in the incident plane, as shown in Fig. 5.13 with $\phi \neq 0$, resulting in increase of coupling loss and decrease of tuning range. The detailed analysis on spatial distribution of planar grating diffraction can be read in Ref. [45].

(3) **Anti-reflective Coating of LD Chip**

The end facet of LD chip facing the external cavity should be coated with anti-reflective (AR) film to get rid of the influence of chip mode to external cavity mode. According to the basic principle of multiple dielectric films, the optical thickness should equal to $\lambda/4$, and the index of the film material should meet the value of $n = \sqrt{n_{\text{LD}}}$. It is important to precisely measure and control the residue reflectance R_2. Since the film is evaporated on LD chip, which is an active medium, conventional method of reflection measurement is not applicable. Instead, it is necessary to utilize the characteristics of LD emitting spectrum for monitoring R_2. The light emitting spectrum of LD below threshold is modulated by the roundtrip propagation between two facets; the modulation depth is a function of their reflectance. The reflectance R_1 of facet without coating is determined by the index of semiconductor, and is mostly about 32%. According to the theory of traveling wave amplifier, the modulation depth is

$$M = \frac{I_{\max} - I_{\min}}{I_{\max} + I_{\min}} = \frac{2\sqrt{R_1 R_2} G_0}{1 + R_1 R_2 G_0^2} \rightarrow \frac{2\sqrt{R_2/R_1}}{1 + R_2/R_1}, \tag{5.48}$$

where $G_0 = \exp(g_{\text{net}} L)$ is the single trip gain of LD. The modulation depth is thus varying with the injected current. Since the LD threshold before coating satisfies $R_1 G_0 = 1$, the residue reflectance after coating can then be calculated by measured modulation depth under the same injected current by the last formula of (5.48) with a known R_1.

(4) **Dependence of Polarization on LD P-N Junction Orientation**

Semiconductor laser has strong polarization selectivity; generally, the electric field of its TE mode is parallel to the P-N junction. Due to the polarization dependence of Fresnel reflection, the diffraction efficiency of planar grating is dependent on the polarization of incident beam. Therefore, the polarization of ECDL output will vary with the P-N junction orientation respect to the beam axis. Different polarization states, from linear, to circular, and to elliptical polarizations can be obtained by rotating LD chip around its beam axis, as reported in Ref. [46]. In addition, since the output beam of LD chip is a long ellipse, the shape of projected area on grating plane is also varying with the junction rotation. It is necessary to adjust carefully the position and orientation of LD chip to avoid changes of the diffraction efficiency and the effect of linewidth reduction.

(5) **Mechanical Stability**

Compared with the monolithic integrated semiconductor lasers, more attentions must be paid to the stability of ECDL. The related factors include:

(1) Robustness. Monolithic structure design with components as less as possible is preferred.
(2) Thermal stability. The material should be with thermal expansion coefficient (TEC) as low as possible; and TEC of different parts should be close with each other. Usually, the LD chip should be cooled and with temperature stabilization measures.
(3) Controllability when LD working point changes. Applications often require an adjustable output power. Long term uses usually require working point adjusted to compensate the degradation of laser's performances due to aging. Detailed analyses and schemes can be read in references.

5.2.4 Various Cavity Structures and Tuning Schemes

(1) **Tuning with Passive Components**

Passive components can be inserted in the external cavity and used as a tuning element. Figure 5.14 gives an example, where the cavity length is varied by rotating an optical plate or a wedge [47]. Reference [48] proposed another method, where the collimating lens of LD beam is translated transversely, and the beam direction can be deflected.

(2) **Tuning with Electro-optic and Acousto-optic Devices**

The electro-optic (EO) and acousto-optic (AO) effects are also utilized to deflect optical beams, instead of mechanical movements. Figure 5.15a shows a scheme with a liquid crystal (LC) cell [49, 50]. The index of LC is a function of the applied electric field, which varies linearly with the position due to its wedge-type and with the applied voltage. The beam deflection is thus varied by the voltage, and the laser wavelength is tuned.

The structure of Fig. 5.15b uses a PLZT plate with triangular electrodes on its upper and bottom surfaces [51]. PLZT is a transparent EO ceramic. When a voltage is applied on the electrodes, the index difference is generated between the two triangular parts, causing refraction at the interface, and the refracted angle can be adjusted by the voltage. The wave plates (WP) in the figure are used to obtain optimal tuning efficiency. Detailed principle and characteristics of EO and AO tuning will be discussed in Chap. 7, together with their dynamic performances.

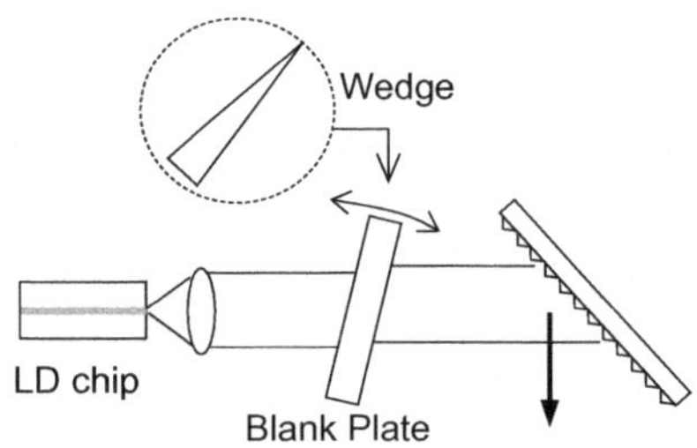

Fig. 5.14 Littrow ECDL with a passive element inserted in cavity

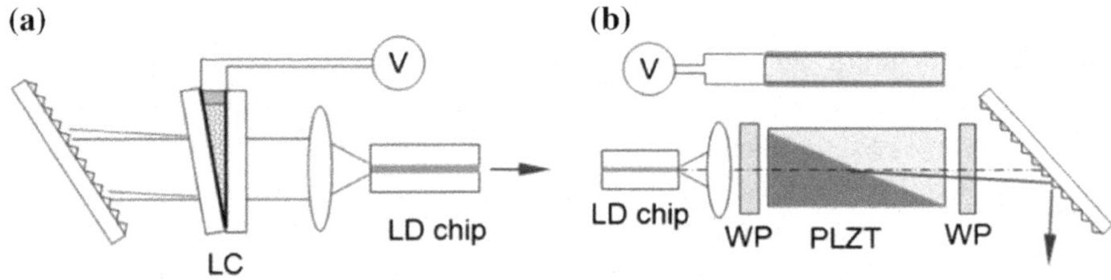

Fig. 5.15 **a** Tuning with a wedge-type LC cell. **b** Tuning with a PLZT deflector

(3) **Extension of Effective Cavity Length**

One of the purposes of ECDL is to extend the photon lifetime for linewidth narrowing by extending the cavity length. Apart from linear cavities, different configurations are proposed. Figure 5.16a shows a folded cavity [52], where refractions, total reflections and total internal reflections (TIR) occur at three surfaces of the prism. The photon lifetime is extended by the multiple roundtrips, as shown by the arrows in the figure. Reference [53] reported a modified Littrow structure by using the second order diffraction of blaze grating, as shown in Fig. 5.16b, where the multiple roundtrips occur between the grating and the output mirror. The finesse of the laser was measured as high as 1855.

(4) **High Power ECDL**

For high output powers, required often in applications, different designs of ECDL structure have been proposed. Two configurations are usually adopted. One is the master oscillator and power amplifier (MOPA) module; the other uses broad area LD chip or LD array bar, instead of a single transverse mode chip. The LD chip in Fig. 5.17a is composed of a narrow stripe section as oscillator and a tapered

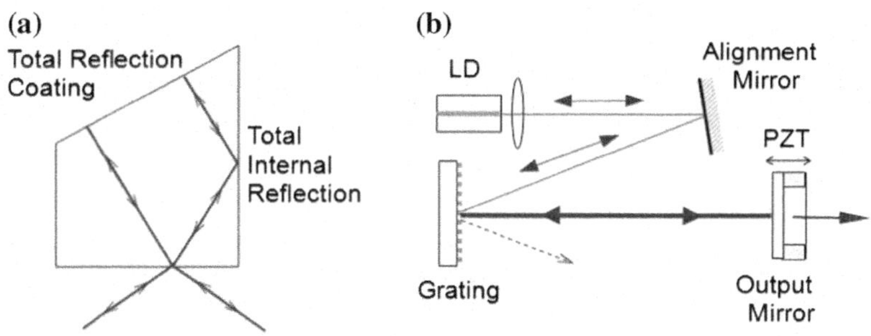

Fig. 5.16 **a** Folded cavity for extension of group delay. **b** Littrow ECDL with second order diffraction

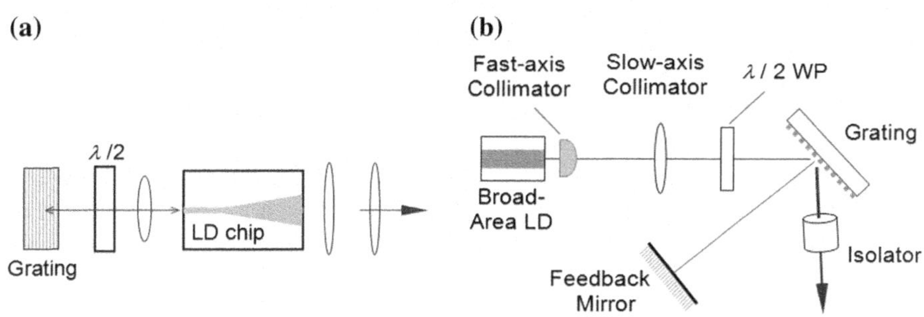

Fig. 5.17 Higher power ECDL, **a** with tapered waveguide amplifier; **b** with a broad area LD

wide stripe as amplifier [54, 55]; and Littrow grating is arranged at the left end side. The incident plane to the grating is now perpendicular to the paper, and a $\lambda/2$ wave plate is used to get higher feedback. Figure 5.17b is a Littman ECDL with a broad area LD chip; its output beam is collimated by a cylindrical lens in its fast axis and by a spherical lens in its slow axis. The wave plate plays the same role as that that in figure (a). The output power up to 25 W was obtained from this ECDL [56].

The external cavity semiconductor lasers are still being developed to meet the requirement of different applications. Combined with technology of micro-electro-mechanical system (MEMS) or micro-opto-electro-mechanical system (MOEMS), ECDL with advantages of integration and multiple functions is developed [57]. Performances of ECDL are improved further with DFB-LD and VCSEL chips used to replace the ordinary F-P chips [58].

5.3 Bragg Grating External Cavity Diode Laser

The stability, reliability, and compactness of the planar grating ECDL are considered not as good as monolithically integrated lasers. ECDL incorporated with Bragg gratings are then developed, which has advantages of stability better than the former, and flexibility and narrow linewidth better than the monolithically integrated laser. Three kinds Bragg gratings have been successfully used in ECDL: fiber Bragg grating (FBG), waveguide Bragg grating (WBG), and volume Bragg grating (VBG).

5.3.1 Fiber Bragg Grating ECDL

(1) **Characteristics of Fiber Bragg Gratings**

It was discovered in 1978 that the index of silica fiber would be changed when it was illuminated from its end by an argon laser beam [59]. It was found later that lateral illumination of ultraviolet (UV) laser on fiber generated the same effect; and holographic technique and exposure with phase mask were developed to form a periodical index modulation in fiber core along its axis [60–62]. The fiber grating was then developed based on the photosensitivity. Its mechanism is widely recognized as the interactions of UV photons with the various defects in germanium-doped silica. The bonds of Ge–O, Ge–O–Si and so on are changed under UV irradiation, and the optical loss of fiber changes simultaneously, so does the refractive index according to the universal Kramers–Kronig relation [63]. The photosensitivity is dependent of the wavelength of UV laser; 248 and 193 nm eximer lasers are proved most effective, and the UV-induced index increment up to 10^{-4}–10^{-3} in near infrared range are achieved.

The periodic index modulation causes coupling between the forward and backward propagating fundamental modes; and causes also coupling between the forward fundamental mode and forward cladding modes. The former coupling obeys Bragg diffraction equation; i.e., the forward mode with wavelength peaked at $\lambda_B = 2n_{\text{eff}}\,\Lambda$ will be reflected by the grating with period of Λ. It is called fiber Bragg grating (FBG). The coupling with cladding modes occurs in case of grating period as long as hundreds micrometers, called long period fiber grating (LPFG), which is used mostly in optical sensors. The characteristics of FBG is described by the same formulas as (4.32–4.38) given in Sect. 4.2.2 for waveguide grating used in DBR-LD [62, 64].

For uses in ECDL field, the phase shift and group delay of FBG play important roles. The phase factor of FBG is the same as (4.36), which is a function of frequency. When FBG is used to build an external cavity laser, factors A and B in formula (5.27) are concerned. For an ideal FBG, the group delay (GD) is expressed as

$$\begin{aligned}\tau_g = \frac{\partial\phi}{\partial\omega} &= \frac{L}{v_g}\frac{\delta^2 - \kappa^2\sinh 2\sigma L/(2\sigma L)}{\delta^2 - \kappa^2\cosh^2\sigma L} \quad (\sigma = \sqrt{\kappa^2 - \delta^2}) \\ &= \frac{L}{v_g}\frac{\delta^2 - \kappa^2\sin 2sL/(2sL)}{\delta^2 - \kappa^2\cos^2 sL} \quad (s = \sqrt{\delta^2 - \kappa^2})\end{aligned}, \tag{5.49}$$

where $\delta = \beta - \beta_B = 2\pi n/\lambda - \pi/\Lambda$; $\kappa = n_1 k_0/2$ is the coupling coefficient; n and n_1 are the fiber index and its modulation amplitude induced by UV exposure. Factor B of FBG is proportional to

$$\begin{aligned}\frac{\partial|r|}{\partial\omega} &= \tau_g\frac{\delta\kappa(\sigma L\cosh\sigma L - \sinh\sigma L)}{L(\kappa^2\cosh^2\sigma L - \delta^2)^{3/2}} \quad (\delta \le \kappa) \\ &= \tau_g\frac{\delta\kappa(sL\cos sL - \sin sL)}{L(\delta^2 - \kappa^2\cos^2 sL)^{3/2}} \quad (\delta \ge \kappa)\end{aligned}. \tag{5.50}$$

Figure 5.18 gives calculated spectra of related parameters, where $\partial|r|/\partial\lambda$ and τ_g are in arbitrary units. Because of $\partial/\partial\omega \propto -\partial/\partial\lambda$, the best effect of linewidth

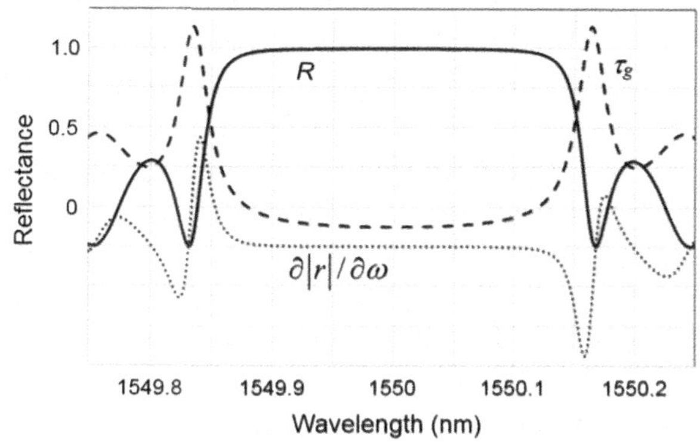

Fig. 5.18 Calculated spectra of reflectance τ_g and $\partial|r|/\partial\omega$ of FBG

reduction can be obtained at the longer wavelength edge, where both of A and B are positive, with higher linewidth reduction factor (1 + A + B).

Similar to the phase-shifted DFB laser, if some phase shift is introduced in FBG, a narrow transmission peak will occur in its reflection band, which is called phase-shifted FBG (PS-FBG). It can be regarded as a cavity with length toward zero, and with wavelength selective mirrors. It is modeled by two uniform gratings joined with a phase shift in between. The coupled mode equations (CME) of grating can be expressed as a matrix equation; and the transfer matrix of phase-shifted FBG is reduced to a product of three matrixes, expressed as [62, 64]

$$\begin{pmatrix} E_T \\ E_B \end{pmatrix} = \begin{pmatrix} T_{11}^{(2)} & T_{12}^{(2)} \\ T_{21}^{(2)} & T_{22}^{(2)} \end{pmatrix} \begin{pmatrix} \mathrm{e}^{j\theta} & 0 \\ 0 & \mathrm{e}^{-j\theta} \end{pmatrix} \begin{pmatrix} T_{11}^{(1)} & T_{12}^{(1)} \\ T_{21}^{(1)} & T_{22}^{(1)} \end{pmatrix} \begin{pmatrix} E_{\mathrm{in}} \\ E_R \end{pmatrix}, \tag{5.51}$$

where θ is the phase shift, E_{in}, E_{R}, and E_{T} are the input, reflected, and transmitted optical fields, and $E_{\mathrm{B}} = 0$ for the case of no backward input. In an ideal case with $\theta = \pi/2$ at the middle point, the transmission spectrum of PS-FBG can be deduced as [64]:

$$T = \frac{1}{1 + 4(\delta^2\kappa^2/\sigma^4)\sinh^4 \sigma l} \approx \frac{1}{1 + 4(\delta^2/\kappa^2)\sinh^4 \kappa l}, \tag{5.52}$$

where $l = L/2$ is half of the whole length of FBG. $T = 1$ at Bragg wavelength is obtained, and the transmission linewidth is $\delta_{1/2} \approx \kappa \mathrm{e}^{-\kappa L}/2$, which is very narrow for a higher κL.

In case of $\Delta\theta = \theta - \pi/2 \neq 0$, the transmission peak will shift from Bragg wavelength to $\beta_{\mathrm{peak}} \approx \beta_{\mathrm{B}} - \kappa \tan \Delta\theta$. If the phase shift position is not at the middle point, the transmission peak is still located at Bragg wavelength, but the peak transmission is lowered. Figure 5.19 shows calculated transmission spectra for phase shift of $\pi/2$ and $\pi/3$.

The narrow linewidth of PS-FBG means high slope at the edges and high contribution to the linewidth reduction factor B; and a high group delay occurs

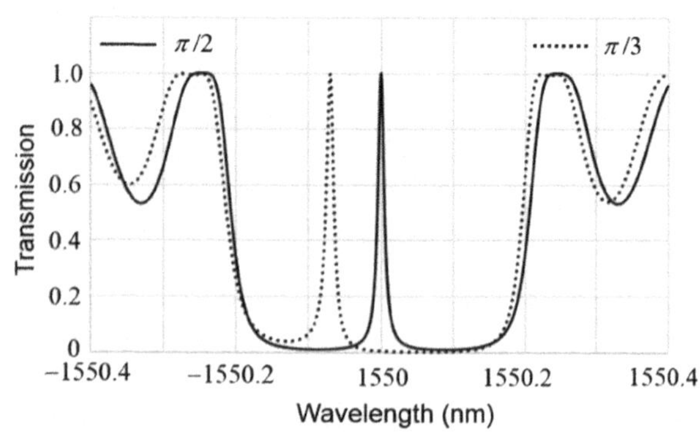

Fig. 5.19 Transmission spectra of phase-shifted grating

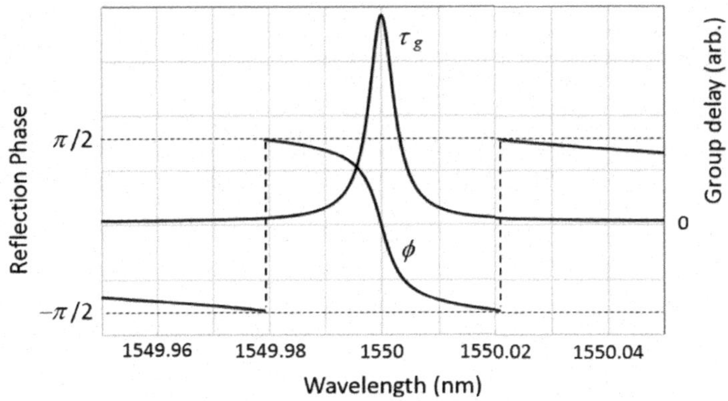

Fig. 5.20 Calculated reflection phase and group delay of phase-shifted FBG

there. For the ideal case with $\theta = \pi/2$ at the middle point, the propagation phase of transmission and reflection are deduced from (5.51), respectively, as

$$\phi_t = \tan^{-1}\frac{\kappa^2 - \delta^2\cosh 2\sigma l}{\delta\sigma\sinh 2\sigma l}, \tag{5.53a}$$

$$\phi_r = \tan^{-1}\frac{\delta\sigma\sinh 2\sigma l}{\kappa^2 - \delta^2\cosh 2\sigma l}. \tag{5.53b}$$

Group delays are then deduced from (5.53a) and (5.53b). A maximum group delay occurs at Bragg wavelength both for transmission and reflection. Since the transmission peak is much narrower than the reflection band of a uniform FBG, much larger A and B factors result. Figure 5.20 gives calculated spectra of ϕ_r and τ_g of a $\pi/2$ PS-FBG, where τ_g is in an arbitrary unit. It is shown that the larger the grating length and the coupling coefficient, the narrower the transmission peak is.

It is shown theoretically and experimentally that strong side lobes appear out of the main reflection band in the spectrum of FBG with uniform index modulation, attributed to the square aperture of the grating. To suppress the side lobes, apodization is developed; that is, the index modulation is designed with its amplitude (and/or other parameters) declined gradually near two ends of grating in some function of the position, such as a cosine function or a SINC function, instead of a steep interrupt in uniform FBGs. The spectrum of PSFBG can also be modified by apodization [62, 64].

(2) **FBG-ECDL**

The optical fiber communication and fiber sensor are the most important applications of semiconductor laser. For the applications, the fiber coupled LD has been well developed and manufactured commercially. The fabrication of FBG-ECDL is a mature technique, so long as the grating is imprinted inside the fiber pigtail [65–68], as shown in Fig. 5.21.

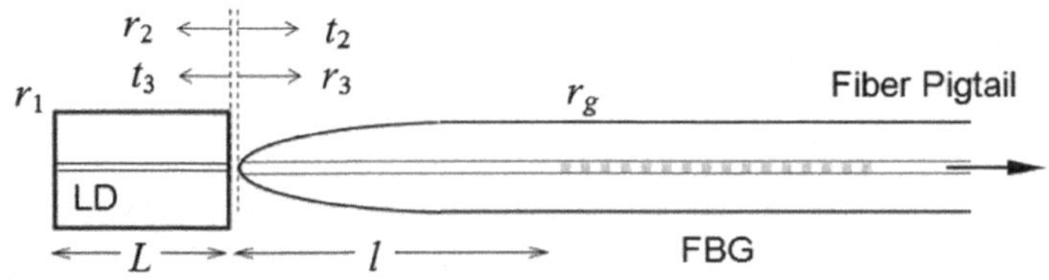

Fig. 5.21 Schematic structure of FBG-ECDL

Compared with the solitary LD the DBG-ECDL has remarkable advantages.

(1) It is easy to make an FBG with linewidth less than the longitudinal mode spacing of chip LD, good for high SMSR.
(2) The temperature coefficient of silica index is roughly one tenth of semiconductor index. The thermal stability of lasing wavelength is enhanced greatly.
(3) It is possible to design and screen the LD chip and FBG separately, so that the performances of ECDL can be optimized.
(4) FBG-ECDL is very convenient to be used in fiber systems.

One of the key technologies is the coupling between LD chip and fiber. The end of the fiber has to be processed into a micro-lens, as shown in Fig. 5.21; many designs and fabrication techniques of the fiber end have been developed. However, the coupling loss is inevitable. Phenomenally, the reflections and transmissions to chip side and to fiber are denoted by symbols depicted in the figure; and their relations are written as $t_{2,3}^2 = \eta_{2,3}(1 - r_{2,3}^2)$ with the coupling efficiency of $\eta_{2,3}$. The effective reflectance to the chip can be deduced as

$$r_{\text{eff}} = \frac{r_2 + [\eta + (1-\eta)r_2^2]|r_g|e^{j(2nkl+\phi_g)}}{1 + r_2|r_g|e^{j(2nkl+\phi_g)}}, \tag{5.54}$$

where $\eta = t_2 t_3/(1 - |r_2 r_3|)$ is the composite coupling efficiency, n is the fiber index. The threshold conditions of FBG-ECDL are now written as

$$g = \alpha_c + L^{-1}\ln(1/r_1|r_{\text{eff}}|), \tag{5.55}$$

$$2n_{\text{LD}}kL + \text{Arg}(r_{\text{eff}}) = 2m\pi. \tag{5.56}$$

For the chip facet AR-coated sufficiently with $r_2 \approx 0$, $r_{\text{eff}} \approx \eta|r_g|\exp[j(2nkl + \phi_g)]$; then the linewidth reduction factors are expressed as $A = \tau_{\text{rt}}^{-1}(\partial\phi_g/\partial\omega + 2n_g l/c)$ and $B = \eta\alpha\tau_{\text{rt}}^{-1}(\partial\ln|r_g|/\partial\omega)$, where n_g is the group index of fiber. A longer length l is beneficial for a larger factor A; however, it will reduce the mode spacing, increase possibility of mode hopping. A careful design of the parameters is necessary.

In order to remove the influence of residue reflection of chip facet as much as possible, a curved stripe LD has been developed, as shown in Fig. 5.22a.

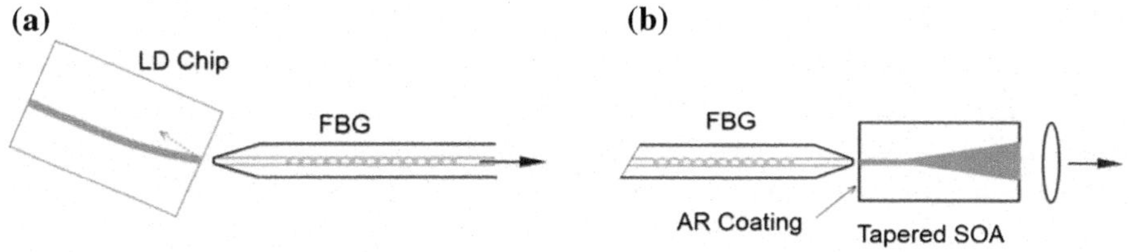

Fig. 5.22 **a** FBG-ECDL with a curved waveguide chip for internal mode suppression. **b** FBG-ECDL with a tapered SOA chip for high power output

The waveguide is perpendicular to the cleaved facet on the left side, and is angled to the facet on the right. Any residue reflected wave will then deviate from the axial direction of waveguide, diffracted out of its lateral boundary. The chip modes are thus suppressed further. It is necessary to optimize the inclined angle for reducing the additional loss in the curved waveguide. Similar to the Littrow ECDL, a MOPA configuration can be used for a high output power FBG-ECDL, as shown in Fig. 5.22b, where a tapered SOA chip is adopted as the active part, and FBG is coupled with the narrow stripe waveguide [66].

The increase of fiber coupling efficiency relies on the quality of micro-lens at fiber tip. A spherical lens is usually formed by arc discharge. The taper angle and a cylindrical micro-lens can be processed by precisely lapping. A shorter fiber length l from the lensed end to FBG is better for larger mode spacing. However, its minimum is limited by the processing of fiber lens, because the UV-induced index may vanish at high temperature near the arc.

The ECDL of Fig. 5.21 is usually called a linear structure. Besides, ring structures are also used for PS-FBG based ECDL, as shown in Fig. 5.23, where light filtered by PSFBG is injected back to the gain medium. Structure (a) uses an optical circulator (OC) to build the ring with a LD chip; structure (b) uses a semiconductor optical amplifier (SOA) chip coupled with two fiber pigtails. The fiber couplers (FC) are used for the outputs, and an optical isolator is inserted to ensure one-direction propagation.

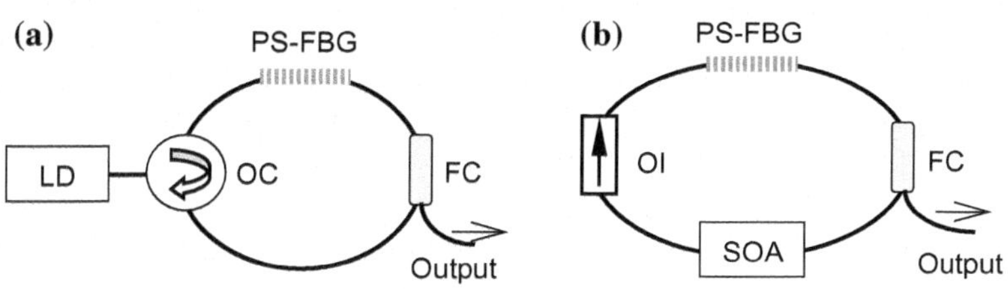

Fig. 5.23 Structures of phase shifted FBG ECDL, **a** with LD chip; **b** with SOA chip

(3) FBG-ECDL for Different Applications

Apart from narrow linewidth applications, FBG are also used to meet the requirement of central wavelength stabilization of high power LD, which may work in continuous wave (cw) or in quasi-cw operations. High power 980 and 1480 nm LD are pump sources widely used in fiber optical amplifiers and fiber lasers. Their wavelengths have to match with the absorption line of erbium-doped silica fiber in varying environment, and in long-term operations with possible aging of LD. The frequency stabilization can be realized by a weak feedback from FBG written in its pigtail. For pump sources of fiber amplifiers and lasers, whose linewidth is not necessary to be very narrow, the FBG-ECDL may work in multiple external cavity modes, but with the LD chip mode locked at the peak wavelength of FBG. On the other hand, its output power should not be decreased much by the FBG. In such cases, a weak reflective FBG is preferred, and the phase factor $2n_fkl$ of external cavity, is smoothed in spectrum; the effective reflectance of external mirror can be written as

$$\overline{R}_{\text{eff}} = \frac{R_2(1-\eta_e R_g) + \eta_e^2 R_g}{1 - R_2 R_g} \tag{5.57}$$

where $\eta_e = \eta + (1 - \eta)R_2$. R_g is small for such applications, usually less than 5%; then the effective reflectance is near R_2 of the LD chip facet. The FBG provides a weak feedback with wavelength selectivity to lock the central frequency of LD.

FBG-ECDLs are also used as the emitter of optical fiber communication system, where the laser is required working in a stable single longitudinal mode under high frequency intensity modulations. The response of FBG-ECDL will be slowed by a longer photon lifetime; therefore, the length of passive fiber section l should be decreased as much as possible.

By utilizing cross-gain modulation of LD active medium, FBG-ECDL has a function of wavelength conversion [69]. When the LD chip of FBG-ECDL is injected by a pulse modulated optical signal with wavelength (λ_0) different from the Bragg wavelength (λ_B) of FBG, the output of ECDL will be modulated by the same waveform of injected signal by so-called cross-gain modulation; as a result, the signal at λ_0 is converted to signal at λ_B. The wavelength selectivity of FBG is utilized also to develop other devices, such as mode locked ECDL [70], wavelength switch [71], and beat frequency output [72] of double wavelength mode locked laser. These results demonstrated characteristics of multiple function and applicability of FBG-ECDL.

Based on the photoelastic effect, thermo-optic effect, and thermal expansion effect, FBG-ECDL can be tuned by strains and by temperature of FBG, expressed as

$$\frac{\Delta\lambda_B}{\lambda_B} = \left(\alpha_T + \frac{1}{n}\frac{\partial n}{\partial T}\right)\Delta T + (1+\gamma)\Delta e, \tag{5.58}$$

where α_T is the thermal expansion coefficient of fiber, e is its strain, and γ is the effective photoelastic coefficient related to the photoelastic coefficient of fiber. It is $\gamma \approx -0.22$ for silica fiber [64]. Using micro heater and thermoelectric cooler, temperature tuning is easily realized. The laser can be tuned by strains applied to FBG, which is stuck on a piezoelectric transducer (PZT) or other elastic element, such as a bendable cantilever. PZT driven by electric voltage has high response speed, up to kHz even more.

5.3.2 Waveguide Bragg Gating ECDL

(1) **Waveguide Grating Materials**

As the development of the integrated optics, the external cavity laser with planar waveguide Bragg grating (WBG-ECDL) attracts attentions of R&D field and industries. By using an LD chip coupled with a waveguide grating, a hybrid integrated WBG-ECDL is built, with mechanisms similar to the monolithically integrated DBR-LD. Basic materials of LD are III–V compounds, e.g., AlGaAs, InGaAs, InGaAsP, InGaAlP, InGaN, etc., whereas various materials different from the chip can be selected for the waveguide, including semiconductors and dielectrics. Silicon dioxide on silicon or silica substrate (SiO_2/Si), lithium niobate crystal ($LiNbO_3$), and polymers are the most mature materials with best performances, used widely. SiO_2/Si waveguide is usually grown on silica substrate or on silicon with a SiO_2 surface by means of chemical vapor deposition (CVD), or flame hydrolysis deposition (FHD), or Sol-Gel technology. The core layer is doped by GeO_2, etc., to increase its refractive index. Stripe structures are fabricated by photolithography and related processing. Similar to FBG, WBG can be written in its core by UV exposure. $LiNbO_3$ waveguide is fabricated by diffusion or ion-exchange of Ti, Ta, or other element. The Bragg grating on $LiNbO_3$ can be fabricated by photolithography, or by dopant diffusion with a grating mask.

WBG has advantages superior to FBG. It can utilize the functions of dielectric materials, such as the electro-optic effect, acousto-optic effect and birefringence of dielectric materials; polymer has an advantage of low cost. The planar waveguide has polarization selectivity; combined with polarization of LD, polarization state of WBG-ECDL is usually more stable than that of FBG-ECDL. In addition, both of waveguide and LD are fabricated by planar process; their coupling is more stable than coupling between LD and fiber. Such a structure has implanted to FBG-ECDL with FBG fixed on a planar substrate to improve coupling stability [67].

(2) **Design and Characteristics of WBG-ECDL**

The basic characteristics of WBG are the same as those of DBR and FBG; the principle and structure of WBG-ECDL are similar to FBG-ECDL. Figure 5.24a

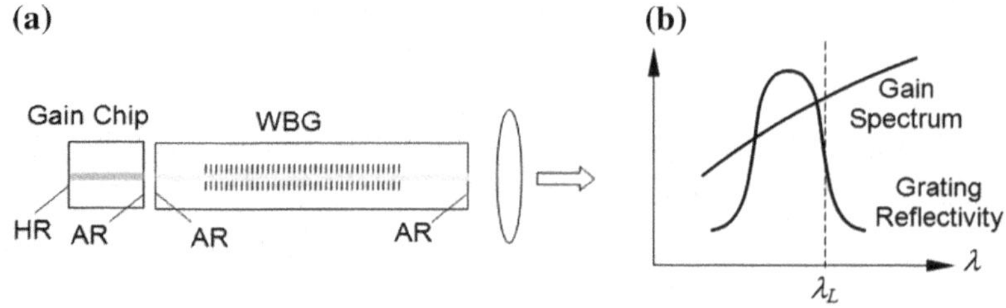

Fig. 5.24 **a** Structure of WBG-ECDL. **b** Spectral relation of LD gain and grating reflectivity

shows a schematic structure of WBG-ECDL [73, 74], where the gain chip is an LD chip or a SOA chip with one cleaved facet high reflective (HR) coated and the other AR-coated. The output of waveguide is collimated by a lens, or coupled to a pigtail fiber.

For a narrow line ECDL, the laser should work at wavelength where the linewidth reduction factors A and B have higher values. For both of FBG and WBG, it is at the longer wavelength edge of grating's spectrum as shown in Fig. 5.18. However, the reflectance there is lower than that at the peak, so it is a problem how to let the laser work with a higher threshold gain. Figure 5.24b gives a solution [74], that is, the laser works at the shorter wavelength side of gain spectrum, where gain is higher than that at FBG peak, to meet the condition of

$$g(\lambda_L) - \frac{1}{L}\ln\frac{1}{R_g(\lambda_L)} > g(\lambda_{\text{peak}}) - \frac{1}{L}\ln\frac{1}{R_g(\lambda_{\text{peak}})}. \tag{5.59}$$

Of cause, the phase condition should be satisfied there at the same time. Careful design and adjustment are necessary to reach such a working point. The spectral relationship of (5.59) is needed also for narrow linewidth FBG-ECDL.

It is noted that an addition merit is provided when working at shorter wavelength side of gain spectrum, that is, the linewidth enhancement factor α is much lower than that at longer wavelength side, as discussed in Sect. 3.2.1. The WBG-ECDL with optimized parameters shows good performance. As Ref. [73] reported, the noise level of WBG-ECDL was measured better than those of DFB-LD, DBR fiber laser, Littman ECDL, and similar to that of nonplanar ring oscillator (NPRO) solid-state laser [75], which is supposed one of the best lasers in coherence. The WBG-ECDL has been developed as a commercial product, and used in practical applications.

For more functions, other optical devices and components can be packaged inside the ECDL modules, such as output power monitor, temperature sensor and TE-cooler, and optical isolator. WBG-ECDL can be tuned by temperature variation, similar to FBG-ECDL.

5.3.3 Volume Bragg Grating ECDL

(1) **Materials and Characteristics of VBG**

The volume Bragg grating (VBG) is fabricated in a special photo-sensitive silicate glass, which consists of aluminum silicate and sodium silicate, and is doped with silver, cerium, bromine, and fluorine, etc., in certain fractions [76–80]. The glass has property of photo-thermo-refractive (PTR) effect, i.e., permanent index change can be generated in the material by UV irradiation and follow-up thermal treatment.

The fabrication of PTR-VBG involves basically two steps: UV exposure first and then annealing at high temperature. The periodical index modulation in volume is generated by UV holography. Mixture of micro crystals containing AgF, NaF, and other compositions is formed in the irradiated regions at high temperature by the combination of photo-induced reaction and thermal reaction. Index modulation amplitude up to 10^{-4}–10^{-3} has been obtained experimentally [76, 78]. The working wavelength of commercial VBG is mostly in 800–1000 nm band, especially suitable for applications of LD-pumped solid-state lasers and other near infrared laser technologies. Figure 5.25 shows a VBG plate and the optical paths of its diffraction.

Characteristics of VBG can be analyzed by Bragg diffraction equation and coupled mode theory, similar to the analysis of FBG. The difference between them is that the incident angle is not limited in normal direction of VBG. The peak wavelength of diffraction is now determined by, $m\lambda = 2n\Lambda\cos\alpha$, where Λ is the period of grating in normal direction, m is the order of diffraction, which is taken to be unity in usual cases, and α is the incident angle. In case of normal incident, the Bragg wavelength is $\lambda_B = 2n\Lambda$, the same as that in FBG and WBG. The reflection spectrum, as a function of $\Delta\lambda = \lambda - \lambda_B$, is deduced to be:

$$\begin{aligned} R(\Delta\lambda) &= \frac{\sinh^2\left(\pi n_1 t\sqrt{1-\Delta\lambda^2/\mu^2}/\lambda\right)}{\cosh^2\left(\pi n_1 t\sqrt{1-\Delta\lambda^2/\mu^2}/\lambda\right)-\Delta\lambda^2/\mu^2} \quad (|\Delta\lambda| \le \mu) \\ &= \frac{\sin^2\left(\pi n_1 t\sqrt{\Delta\lambda^2/\mu^2-1}/\lambda\right)}{\Delta\lambda^2/\mu^2-\cos^2(\pi n_1 t\sqrt{\Delta\lambda^2/\mu^2-1}/\lambda)} \quad (|\Delta\lambda| \ge \mu), \end{aligned} \tag{5.60}$$

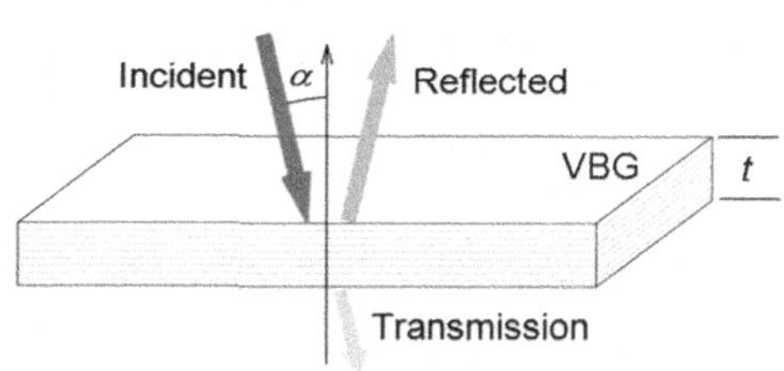

Fig. 5.25 Diffraction paths of a VBG plate

where t is the thickness of grating, $\mu = n_1\lambda_B/2n$ is the coupling coefficient in wavelength domain, n_1 is the amplitude of index modulation. The relation to coupling coefficients in wave vector domain is $\mu = \kappa\lambda_B/2\pi n$. The spectrum of reflection is similar to Fig. 4.8 of FBG. The peak reflectance is obtained from (5.60):

$$R_0 = \tanh^2(\pi n_1 t/\lambda_B) = \tanh^2(\kappa t). \tag{5.61}$$

The spacing between two first zero points is $\Delta\lambda_1 = \lambda_B\sqrt{\lambda_B^2 + n_1^2 t^2}/t$. It is seen that the linewidth of main lobe is inversely proportional to the number of grating grooves in its thickness: $\propto \lambda/t \propto \Lambda/t = 1/N$. When the material's loss is neglected, the transmission is $T = 1 - R$. The field reflectivity of VBG can be approximated as Lorentzian function, with a full width at half maximum linewidth (FWHM) $\Delta\lambda_H$:

$$r_{\mathrm{VBG}} = \frac{\sqrt{R_0}}{1 + j(\lambda - \lambda_B)/\Delta\lambda_H}. \tag{5.62}$$

Figure 5.26 shows a measured VBG reflection spectrum [81], giving its peak wavelength of 807.8 nm, FWHM linewidth of 0.4 nm, and peak reflectance of 80%.

Compared with FBG and WBG, one of the advantages of VBG is the possibility of incident angles variation. It brings about flexibility of its uses and more functions [79]. The peak wavelength, coupling coefficient, and peak reflectance vary with the incident angle, $\lambda_B = 2n\Lambda\cos\alpha$; $\mu = n_1\lambda_B/2n = \mu_0\cos\alpha$, and $R_0 = \tanh^2(\pi n_1 t/\lambda_B)$. The reflection spectrum has the same form as (5.60). It is noted the peak reflectance increases with angle α, similar to Fresnel reflection at a medium surface. Detailed analysis of the variations of reflection spectrum and diffraction coefficient with incident angle is given in Ref. [82].

The optical wave in FBG and WBG is usually the fundamental mode of waveguide, whereas the incident beam of VBG may have different beam shape.

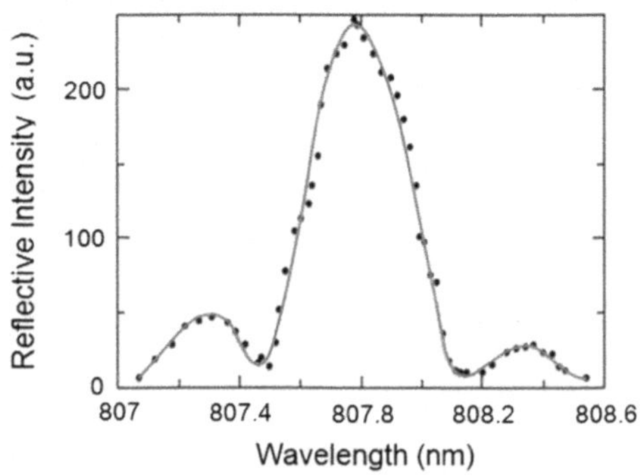

Fig. 5.26 A measure reflection spectrum of VBG

The formulas given above are deduced for a planar wave. In practice, the transverse dimension of incident beam is much narrower than the planar wave approximation. The effective reflection spectrum should be derived by integration with a weighting factor of angular distribution. Reference [82] presented the results of analysis of a Gaussian beam, and the influence of polarization state of incident beam.

As a frequency selective element and a frequency discriminator, the thermal characteristics of VBG are surely concerned. It was reported that the peak wavelength temperature coefficient of typical VBG is 0.01 nm/°C [83]. Typically the longitudinal mode temperature coefficient of 800 nm LD is 0.06 nm/°C, the gain peak temperature coefficient is 0.2–0.3 nm/°C; therefore, the role of VBG in frequency selectivity and stabilization of LD is remarkable.

(2) **VBG External Cavity LD**

The typical structure of VBG-ECDL is given in Fig. 5.27, where the VBG is used as the output mirror. According to (5.1a), (5.1b) and (5.1c), the effective reflectance of LD chip is expressed as

$$\hat{r}_{\text{eff}} = \frac{r_2 + r_{\text{VBG}} e^{j2kl}}{1 + r_2 r_{\text{VBG}} e^{j2kl}}. \tag{5.63}$$

One of purposes of VBG-ECDL is to realize the single longitudinal mode operation. It is reqiuired that the mode spacing $\Delta\lambda_{\text{ECDL}}$ is larger than the linewidth of VBG:

$$\Delta\lambda_{\text{ECDL}} = \lambda_{\text{B}}^2/(2nL + 2l) > \Delta\lambda_H. \tag{5.64}$$

Therefore, the external cavity length l should be short, and the thickness of VBG should be large. The LD output beam must be collimated by a lens; therefore, the reduction of l is limited in practice. It is thus necessary to adjust LD pump current, and/or temperatures of LD and VBG, together with adjusting l precisely, to enhance the dominance of main mode.

The mode competition also depends on the LD gain spectrum, expressed by Lorenzian function approximately, as discussed in Chapter 2. To maintain the laser working at Bragg wavelength, it is required that the spacing between net gain and threshold gain at λ_{B} should be smaller than that at the gain peak λ_p:

$$g_{\text{th}}^{[\text{B}]}(\lambda_{\text{B}}) - g(\lambda_{\text{B}}) < g_{\text{th}}(\lambda_p) - g(\lambda_p). \tag{5.65}$$

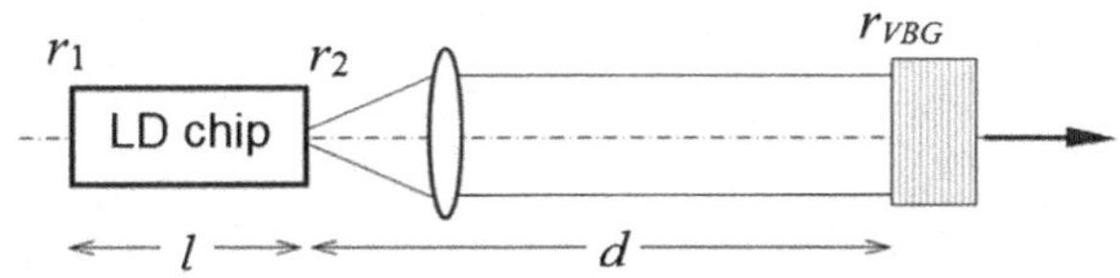

Fig. 5.27 Schematic structure of VBG-ECDL

where $g_{th}^{[B]}(\lambda) = \alpha_c + (1/L)\ln(1/R_1 R_{eff})$, and $g_{th}(\lambda) = \alpha_c + (1/2L)\ln(1/R_1 R_2)$. The required effective reflectance is thus deduced to be

$$R_{eff} = |r_{eff}|^2 \geq R_2 \exp\frac{2g_N NL}{1 + (\lambda - \lambda_p)^2/\Delta\lambda_g^2}. \tag{5.66}$$

In LD-pumped solid-state lasers and fiber lasers, the wavelength of LD must be locked at the absorption line of rare-earth ions. For example, 808 nm line of Nd-YAG, 980 nm line of Er/Yb-doped fiber (EDF/YDF). The wavelength of free-running LD often varies with temperature, with injection current variation, and with device aging for long tern operation. The wavelength locking of high power LD and LD array by VBG is proved an effective method [81, 83–85]. For such applications, the ECDL may work in multiple longitudinal modes, so long as the fluctuation of central wavelength of envelop spectrum is less than the linewidth of ion absorption. The phase factors in formula (5.63) can now be averaged, and the effective reflectance is rewritten as

$$|r_{eff}| = \frac{r_2 + \eta_c|r_{VBG}(\lambda)|}{1 + \eta_c r_2|r_{VBG}(\lambda)|}. \tag{5.67}$$

where η_c is the coupling coefficient between VBG and LD chip. A lower r_2 and a higher η_c are surely beneficial for a stronger effect of VBG. On the other hand, the peak reflectance of VBG is limited by the output power requirement. Design of parameters should take the external differential efficiency and the ratio of output powers from two ends into account, expressed as

$$\eta_d = \eta_{in}\frac{(1 - R_{eff})\sqrt{R_1}}{(1 - R_{eff})\sqrt{R_1} + (1 - R_1)\sqrt{R_{eff}}}; \tag{5.68}$$

$$\frac{P_{front}}{P_{rear}} = \frac{(1 - R_{eff})\sqrt{R_1}}{(1 - R_1)\sqrt{R_{eff}}}. \tag{5.69}$$

Figure 5.28a gives a spectrum of VBG-ECDL with a broad area LD chip, measured by an optical spectrum analyzer. Figure 5.28b is signals measured by a scanning Fabry-Perot etalon for precise data [84], showing a high SMSR and linewidth of 19 MHz, much narrower than the original linewidth of broad area LD.

VBG has been used widely for wavelength locking of high power diode lasers. Figure 5.29 shows the spectra of LD output at different temperatures of LD heat sink, showing that the lasing wavelength of VBG-ECDL keeps a constant within the limitation of apparatus precision in temperature range of 22–35 °C, while the wavelength of free-running LD varies about 2.5 nm [81].

Various configurations of ECDL can be designed by utilizing the spatial diffraction characteristics of VBG [77]. Fox-Smith type cavity lasers can be built by using VBG as a beam splitter with narrow line spectral characteristics, as shown in

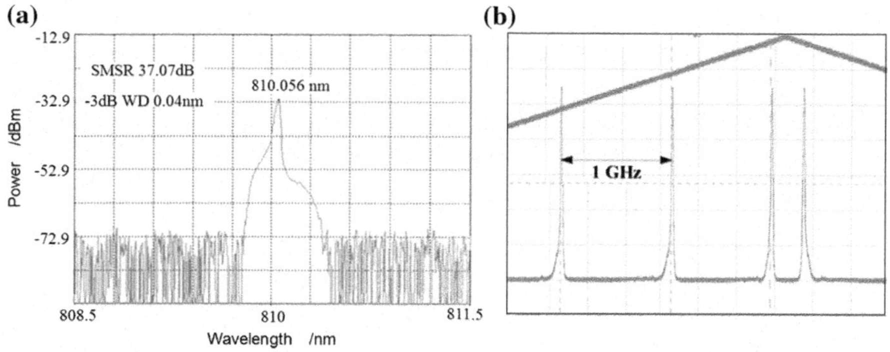

Fig. 5.28 **a** Spectrum of VBG-ECDL output; **b** Signal of scanning FPI. *Thicker line* scanning voltage; *Thinner line* F-P output. Reprinted from Ref. [84] with permission

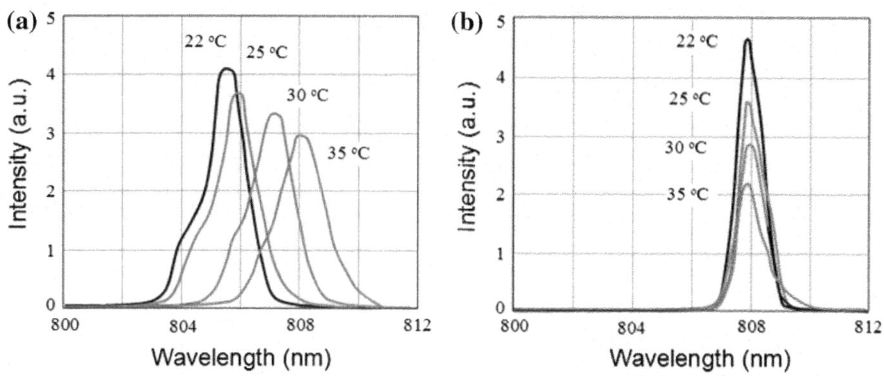

Fig. 5.29 Laser spectrum at different temperatures: **a** Free-running LD; **b** VBG-ECDL. Reprinted from Ref. [81] with permission

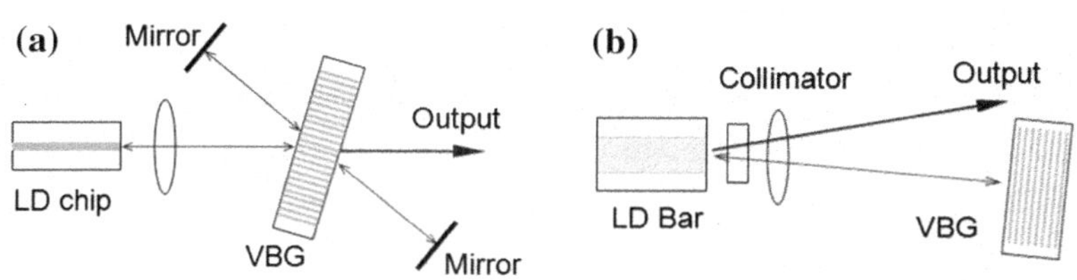

Fig. 5.30 **a** Fox-Smith type cavity with VBG. **b** VBG for beam quality improvement

Fig. 5.30a. The beam quality of broad area LD and LD array can also be improved by utilizing the spatial diffraction of VBG, as shown in Fig. 5.30b, where the VBG not only plays a role of the wavelength locking, but also enhances the oscillation of main spatial lobe, and suppresses the side lobes, resulting in beam divergence reduction. The different array elements, or beams from different position of broad area laser bar, will couple with each other by feedback from VBG, and the lateral phase locking is thus enhanced.

5.4 Diode Laser with Cavity Feedback

Apart from the gratings discussed in Sects. 5.2 and 5.3, Fabry-Perot (F-P) cavity and ring cavity are also used to build ECDL, since both of them have properties of narrow line resonances.

5.4.1 *Feedback from Fabry-Perot Cavity*

(1) **Basic Characteristics of Fabry-Perot Cavity**

The basic theory of F-P cavity can be found in many textbooks [5, 63]. It is composed of two planar mirrors, as shown in Fig. 5.31a, with field reflectance of $r_1 = \sqrt{R_1}$ and $r_2 = \sqrt{R_2}$, and transmission of $t_1 = \sqrt{1 - R_1}$ and $t_2 = \sqrt{1 - R_2}$, if the mirrors are lossless.

The property of F-P is based on the model of multiple beam interference. The field reflection and transmission of F-P cavity are deduced to be

$$\begin{aligned} r_{\mathrm{FP}} &= \frac{-r_1 + r_2 e^{j\Delta}}{1 - r_1 r_2 e^{j\Delta}} = |r_{\mathrm{FP}}| \exp(j\phi_r) \\ t_{\mathrm{FP}} &= \frac{t_1 t_2 e^{j\Delta/2}}{1 - r_1 r_2 e^{j\Delta}} = |t_{\mathrm{FP}}| \exp(j\phi_t) \end{aligned}, \tag{5.70}$$

where $\Delta = 4\pi nL\cos\theta/\lambda$ is the phase shift between the adjacent beams. The intensity reflection and transmission are expressed as

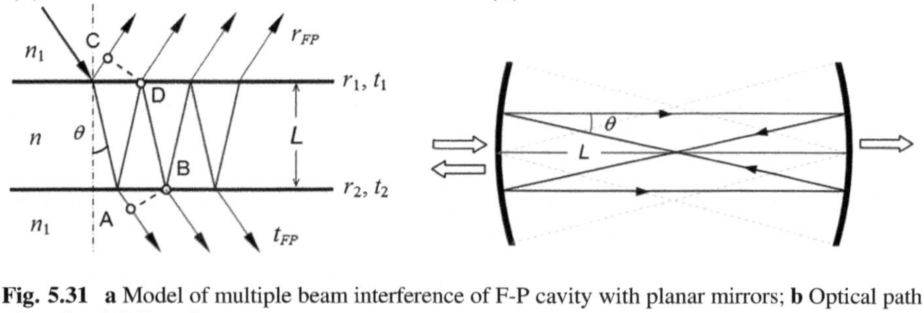

Fig. 5.31 **a** Model of multiple beam interference of F-P cavity with planar mirrors; **b** Optical path of confocal F-P cavity

$$\begin{aligned} R_{\mathrm{FP}} &= \frac{R_1 + R_2 - 2\sqrt{R_1 R_2}\cos\Delta}{1 + R_1 R_2 - 2\sqrt{R_1 R_2}\cos\Delta} = \frac{R_0 + s^2\sin^2(\Delta/2)}{1 + s^2\sin^2(\Delta/2)} \\ T_{\mathrm{FP}} &= \frac{T_1 T_2}{1 + R_1 R_2 - 2\sqrt{R_1 R_2}\cos\Delta} = \frac{T_0}{1 + s^2\sin^2(\Delta/2)} \end{aligned}, \tag{5.71}$$

where $R_0 = \frac{(\sqrt{R_1}-\sqrt{R_2})^2}{(1-\sqrt{R_1 R_2})^2}$, $T_0 = \frac{T_1 T_2}{(1-\sqrt{R_1 R_2})^2}$, and $s^2 = \frac{4\sqrt{R_1 R_2}}{(1-\sqrt{R_1 R_2})^2}$. For lossless mirrors, $T_{1,2} + R_{1,2} = 1$, so that $T_{\mathrm{FP}} + R_{\mathrm{FP}} = 1$. For symmetric cavity with $R_1 = R_2 = R$, $R_0 = 0$, $T_0 = 1$, and $s = 2\sqrt{R}/(1-R)$. A series of resonances occurs at $\Delta = 2m\pi$ with narrow transmission peaks. The resonant wavelengths and frequencies F-P are expressed as

$$\begin{aligned} \lambda_m &= 2nL\cos\theta/m \\ \nu_m &= mc/(2nL\cos\theta) \end{aligned}. \tag{5.72}$$

The spacing between adjacent resonances is called the free spectral range (FSR) of F-P:

$$\begin{aligned} \lambda_{\mathrm{FSR}} &= \lambda^2/(2nL\cos\theta) \\ \nu_{\mathrm{FSR}} &= c/(2nL\cos\theta) \end{aligned}. \tag{5.73}$$

The line shape near one of the resonances λ_m is then written as

$$\begin{aligned} T_{\mathrm{FP}} &= \frac{1}{1 + s^2\sin^2(m\pi\lambda_m/\lambda)} \approx \frac{1}{1 + (\lambda - \lambda_m)^2/\delta\lambda_{FP}^2} \\ \delta\lambda_{\mathrm{FP}} &= \frac{\lambda_m}{m\pi}\sin^{-1}\frac{1}{s} = \frac{\lambda_{\mathrm{FSR}}}{\pi}\sin^{-1}\frac{1}{s} \approx \frac{\lambda_{\mathrm{FSR}}}{\pi s} \end{aligned}, \tag{5.74}$$

It is seen that the line shape is in Lorentzian basically. The performance of F-P cavity is characterized by its finesse, defined as the ratio of FSR to FWHM linewidth; for a symmetric F-P with $R \approx 1$, it is expressed as

$$F = \frac{\lambda_{\mathrm{FSR}}}{\delta\lambda_{\mathrm{FP}}} = \frac{\pi\sqrt{R}}{1-R}. \tag{5.75}$$

The resonator is also characterized by its quality factor Q, defined as $Q = \nu/\delta\nu$, which is proportional to the finesse of F-P. Figure 5.32a shows a calculated reflection spectrum of a lossless symmetric F-P with $R = 0.99$.

If the medium in cavity is with loss α, including the loss caused by beam divergence, the resonance positions and FSR are the same as expressed by (5.72) and (5.73), whereas the finesse is expressed as $F = \pi\sqrt{\rho R}/(1-\rho R)$ with $\rho = \mathrm{e}^{-\alpha L}$, lower than that with lossless medium.

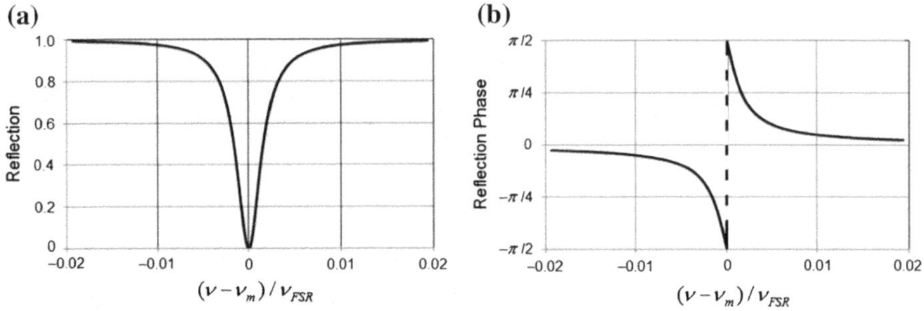

Fig. 5.32 **a** Reflection spectrum of F-P cavity; **b** Phase shift of F-P reflection

The F-P mirrors have to be adjusted and kept highly parallel. A minor deviation will cause optical loss and decrease of finesse. The divergence of incident beam must also be low enough, or otherwise the resolution of spectrum analysis will be lowered. To mitigate such a requirement and to improve the applicability, F-P cavities with spherical mirrors are developed. Among them the confocal F-P cavity is used widely, in which the cavity length L equals the curvature radius of the spherical mirror, as shown in Fig. 5.32b. Compared with the planar mirror F-P, the roundtrip path and the phase difference between the adjacent beams is doubled. For paraxial rays with $\theta \approx 0$ and for cavity medium of $n = 1$, the phase difference Δ and frequency FSR are modified as

$$\Delta = 4kL\frac{2\cos\theta - 1}{\cos 2\theta}\cos^2\theta \approx 4kL(1+\theta^2/2) \approx 4kL; \tag{5.76}$$

$$\nu_{\rm FSR} = c/4L. \tag{5.77}$$

The formulas for planar mirror F-P are applicable to the confocal F-P cavity, so long as the phase factor is replaced by expression (5.76).

(2) **Phase Shift and Group Delay of F-P Cavity**

The optical wave will experience phase changes when reflected and transmitted from F-P cavity. For a lossless F-P, the phase factors are deduced from (5.70), expressed as:

$$\begin{aligned}\phi_t &= \arctan\left(\frac{1+r_1r_2}{1-r_1r_2}\tan\frac{\Delta}{2}\right) \to \arctan\left(\frac{1+R}{1-R}\tan\frac{\Delta}{2}\right) \\ \phi_r &= \arctan\frac{a\sin\Delta}{\cos\Delta - b} \to -\arctan\left(\frac{1-R}{1+R}\cot\frac{\Delta}{2}\right)\end{aligned}, \tag{5.78}$$

where $a = (1-R_1)/(1+R_1)$, $b = r_1(1+R_2)/[r_2(1+R_1)]$. The last expressions of (5.78) after arrows are for the symmetric F-P with $r_1 = r_2$. Figure 5.32b shows a calculated reflection phase of a symmetric F-P cavity with $R = 0.99$. The phase shifts cause group delays, deduced as

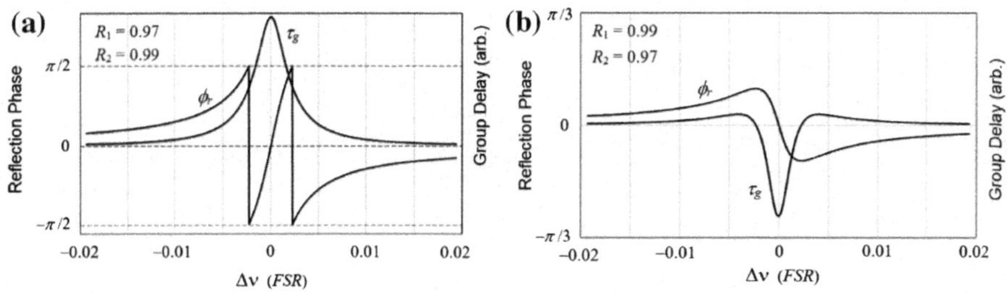

Fig. 5.33 Calculated phase and delay spectra of asymmetric F-P. **a** $R_1 < R_2$; **b** $R_1 > R_2$

$$\tau_t = \frac{\partial \phi_t}{\partial \omega} = \frac{\tau_c}{2} \frac{\sqrt{1+s^2}}{1+s^2 \sin^2(\Delta/2)}, \tag{5.79a}$$

$$\tau_r = \frac{\partial \phi_r}{\partial \omega} = \frac{a\tau_c(1-b\cos\Delta)}{(\cos\Delta - b)^2 + a^2 \sin^2\Delta} \to \frac{\tau_c}{2} \frac{\sqrt{1+s^2}}{1+s^2 \sin^2(\Delta/2)}, \tag{5.79b}$$

where the last expression of (5.79a) and (5.79b) is for a symmetric F-P. Highest group delays are obtained at resonances: $\tau_{t0} = \tau_{r0} = \frac{\tau_c}{2}\frac{1+R}{1-R}$ for $R_1 = R_2 \approx 1$. However, ϕ_r and τ_r are different from ϕ_t and τ_t for the case of $R_1 \neq R_2$; in addition, ϕ_r and τ_r are different between cases of $R_1 < R_2$ and $R_1 > R_2$. The peak group delay of a F-P with $R_1 < R_2$ is much higher than that of the symmetric F-P with $R \approx \sqrt{R_1 R_2}$; and the peak group delay for $R_1 > R_2$ is negative. Figure 5.33 shows calculated curves as examples.

The differential of field reflectance over frequency, related to linewidth reduction factor B, is obtained from (5.71):

$$\frac{\mathrm{d}|r_{\mathrm{FP}}|}{\mathrm{d}\omega} = \frac{\tau_c}{2} \frac{\mathrm{sgn}(\Delta)\cos(\Delta/2)}{[1+s^2 \sin^2(\Delta/2)]^{3/2}}. \tag{5.80}$$

Figure 5.34 shows calculated spectra of reflectance, group delay and differential of $|r_{\mathrm{FP}}|$ for the case of symmetric F-P.

(3) **Applications of F-P Cavity in ECDL**

To utilize the characteristics of F-P cavity in ECDL, several schemes were reported in journals. One of them is to insert an F-P inside the extended cavity [86, 87], as shown in Fig. 5.35a. One of the external cavity modes is tuned by adjusting the position of output mirror, attached on a PZT, to match with one of the F-P resonant peaks. The side modes are thus suppressed effectively, and the linewidth of main mode is narrowed greatly. The F-P resonant peak can also be tuned by rotation as marked in the figure. In addition, the output mirror is placed at the focus point of lenses to mitigate the precision requirement of adjusting. The thin film filter (TFF) can play similar role to that of F-P filter, as shown in the figure by

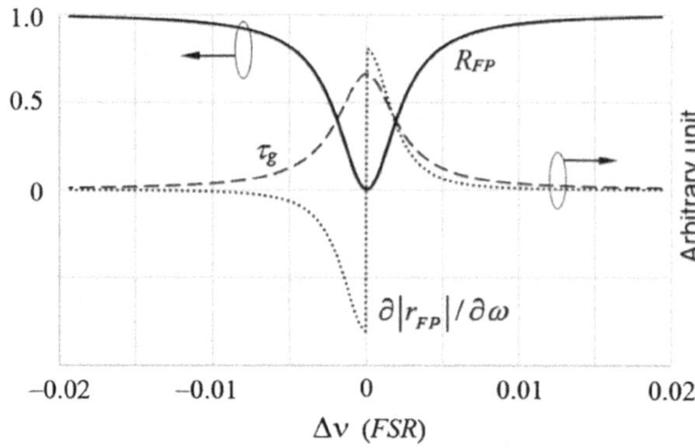

Fig. 5.34 Calculated characteristics of a symmetric Fabry-Perot resonator

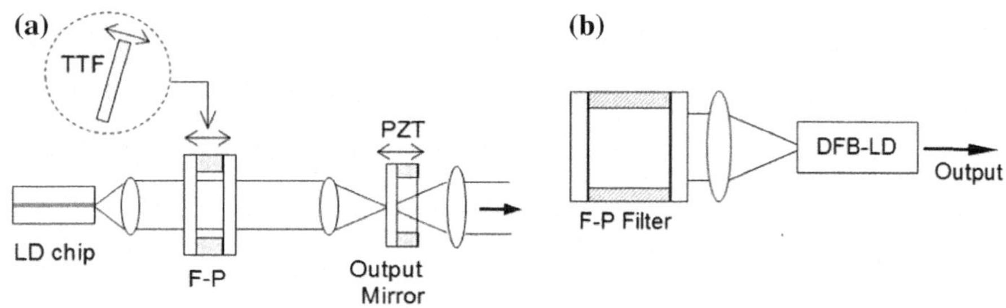

Fig. 5.35 **a** ECDL with a tunable F-P cavity; **b** ECDL with feedback from F-P filter

circled insert. Although the linewidth of TFF is much larger than F-P, it is attractive for its low cost and used in cases with lower linewidth requirements.

Another configuration is to use F-P cavity as a feedback element [88, 89], as shown in Fig. 5.35b. Different from the ECDL introduced above, the active element is a DFB-LD, so that the lasing wavelength is determined by DFB itself. When the wavelength is tuned to locate at the edge of F-P cavity reflection spectrum, the linewidth will be greatly suppressed by the linewidth reduction factor A and B of F-P cavity.

Frequency of a DFB diode laser is stabilized in Ref. [90] by using a mesoscopic monolithic confocal F-P cavity. Such a design improves greatly the stability of cavity. Moreover, an asymmetric cavity is used, i.e., with surface reflection of 0.97 and 0.9997, respectively. The effective reflective delay is thus increased greatly, as illustrated by formula (5.78) and Fig. 5.33. The Lorentzian linewidth down to 15.7 Hz was obtained using the scheme.

5.4.2 Feedback from Ring Cavity

The ring cavity has similar characteristics to F-P cavity with sharp resonance peaks. It can be constructed by various structures and with different components.

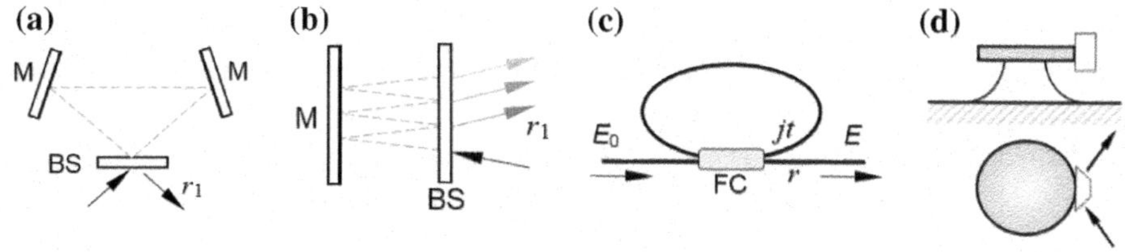

Fig. 5.36 Typical ring cavities. **a** Composed of planar mirrors (M) and beam splitter (BS); **b** G-T cavity; **c** fiber ring cavity with fiber coupler (FC); **d** Whispering gallery mode cavity

Figure 5.36 shows different schemes of ring cavities. Ring cavity (a) is composed of two totally reflective mirrors (M) and a beam splitter (BS). Cavity (b) is similar to F-P cavity, but one of the reflectors is a totally reflective mirror. Such a cavity is called Gires-Tournois (GT) interferometer, after the names of its inventors. Its characteristics are similar to an asymmetric F-P. Ring cavity (c) uses a fiber coupler (FC) with its two ports connected. Figure (d) is a whispering gallery mode (WGM) cavity.

Characteristics of ring cavities (a) and (b) can be deduced by referring formulas of an asymmetric F-P, with $r_2 = 1$ and $t_2 = 0$:

$$\begin{aligned} r_{GT} &= \frac{-r_1 + e^{j\Delta}}{1 - r_1 e^{j\Delta}} = \frac{(1+r_1^2)\cos\Delta - 2r_1 + j(1-r_1^2)\sin\Delta}{1 + r_1^2 - 2r_1\cos\Delta} = e^{j\phi_{GT}} \\ t_{GT} &= 0 \end{aligned}, \tag{5.81}$$

where $\Delta = 4\pi nL\cos\theta/\lambda$, the same as that of F-P cavity. It is seen that $|r_{\mathrm{GT}}| = 1$; i.e., G-T cavity is an all-pass filter when the losses of medium and reflectors are neglected. Its phase shift is

$$\phi_{\mathrm{GT}} = \arctan\frac{(1-r_1^2)\sin\Delta}{(1+r_1^2)\cos\Delta - 2r_1}. \tag{5.82}$$

Structure of Fig. 5.36c uses a 2 × 2 fiber coupler with intensity split ratio of r^2: t^2. It is also an all-pass filter for a lossless ring; its transmission is deduced as

$$\frac{E}{E_0} = \frac{r - e^{j\beta L}}{1 - re^{j\beta L}} = e^{j\phi_{\mathrm{FR}}}, \tag{5.83a}$$

$$\phi_{\mathrm{FR}} = \arctan\frac{(1-r^2)\sin\Delta}{(1+r^2)\cos\Delta - 2r}, \tag{5.83b}$$

where $\Delta = \beta L = n_f kL$, β is the propagation constant of fiber, L is the length of fiber ring. The fiber ring has the same characteristics as G-T cavity. Their group delay is deduced as

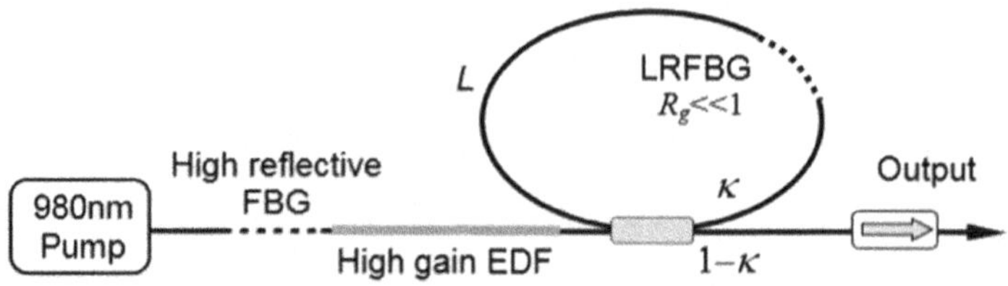

Fig. 5.37 A narrow linewidth fiber laser with a slow-light fiber ring as its output mirror

$$\tau_g = \frac{\partial \phi}{\partial \omega} = \tau_c \frac{1 - r^2}{1 + r^2 - 2r\cos\Delta} \tag{5.84}$$

where $\tau_c = nL/c$ is the single trip delay in the cavity. At resonant peaks with $\Delta = 2m\pi$, the group delay reaches its maxima: $\tau_g = \tau_c(1 + r)/(1 - r)$, which is toward infinitive when $r \sim 1$. For practical applications in laser devices, the loss of fiber coupler may not be neglected because the beam will pass through it multiple times, i.e., $t^2 + r^2 = a < 1$. The transmissions of field and intensity are then rewritten as

$$\begin{aligned} E/E_0 &= (r - ae^{j\beta L})/(1 - re^{j\beta L}) \\ I/I_0 &= (a - r)^2/(1 - r)^2 \end{aligned} \tag{5.85}$$

Similar to a fiber ring, a waveguide ring can be built with a waveguide coupler by the integrated optics technology [91], giving similar characteristics. Various micro cavities have been proposed and fabricated as reviewed in Ref. [92]. Among them, the whispering gallery mode (WGM) cavity is attractive. The acoustic whispering gallery wave was found very earlier in a dome or a round wall construction. The effect in optics is utilized as a miniature ring cavity in microcavity semiconductor lasers, as shown in Fig. 5.36d. The optical wave in WGM cavity is concentrated along the boundary, and will resonate the same as an ordinary ring. By using a prism coupler as input and output component, it plays a role of narrow band filter. A whispering gallery mode cavity made of dielectric crystal, which has advantages of low loss, high Q, and high stability, is used as a resonator to stabilize LD frequency is demonstrated in Ref. [93]. The cavity is integrated heterogeneously with LD chip with the stability improved greatly. The integral linewidth down to 30 Hz and instantaneous linewidth of sub-Hz is obtained.

The fiber ring shown in Fig. 5.36c cannot be used as a laser cavity mirror since it is an all-pass filter. If a very low reflective fiber Bragg grating is inserted in the fiber loop, it will show backward reflection. Moreover, multiple roundtrips make strong resonances; the reflection linewidth is suppressed and the effective group delay is increased greatly. Reference [94] built a fiber laser with such a composite ring as cavity mirror, as shown in Fig. 5.37, showing low phase noise and low intensity noise. The intrinsic Lorentzian linewidth of 10 Hz was estimated, and the linewidth in longer integration was measured to be 150 Hz. The characteristics of slow-light effect of the composite ring and its optimization are studied in detailed in Ref. [95].

References

1. Pertermann K (1988) Laser diode modulation and noise. Kluwer Academic Publishers
2. Ohtsu M (1992) Highly coherent semiconductor lasers. Artech House
3. Ye C (2004) Tunable external cavity diode lasers. World Scientific Publishing Corporation
4. Ohtsubo J (2013) Semiconductor lasers—stability, instability and chaos, 3rd edn. Springer, Berlin
5. Born M, Wolf E (1999) Principles of optics, 7th edn. Cambridge University Press
6. Lang R, Kobayashi K (1980) External optical feedback effects on semiconductor injection laser properties. IEEE J Quantum Electron 16(3):347–355
7. Schawlow AL, Townes CH (1958) Infrared and optical masers. Phys Rev 112:1940–1949
8. Henry CH (1983) Theory of the phase noise and power spectrum of a single mode injection laser. IEEE J Quantum Electron 19(9):1391–1397
9. Spano P, Piazzolla S, Tamburrini M (1984) Theory of noise in semiconductor lasers in the presence of optical feedback. IEEE J Quantum Electron 20(4):350–357
10. Agrawal G (1984) Line narrowing in a single-mode injection laser due to external optical feedback. IEEE J Quantum Electron 20(5):468–471
11. Kitaoka Y, Sato H, Mizuuchi Q et al (1996) Intensity noise of laser diodes with optical feedback. IEEE J Quantum Electron 32(5):822–828
12. Detoma E, Tromborg B, Montrosset I (2004) Frequency and time domain analysis of an external cavity laser with strong filtered optical feedback. Proc SPIE 5452:283–290
13. Fischer APA, Andersen OK, Yousefi M et al (2000) Experimental and Theoretical study of filtered optical feedback in a semiconductor laser. IEEE J Quantum Electron 32(3):375–384
14. Sun H, Menhart S, Adams A (1994) Calculation of spectral linewidth reduction of external-cavity strong-feedback semiconductor lasers. Appl Opt 33(21):4771–4775
15. Hau LV, Harris SE, Dutton Z et al (1999) Light speed reduction to 17 metres per second in an ultracold atomic gas. Nature 397:594–598
16. Khurgin JB (2010) Slow light in various media: a tutorial. Adv Opt Photonics 2(3):287–318
17. Patzak E, Sugimura A, Saito S et al (1983) Semiconductor laser linewidth in optical feedback configurations. Electron Lett 19(24):1026–1027
18. Kazarinov RF, Henry CH (1987) The relation of line narrowing and chirp reduction resulting from the coupling of a semiconductor laser to a passive resonator. J Quantum Electron 23 (9):1401–1409
19. Tromborg B, Olesen H, Pan X et al (1987) Transmission line description of optical feedback and injection locking for Fabry-Perot and DFB lasers. J Quantum Electron 23(11):1875–1889
20. Agrawal GP, Henry CH (1988) Modulation performance of a semiconductor laser coupling to an external high-Q resonator. J Quantum Electron 24(2):134–142
21. Kobayashi S, Kimuw T (1981) Injection locking in AlGaAs semiconductor laser. IEEE J Quantum Electron 17(5):681–689
22. Lang R (1982) Injection locking properties of a semiconductor laser. IEEE J Quantum Electron 18(6):76–983
23. Spano P, Piazzola S, Tamburrini M (1986) Frequency and intensity noise in injection-locked semiconductor lasers: theory and experiments. IEEE J Quantum Electron 22(3):427–435
24. Gordon R (2006) Fabry-Perot semiconductor laser injection locking. IEEE J Quantum Electron 42(4):353–356

25. Lau EK, Sung HK, Wu MC (2007) Scaling of resonance frequency for strong injection-locked lasers. Opt Lett 32(23):3373–3375
26. Lau EK, Sung HK, Wu MC (2008) Frequency response enhancement of optical injection-locked lasers. IEEE J Quantum Electron 44(1):90–99
27. Moon HS, Park SE, Park YH et al (2006) Passive atomic frequency standard based on coherent population trapping in 87Rb using injection-locked lasers. J Opt Soc Am B 23 (11):2393–2397
28. Hong Y, Shore KA (1999) Locking characteristics of a side-mode injected semiconductor laser. IEEE J Quantum Electron 35(11):1713–1717
29. van Voorst PD, Offerhaus HL, Boller KJ (2006) Single-frequency operation of a broad-area laser diode by injection locking of a complex spatial mode via a double phase conjugate mirror. Opt Lett 31(8):1061–1063
30. Wyatt R (1985) Spectral linewidth of external cavity semiconductor lasers with strong, frequency-selective feedback. Electron Lett 21(15):658–659
31. de Labachelerie M, Passedat G (1993) Mode-hop suppression of Littrow grating-tuned lasers. Appl Opt 32(3):269–274
32. Trutna WR Jr, Stokes LF (1993) Continuously tuned external cavity semiconductor laser. J Lightwave Tech 11(8):1279–1286
33. Nilse L, Davies HJ, Adams CS (1999) Synchronous tuning of extended cavity diode lasers: the case for an optimum pivot point. Appl Opt 38(3):548–553
34. Favre F, Le Cuen D, Simon JC et al (1986) External-cavity semiconductor laser with 15 nm continuous tuning range. Electron Lett 22(15):795–796
35. Hult J, Burns IS, Kaminski CF (2005) Wide-bandwidth mode-hop-free tuning of extended-cavity GaN diode lasers. Appl Opt 44(18):3675–3685
36. Hawthorn CJ, Weber KP, Scholten RE (2001) Littrow configuration tunable external cavity diode laser with fixed direction output beam. Rev Sci Instrum 72(12):4477–4479
37. de Labachelerie M, Latrasse C, Diomande K et al (1991) A 1.5 mm absolutely stabilized extended-cavity semiconductor laser. IEEE Trans Instrum Measur 40(2):185–190
38. Liu K, Littman MG (1981) Novel geometry for single-mode scanning of tunable lasers. Opt Lett 6(3):117–118
39. McNicholl P, Metcalf HJ (1985) Synchronous cavity mode and feedback wavelength scanning in dye laser oscillators with gratings. Appl Opt 24(17):2757–2761
40. Littman MG, Metcalf HJ (1987) Spectrally narrow pulsed dye laser without beam expander. Appl Opt 17(14):2224–2227
41. Godard A, Pauliat G, Roosen G et al (2002) Side-mode gain in grating-tuned extended-cavity semiconductor lasers: investigation of stable single-mode operation conditions. IEEE J Quantum Electron 38(4):390–401
42. Repasky KS, Nehrir AR, Hawthorne JT et al (2006) Extending the continuous tuning range of an external-cavity diode laser. Appl Opt 45(35):9013–9020
43. Führer T, Walther T (2008) Extension of the mode-hop-free tuning range of an external cavity diode laser based on a model of the mode-hop dynamics. Opt Lett 33(4):372–374
44. Wei F, Chen D, Xin G et al (2013) A compact and rugged tunable external cavity diode laser with Littman-Metcalf configuration. Chin J Lasers 40(11):1102012(1–7) (in Chinese)
45. Moharam MG, Gaylord TK (1983) Three-dimensional vector coupled-wave analysis of planar-grating diffraction. J Opt Soc Am 73(9):1105–1112
46. Chen D, Fang Z, Cai H et al (2009) Polarization characteristics of an external cavity diode laser with Littman-Metcalf configuration. IEEE Photonics Technol Lett 21(14):984–986
47. Filimonov S, Borysow J (1995) Long-range tunable diode laser. Appl Opt 34(3):438–443
48. Okamura H (2010) Shift lens external-cavity diode laser for broad wavelength tuning and switching. Opt Lett 35(8):1175–1177
49. Ménager L, Cabaret L, Lorgeré I et al (2000) Diode laser extended cavity for broad-range fast ramping. Opt Lett 25(17):1246–1248
50. Wang P, Seah LK, Murukeshan VM et al (2006) External-cavity wavelength tunable laser with an electro-optic deflector. Appl Opt 45(34):8772–8776

51. Wei F, Sun Y, Chen D et al (2011) Tunable external cavity diode laser with a PLZT electrooptic ceramic deflector. IEEE Photonics Technol Lett 23(5):296–298
52. Zhao Y, Peng Y, Yang T et al (2011) External cavity diode laser with kilohertz linewidth by a monolithic folded Fabry-Perot cavity optical feedback. Opt Lett 36(1):34–36
53. Britzger M, Khalaidovski A, Hemb B et al (2012) External-cavity diode laser in second-order Littrow configuration. Opt Lett 37(15):3117–3119
54. Chi M, Erbert G, Sumpf B et al (2010) Tunable high-power narrow-spectrum external-cavity diode laser based on tapered amplifier at 668 nm. Opt Lett 35(10):1545–1547
55. Jensen OB, Sumpf B, Erbert G et al (2011) Widely tunable high-power tapered diode laser at 1060 nm. IEEE Photonics Technol Lett 23(21):1624–1626
56. Sell JF, Miller W, Wright D et al (2009) Frequency narrowing of a 25 W broad area diode laser. Appl Phys Lett 94:051115(1–3)
57. Liu AQ, Zhang XM, Tang DY et al (2004) Tunable laser using micromachined grating with continuous wavelength tuning. Appl Phys Lett 85(17):3684–3686
58. Laurain A, Myara M, Beaudoin G et al (2009) High power single-frequency continuously-tunable compact extended-cavity semiconductor laser. Opt Express 17 (12):9503–9508
59. Hill KO, Fujii Y, Johnson DC et al (1978) Photosensitivity in optical fiber waveguides: application to reflection filter fabrication. Appl Phys Lett 32(10):647–649
60. Meltz G, Morey WW, Glenn WH (1989) Formation of Bragg gratings in optical fibers by a transverse holographic method. Opt Lett 14(15):823–825
61. Hill KO, Meltz G (1997) Fiber Bragg grating technology fundamentals and overview. J Lightwave Technol 15(8):1263–1276
62. Kashyap R (1999) Fiber Bragg gratings. Academic Press
63. Saleh BEA, Teich MC (2007) Fundamentals of photonics. Wiley
64. Fang Z, Chin K, Qu R et al (2012) Fundamentals of optical fiber sensors. Wiley
65. Bird DM, Armitage JR, Kashyap R et al (1991) Narrow line semiconductor laser using fiber grating. Electron Lett 27(13):1115–1116
66. Morton PA, Mizrahi V, TanbunEk T et al (1994) Stable single mode hybrid laser with high power and narrow linewidth. Appl Phys Lett 64(20):2634–2636
67. Goyal AK, Gavrilovic P, Po H (1998) 1.35 W of stable single-frequency emission from an external-cavity tapered oscillator utilizing fiber Bragg grating feedback. Appl Phys Lett 73 (5):575–577
68. Loh W, O'Donnell FJ, Plant JJ et al (2011) Packaged, high-Power, narrow-linewidth slab-coupled optical waveguide external cavity laser (SCOWECL). IEEE Photonics Technol Lett 23(14):974–976
69. Chen G, Qu R, Zhao H et al (1998) Wavelength conversion in a DBR laser with fiber Bragg rating external cavity. ACTA Opt Sin 18(3):257–261 (in Chinese)
70. Zhang W, Ding H, Zhao H et al (1996) An active mode-locked semiconductor laser with a single-mode fiber phase-grating external cavity. ACTA Opt Sin 16(12):1681–1683 (in Chinese)
71. Ding H, Zhao H, Zhang W et al (1998) Electrical wavelength switching of an active mode-locked fiber Bragg grating incorporated fiber ring laser. Chin J Lasers A 25(5):397–400 (in Chinese)
72. Barry LP, Dudley JM, Thomsen BC et al (1998) Frequency-resolved optical gating measurement of 1.4THz beat frequencies from dual wavelength self-seeded gain switched laser diode. Electron Lett 34(10):988–990
73. Numata K, Camp J, Krainak MA et al (2010) Performance of planar-waveguide external cavity laser for precision measurements. Opt Express 18(22):22781–22788
74. Numata K, Alalusi M, Stolpner L et al (2014) Characteristics of the single-longitudinal mode planar-waveguide external cavity diode laser at 1064 nm. Opt Lett 39(7):2101–2104
75. Kane TJ, Byer RL (1985) Monolithic, unidirectional single-mode Nd:YAG ring laser. Opt Lett 10(2):65–67

76. Efimov OM, Glebov LB, Glebova LN et al (1999) High-efficiency Bragg gratings in photothermorefractive glass. Appl Opt 38(4):619–627
77. Volodin BL, Dolgy SV, Melnik ED et al (2004) Wavelength stabilization and spectrum narrowing of high-power multimode laser diodes and arrays by use of volume Bragg gratings. Opt Lett 29(16):1891–1893
78. Venus GB, Sevian A, Smirnov VI et al (2005) High-brightness narrow- line laser diode source with volume Bragg-grating feedback. Proc SPIE 5711:166–176
79. Lumeau J, Glebov LB, Smirnov V (2006) Tunable narrowband filter based on a combination of Fabry-Perot etalon and volume Bragg grating. Opt Lett 31(16):2417–2419
80. Meng LS, Nizamov B, Madasamy P et al (2006) High power 7-GHz bandwidth external-cavity diode laser array and its use in optically pumping singlet delta oxygen. Opt Express 14(22):10469–10474
81. Xin G, Cheng C, Qu R et al (2007) Study of spectral characteristics of external-cavity semiconductor laser with a volume Bragg grating. ACTA Opt Sin 27(10):1821–1826 (in Chinese)
82. Ciapurin IV, Glebov LB, Smirnov VI (2005) Modeling of Gaussian beam diffraction on volume Bragg gratings in PTR glass. Proc SPIE 5742:183–194
83. Cheng C, Xin G, Feng H et al (2008) Temperature characteristics of volume Bragg grating external cavity semiconductor laser working at continuous wave. Chin J Lasers 35(1):27–30 (in Chinese)
84. Shen L, Ye Q, Cai H et al (2011) Mode-hop-free electro-optically tuned external-cavity diode laser using volume Bragg grating and PLZT ceramic. Opt Express 19(18):17244–17249
85. Chuang HC, Jimenez-Martinez R, Braun S et al (2008) Tunable external cavity diode laser using a micromachined silicon flexure and a volume holographic reflection grating for applications in atomic optics. J Micro/Nanolith MEMS MOEMS 7(2):021010(1–13)
86. Allard F, Maksimovic I, Abgrall M et al (2004) Automatic system to control the operation of an extended cavity diode laser. Rev Sci Instrum 75(1):54–58
87. Baillard X, Gauguet A, Bize S et al (2006) Interference-filter-stabilized external-cavity diode lasers. Opt Commun 266:609–613
88. Aoyama K, Yoshioka R, Yokota N et al (2014) Experimental demonstration of linewidth reduction of laser diode by compact coherent optical negative feedback system. Appl Phys Express 7:122701
89. Aoyama K, Yoshioka R, Yokota N et al (2015) Optical negative feedback for linewidth reduction of semiconductor lasers. IEEE Photonics Technol Lett 27(4):340–343
90. Lewoczko-Adamczyk W, Pyrlik C, Häger J et al (2015) Ultra-narrow linewidth DFB-laser with optical feedback from a monolithic confocal Fabry-Perot cavity. Opt Express 23 (8):9705–9709
91. Liu B, Shakouri A, Bowers JE (2001) Passive microring-resonator-coupled lasers. Appl Phys Lett 79(22):3561–3563
92. Vahala KJ (2003) Optical microcavities. Nature 424:839–846
93. Ling W, Ilchenko VS, Eliyahu D et al (2015) Ultralow noise miniature external cavity semiconductor laser. Nat Commun 8371:1–6
94. Pan Z, Ye Q, Cai H et al (2014) Fiber ring with long delay used as a cavity mirror for narrowing fiber laser. IEEE Photonics Technol Lett 26(16):1621–1624
95. Ye Q, Pan Z, Wang Z et al (2015) Novel slow-light reflector composed of a fiber ring resonator and low-reflectivity fiber Bragg grating. J Lightwave Technol 33(14):3016–3022

Chapter 6
Frequency Stabilization of Semiconductor Lasers

6.1 Introduction to Saturated Absorption Spectroscopy

Based on the technologies introduced in Chaps. 4 and 5, the semiconductor laser can work in a single longitudinal mode with high side mode suppression ratio (SMSR) and narrow linewidth. However, even if technical measures are adopted for temperature stabilization, vibration isolation, and pump current stabilization, laser's frequency noise and drift may not meet the requirements of applications, where low noises and high frequency stability are required, such as high resolution spectroscopy, high precision metrology, laser cooling of atoms and atomic clock, and high speed optical communications. It is necessary to stabilize the frequency by active feedback.

The laser frequency stabilization involves mainly three parts: a high precision frequency standard as reference, a frequency discriminator to detect the difference between reference frequency and laser's frequency, and an approach of controlling laser's frequency with feedback of error signal. Three frequency stabilization technologies have been developed successfully. One of them uses absorption lines of atoms or molecules as reference. The second uses high precision and high stability resonators, especially the F-P cavity. The method is named PDH method after its inventors, R.V. Pound, R.W.P. Drever, and J.L. Hall. The third uses a laser with precision and stability much higher than the laser to be stabilized, which will be introduced in Chap. 8.

Lasers with frequency stabilized on absorption lines of atoms or molecules are widely used in many important areas, such as spectroscopy, and laser cooling of atoms. One of the milestones in the field is the saturated absorption spectroscopy developed by A.L. Schawlow, one of the laser's inventors, who was awarded Nobel Prize in Physics 1981 due to the contribution [1, 2].

Z. Fang et al., *Single Frequency Semiconductor Lasers*, Optical and Fiber Communications Reports 9, DOI 10.1007/978-981-10-5257-6_6

6.1.1 Classical Theory of Absorption Spectrum

(1) **Classical Harmonic Oscillator Model**

When an optical wave $\boldsymbol{E}$ with frequency ω is propagating in a medium, the electric polarization will generate in it, $\boldsymbol{P} = \chi \boldsymbol{E}$, by two effects, one is the change of charge distribution inside the atoms or molecules induced by optical field, the other is the orientation change of polarized molecules from random to dominantly along with the optical filed direction. In case the incident optical intensity is not so strong that nonlinear optical effects can be neglected, the electric susceptibility can be deduced form movement of electrons pushed by optical field, described by the classical harmonic oscillator model [3–6]:

$$\chi = \frac{Nq^2}{m_e} \frac{1}{\omega_0^2 - \omega^2 - j\omega\gamma}, \tag{6.1}$$

where N is the number of atoms in unit volume, q is the electric charge of electron, m_e is its mass, ω_0 is the intrinsic frequency of electron oscillation, γ is the damping coefficient. The real part and imaginary part of electric susceptibility, $\chi = \chi_r + j\,\chi_i$, are expressed as:

$$\chi_r = \frac{Nq^2}{m_e} \frac{\omega_0^2 - \omega^2}{(\omega_0^2 - \omega^2)^2 + (\omega\gamma)^2}, \tag{6.2}$$

$$\chi_i = \frac{Nq^2}{m_e} \frac{\omega\gamma}{(\omega_0^2 - \omega^2)^2 + (\omega\gamma)^2}. \tag{6.3}$$

They are related with each other, as

$$\chi_r = \chi_i \frac{\omega_0^2 - \omega^2}{\omega\gamma} \approx \chi_i \frac{\omega_0 - \omega}{\gamma/2}. \tag{6.4}$$

The real and imaginary parts of susceptibility obey the universal Kramers–Kronig (K-K) relations, as expressed in Chap. 2 by Formula (2.35). The susceptibility determines the dielectric constant of medium and refractive index with $\chi = \varepsilon/\varepsilon_0 - 1 = n^2 - 1$. In the frequency stabilization by absorption lines, gas cells are often used with the background index of $n_{r0} \sim 1$. The real part and imaginary part of index near the resonant frequency ω_0 are written as:

$$n_r = 1 + \frac{\pi N q^2}{m_e\omega} \frac{\omega_0 - \omega}{(\omega_0 - \omega)^2 + \gamma^2/4}, \tag{6.5}$$

$$n_i = \frac{\pi N q^2}{2m_e\omega} \frac{\gamma}{(\omega_0 - \omega)^2 + \gamma^2/4}. \tag{6.6}$$

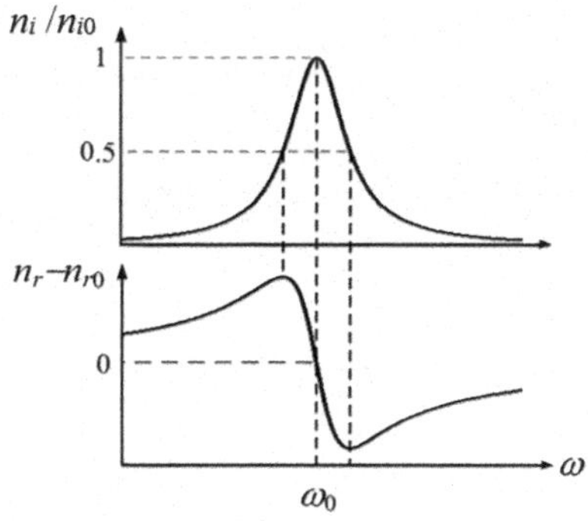

Fig. 6.1 Absorption line and index dispersion spectra

The imaginary part of index is termed the extinction coefficient, which is proportional to absorption coefficient α, $n_i = \alpha/2k$. Formulas (6.5)–(6.6) describe the chromatic dispersion of index, as shown in Fig. 6.1. The index n_r changes asymmetrically respect to the resonant frequency ω_0, where the dispersion reaches its maximum.

The absorption line has a Lorenzian shape; its full width at half maximum (FWHM) is obtained from (6.6): $\Delta\omega_L = \gamma$. Both of the absorption line and the index dispersion curve can be used to acquire error signal for frequency stabilization. Usually, atoms and molecules have several absorption lines, which can be utilized for different frequency bands.

(2) **Absorption Saturation by Optical Power**

In the process of optical absorption the particle (atom or molecule) absorbs the energy of a photon, and its electron transits from the ground state to an excited state. The electron at the excited state will transit back stochastically in a short time, the average of which is the spontaneous relaxation time τ. As the optical intensity increases, the population at the ground state will decrease, and the absorption coefficient will decrease. Such a phenomenon is called the absorption saturation. According to Einstein's radiation theory, there exist spontaneous emission and stimulated emission at the same time, the rate equation of population at the ground state is written as

$$\frac{\mathrm{d}N_1}{\mathrm{d}t} = -N_1 B_{12} c\alpha(\omega)\frac{I}{hv} + \frac{N_0 - N_1}{\tau}, \tag{6.7}$$

where B_{12} is Einstein coefficient, c is the light velocity, $\alpha(\omega)$ is the absorption coefficient, N_0 is the total number of particles, $N_2 = N_0 - N_1$ is the number of particles at the excited state level, τ is also termed the longitudinal relaxation time. The stationary solution of rate equation can be deduced as $N_1 = N_0[1 + \tau B_{12} c\alpha(\omega) I/(hv)]^{-1}$. In the semi-classic theory of laser physics, the populations and transitions are expressed as density matrix, and obeys the density matrix equation, which takes the coherent transition into account. The population difference between upper and down levels in the stationary state is deduced as [4–8]:

$$\Delta N = \Delta N_0 \frac{1+(\omega-\omega_0)^2 T_2^2}{1+(\omega-\omega_0)^2 T_2^2 + \Omega_R^2 T_2 \tau}, \tag{6.8}$$

where $T_2 = 1/\gamma$ is the relaxation time of coherent excitation, termed also the transverse relaxation time; $\Omega_R = \mu E_0/\hbar = \mu\sqrt{I}/\hbar$ is Rabi oscillation frequency with electric dipole μ [6–8]. ΔN_0 is the population difference without incident light, which is different from the total atom's number N_0 due to the existence of thermal excitation. The absorption coefficient is deduced from (6.8)

$$\alpha(\omega) = \alpha_0 \frac{(\gamma/2)^2}{(\omega-\omega_0)^2 + (\gamma/2)^2(1+I/I_s)}, \tag{6.9}$$

where α_0 is the peak absorption coefficient for a weak incident light. Therefore the absorption coefficient decreases with increase of the optical power; that is the saturation effect of absorption. At the peak, $\alpha(\omega_0) \propto (1+I/I_s)^{-1}$. I_s is the saturation intensity, expressed as

$$I_s = \frac{c\varepsilon_0\hbar^2}{4\mu^2 T_2 \tau} = \frac{c\varepsilon_0\hbar^2\gamma}{4\mu^2\tau}. \tag{6.10}$$

The saturation intensity is proportional to the inherent linewidth γ, and inversely proportional to the spontaneous relaxation time. The lower the saturation intensity, the easier the saturation effect occurs. It is seen from (6.9) that the linewidth increases with the intensity, called the effect of saturation broadening: $\Delta\omega_{\mathrm{FWHM}} = \gamma\sqrt{1+I/I_s}$.

6.1.2 Doppler Broadening and Saturated Absorption Spectroscopy

(1) **Doppler Broadening of Absorption Line**

In the model above, the electrons at excited state will transit back to the ground state by spontaneous emission and stimulated emission. The spontaneous relaxation time, expressed by damping coefficient in the classical model, describes a natural linewidth. However, other mechanisms exist in the medium, shortening the relaxation time. The downward transition may be caused by collisions of particles with each other, and with walls of gas container. In solid state, the transition may be induced by vibrations of host lattice, i.e., by collision with phonons. The mechanisms shorten the lifetime of electrons at high level and broaden the absorption line.

If the probability of transition by the mechanism is identical to all of the particles, and the transitions of different particles give same contribution to the spectrum of broadened line, such a broadening is called the homogeneous broadening. In such a case it is impossible to distinct which of particles gives contributions to the different spectral composition of broadened line. In saturated absorption, the line declines with the intensity of incident light, whereas its spectral shape keeps unchanged. The absorption line due to homogeneous broadening has a Lorentzian line shape, the same as that due to natural broadening.

It is well known that particles in gas are moving with different directions and different velocities, and the probability of velocities obeys Maxwell–Boltzmann distribution. Similar to the case of acoustic wave, optical Doppler Effect occurs for moving particles, i.e., the optical frequency observed by moving particles changes with their velocities $\boldsymbol{u}$, expressed as

$$\omega' = \omega - \boldsymbol{k} \cdot \boldsymbol{u} \rightarrow \omega(1 \pm u/c). \tag{6.11}$$

The last expression is for the particles with velocity parallel or antiparallel to the optical wave vector $\boldsymbol{k}$; the former causes red-shift as $(1 - u/c)$, and the latter causes blue-shift as $(1 + u/c)$. Therefore, the same particles but moving differently will thus absorb different frequency photons corresponding to their velocities; and the absorption line is thus broadened; particles with different velocities give different contributions to the line shape. Such a mechanism is called the inhomogeneous broadening. Taking Doppler Effect into account, the term $(\omega - \omega_0)$ in (6.8) and (6.9) should be replaced by $(\omega - \boldsymbol{k} \cdot \boldsymbol{u} - \omega_0)$.

The population should be resolved according to Maxwell–Boltzmann distribution, expressed as

$$f_i(u_i) = \frac{1}{u_p\sqrt{\pi}} \exp\frac{-u_i^2}{u_p^2}, \tag{6.12a}$$

$$f(u) = \frac{4u^2}{u_p^3\sqrt{\pi}} \exp\frac{-u^2}{u_p^2}, \tag{6.12b}$$

where $u_i (i = x, y, z)$ is the three components of velocity, and $u = \sqrt{u_x^2 + u_y^2 + u_z^2}$. The velocity with the maximum probability is $u_p = \sqrt{2k_B T/m}$, where k_B is Boltzmann constant, T is temperature in Kelvin, m is the mass of particle.

When an optical wave in z-direction propagates in the gas, the number of particles absorbing photons is dependent on their velocities u_z. Substituting $u_z = (\omega - \omega_0)/k$ into (6.12a, 6.12b), the absorption line is broadened in a Gaussian shape of

$$\alpha(\omega) \propto \frac{1}{\Delta\omega_G} \exp \frac{-(\omega - \omega_0)^2}{(\Delta\omega_G)^2}, \tag{6.13}$$

where $\Delta\omega_G = ku_p = \omega u_p/c \propto \omega\sqrt{T/m}$ is the half width at $1/e$ amplitude, the FWHM linewidth is $\Delta\omega_{1/2} = 2\Delta\omega_G\sqrt{\ln 2}$. With the Lorentzian line of static particles, the composed line is the convolution of two line shapes, i.e., the Voigt line, expressed as

$$\alpha_z(\omega) = \alpha_0 \frac{\gamma^2}{4u_p\sqrt{\pi}} \int_0^\infty \frac{\exp(-u_z^2/u_p^2)du_z}{(\omega - \omega_0 - ku_z)^2 + (\gamma/2)^2(1 + I/I_s)}. \tag{6.14}$$

The inherent absorption line of gas particles is usually very narrow but the measured line is much wider than the inherent linewidth due to Doppler broadening. For example, the Doppler broadened line of rubidium (Rb) vapor is measured around 500 MHz at room temperature, while its inherent linewidth is less than 10 MHz. Inherent lines are buried in a Doppler broadening envelope, impossible to be resolved; the case is even worse for the ultra-fine spectral structures.

Combination of Doppler broadening, collision broadening, and natural line will generate different line shapes in different cases related to gas pressure or particle density and so [9]. This book will not go further in details here.

(2) **Saturated Absorption Spectroscopy**

When the frequency of an optical wave propagating in the gas is $\omega = \omega_0 + ku_z$, the photons will be absorbed by the particles with velocity u_z. The population at the ground state will thus be decreased; the original Gaussian spectrum will be deformed with a hole at the correspondent frequency. Such a hole burning of gain spectrum is termed Lamb dip in laser physics [10, 11]. The idea of saturated absorption spectroscopy is based on the similar effect; that is, to decrease the population at the ground state by injecting a laser beam (pump) strong enough and with correspondent frequency, and to detect the inherent absorption line generated by the particles moving perpendicular to the direction of pump beam by a weak laser beam (probe). The probe beam should be parallel or reverse parallel to the pump beam, so that no more Doppler shift occurs.

When the pump of frequency ω_1 and intensity I_1 is injected into the medium, its population is changed; the population difference between the upper and down levels becomes

$$\Delta N = \Delta N_0 \frac{(\omega_1 - \omega_0 - \boldsymbol{k}_1 \cdot \boldsymbol{u})^2 + (\gamma/2)^2}{(\omega_1 - \omega_0 - \boldsymbol{k}_1 \cdot \boldsymbol{u})^2 + (\gamma/2)^2(1 + I_1/I_s)}. \tag{6.15}$$

The contribution of particles of velocity $\boldsymbol{u}$ to absorption of probe with frequency ω_2 is

$$\alpha_{\boldsymbol{u}}(\omega_1,\omega_2)\propto\frac{\Delta N_0[(\omega_1-\omega_0-\boldsymbol{k}_1\cdot\boldsymbol{u})^2+(\gamma/2)^2]}{(\omega_1-\omega_0-\boldsymbol{k}_1\cdot\boldsymbol{u})^2+(\gamma/2)^2(1+I_1/I_s)}\frac{(\gamma/2)^2}{(\omega_2-\omega_0-\boldsymbol{k}_2\cdot\boldsymbol{u})^2+(\gamma/2)^2}.$$

The absorption coefficient is obtained by an integral along the velocities:

$$\alpha(\omega_1,\omega_2)=\frac{1}{\pi^{3/2}u_p^3}\iiint\alpha_{\boldsymbol{u}}(\omega_1,\omega_2)\mathrm{e}^{-(u_x^2+u_y^2+u_z^2)/u_p^2}\mathrm{d}u_x\mathrm{d}u_y\mathrm{d}u_z. \tag{6.16}$$

Usually the pump and the probe of same frequency are used, and are tuned simultaneously; they propagate in the same or reverse direction. Expression (6.16) is then reduced to [6]:

$$\alpha(\omega)\propto\int_0^\infty\frac{[(\omega-\omega_0-ku_z)^2+(\gamma/2)^2]\mathrm{e}^{-u_z^2/u_p^2}\mathrm{d}u_z}{[(\omega-\omega_0-ku_z)^2+(\gamma/2)^2(1+I_1/I_s)][(\omega-\omega_0\mp ku_z)^2+(\gamma/2)^2]}, \tag{6.17}$$

where sign – of $\mp$ corresponds to unidirectional pump and probe, and sign + for counter directions. In case the inherent linewidth is much smaller than the Doppler broadened linewidth, $\gamma\ll ku_p$, the Lorentzian line shape can be approximated as a delta function [6]:

$$\frac{\gamma}{\pi[(\omega-\omega_0\mp ku_z)^2+\gamma^2]}\approx\delta(\omega-\omega_0\mp ku_z). \tag{6.18}$$

When the probe and pump are in counter directions, $ku_z\approx\omega_0-\omega$, (6.17) is further reduced to

$$\alpha(\omega,I_1)\propto\frac{4(\omega-\omega_0)^2+(\gamma/2)^2}{4(\omega-\omega_0)^2+(\gamma/2)^2(1+I_1/I_s)}\exp\frac{-(\omega-\omega_0)^2}{\Delta\omega_G^2}. \tag{6.19}$$

Therefore the line measured by probe shows a Lorentzian dip in the Doppler broadened Gaussian line profile; and the downward Lorentzian line is the inherent absorption line with effect of intensity broadening. It is necessary to acquire the quantitative relation and to optimize the pump power. Formula (6.19) is rewritten as $\alpha(x)\propto[(x^2+1)/(x^2+a)]\exp(-bx^2)$ with $x=4(\omega-\omega_0)/\gamma, a=1+I_1/I_s$ and $b=\gamma^2/(2\Delta\omega_G^2)$, as depicted in Fig. 6.2.

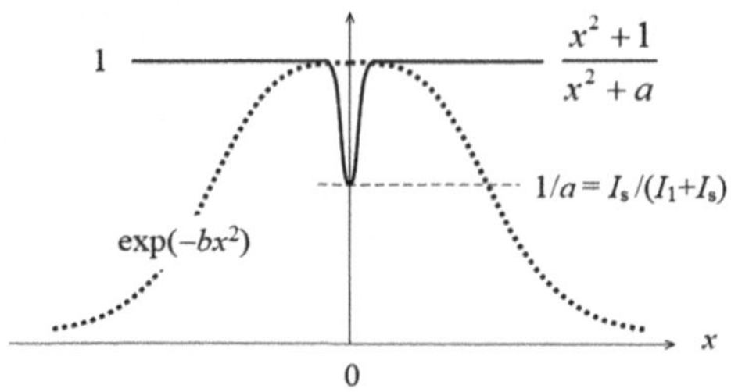

Fig. 6.2 Illustration of Eq. 6.19. $x = 4(\omega - \omega_0)/\gamma, a = 1 + I_1/I_s, b = \gamma^2/(2\Delta\omega_G^2)$

By approximating $\exp(-bx^2) \approx 1$ near the peak, the minimum is obtained as

$$\Delta\alpha(\omega_0) = \alpha(\omega_0, 0) - \alpha(\omega_0, I_1) \propto 1 - \frac{1}{a} = \frac{I_1}{I_1 + I_s}. \tag{6.20}$$

The FWHM linewidth of absorption line (the dip) is just the Doppler-free intrinsic width but broadened by intensity saturation effect, expressed as

$$\Delta\omega_s = (\gamma/2)\sqrt{1 + I_1/I_s} \tag{6.21}$$

The curvature at the peak is obtained as

$$\left|\frac{\partial^2\alpha}{\partial\omega^2}\right|_{\omega=\omega_0} \propto \frac{2I_1 I_s}{(I_1 + I_s)^2}. \tag{6.22}$$

Obviously, a larger curvature is beneficial for a higher sensitivity as a frequency discriminator. Expression (6.22) reaches its maximum at $I = I_s$, and linewidth of $\Delta\omega_s = \gamma/\sqrt{2}$ will be measured.

The above discussion describes the basic principle and main characteristics of saturated absorption. In practice, many complicated factors will affect the experimental results, such as multiple lines may be covered underneath the Doppler profile. The undistinguishable lines can then be separated clearly by the saturated absorption spectroscopy [1, 2].

In the above scheme, the probe beam direction is taken opposite to the pump beam. When two beams propagate in the same direction, (6.17) will be reduced to the same as (6.14), remaining Doppler broadening unchanged. In such a configuration, if the pump and the probe are with different frequencies, Doppler broadening may also be removed under certain conditions.

In hyperfine spectroscopy, the excited state may contain two or more levels, with interval much smaller than the Doppler linewidth. In such a case, crossover (CO) resonance transition occurs often. Suppose two levels exist with lines at ω_{01} and ω_{02}. Under conditions of $\omega_{01} = \omega(1 - u/c)$ and $\omega_{02} = \omega(1 + u/c)$, a pump beam with frequency ω will be absorbed by both particles moving in opposite directions with $u = c(\omega_{02} - \omega_{01})/(\omega_{02} + \omega_{01})$. The resonance occurs at pump frequency $\omega = (\omega_{02} + \omega_{01})/2$. The absorption saturation occurs under a certain pump level, similarly to the case of a single excited level; and a Doppler-free absorption spectrum can thus be obtained by a weak probe beam.

6.2 Frequency Stabilization by Modulation Spectroscopy

The absorption spectrum is usually measured by modulating the optical source wavelength. The method is widely used in frequency stabilization with an absorption line as reference frequency. Spectroscopic systems by modulating tunable semiconductor laser as sources are termed the tunable diode laser spectroscopy (TDLS).

6.2.1 *Spectrum of Frequency Modulated Optical Signals*

The basic concept of modulation spectroscopy is illustrated in Fig. 6.3. A frequency modulated laser is used as probe; the power of transmission passing through the absorbing medium is detected as signal of the absorption spectrum. The technology has advantage of noise suppression, including noises from detector, from optical system and electronics system, especially the $1/f$ noises.

The frequency modulation of a laser beam is usually implemented by modulating the optical phase. Denoting the modulation frequency by Ω and the modulation amplitude by β, the modulated phase and the transient optical frequency are written as $\Delta\Phi = \beta \sin(\Omega t)$ and $\omega(t) = \omega_0 + \beta\Omega \cos(\Omega t)$. The frequency modulated optical field is composed of a series of side components [12, 13]:

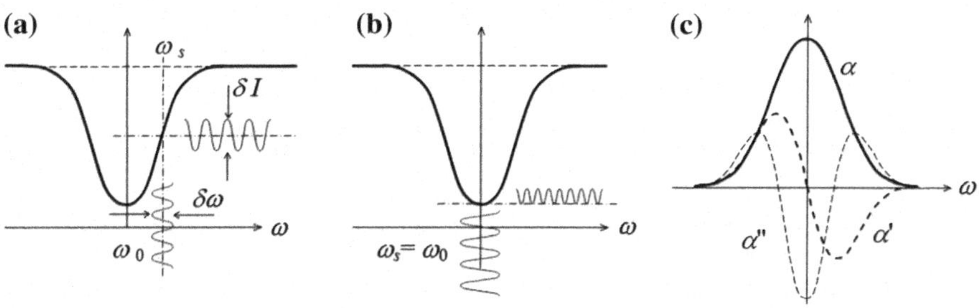

Fig. 6.3 **a** Illustration of modulation spectroscopy; **b** signals on the top of absorption; **c** absorption line shape and its first order and second order differentials

$$E(t) = E_0 e^{j(\omega_0 t + \beta \sin \Omega t)} = E_0 e^{j\omega_0 t} \sum_{n=-\infty}^{\infty} J_n(\beta) e^{jn\Omega t}, \tag{6.23}$$

where J_n (β) is the n-th Bessel function.

The modulation spectroscopy is sorted into two technologies conventionally: one is called the wavelength modulation spectroscopy (WMS) with modulation frequency Ω lower than the absorption linewidth $\Delta\omega$; the other is the frequency modulation spectroscopy (FMS) with $\Omega \approx \Delta\omega$ or $\Omega > \Delta\omega$ [12]. Physically, the wavelength is equivalent to the frequency, no difference between them in principle. However, because the noises of laser and detector, and their frequency response, are functions of modulation frequency, and the materials to be analyzed have different properties, the two technologies have different features and different applicable objects. Generally higher modulation frequency is beneficial for suppression of source noise and detector noise, but higher frequency responses of source and detector are needed. On the other hand, modulation frequency should not be too high to cover two or more absorption lines. In comparison, WMS has an advantage of low cost, since high response devices are not required. In a modified FMS, two different modulation frequencies are used simultaneously, so that the demodulated frequency is reduced to their beat frequency and the response requirement of detection and data processing is lowered [13].

The frequency modulation amplitude $\delta\omega = \beta\Omega$ should be much smaller than the absorption linewidth $\Delta\omega$ to obtain a highly precise measured line, $\delta\omega/\Delta\omega \ll 1$. Therefore small phase modulation amplitude $\beta \ll 1$ is required in FMS. According to the approximations of Bessel function $J_n(\beta) \approx (\beta/2)^n/n!$, the higher order terms in (6.23) can be neglected, written as

$$E_{\text{FMS}}(t) \approx E_0 e^{j\omega_0 t}(1 + j\beta \sin \Omega t). \tag{6.24}$$

In WMS with $\Omega < \Delta\omega$, larger phase modulation amplitude is used with $\beta \sim 1$ or $\beta > 1$ to maintain $\delta\omega$ and detected signal large enough.

The spectrum is actually measured by detecting the signal transmitted from an absorption cell with length of l. In case nonlinear optical effects is neglected, transmission of the cell obeys Beer–Lambert law:

$$I(\omega) = I_{\text{in}} \exp[-\alpha(\omega)l] = I_{\text{in}} \exp[-z(\omega)], \tag{6.25}$$

where $z = \alpha(\omega)l$ is the optical depth. Therefore the line shape of transmission is different from that of absorption line. For an absorption line with Lorentzian profile the optical depth is written as

$$z = \frac{\alpha_p l}{1 + (\omega_0 + \delta\omega \cos \Omega t - \omega_p)^2/(\Delta\omega)^2} = \frac{z_0}{1 + (x + m \cos \Omega t)^2}, \tag{6.26}$$

where α_p, ω_p and $\Delta\omega$ are the peak absorption and peak frequency and absorption linewidth; $x = (\omega_0 - \omega_p)/\Delta\omega$ and $m = \delta\omega/\Delta\omega$. In the measurement laser frequency ω_0 is linearly scanned around the absorption line ω_p with a speed much lower than the modulation frequency. The optical depth is expanded as a Fourier series [14–20]:

$$z = \frac{z_0}{1 + (x + m \cos \Omega t)^2} = z_0 \sum_{n=0}^{\infty} S_n \cos(n\Omega t)$$
$$S_n = \left\{ \frac{j^n [(1 + jx) - \sqrt{(1 + jx)^2 + m^2}]^n}{m^n \sqrt{(1 + jx)^2 + m^2}} + c.c. \right\} \left(1 - \frac{\delta_{n0}}{2}\right). \tag{6.27}$$

Approximations of the first three coefficients S_n at the line peak are given as

$$\begin{aligned} S_0 &\approx \frac{1}{(1 + m^2)^{1/2}} - \frac{x^2}{2(1 + m^2)^{5/2}} \\ S_1 &\approx -2mx/(1 + m^2)^{3/2} \\ S_2 &\approx \frac{-2(\sqrt{1 + m^2} - 1)^2}{m^2 \sqrt{1 + m^2}} + \frac{2(1 + 3m^2)}{m^2 (1 + m^2)^{5/2}} x^2. \end{aligned} \tag{6.28}$$

Under condition of $m \ll 1$, the transmission of absorption cell is reduced to:

$$\begin{aligned} e^{-z} &= e^{-z_0 S_0} \sum_{k=0}^{\infty} \frac{1}{k!} \left(\sum_{n=0}^{\infty} S_n \cos(n\Omega t) \right)^k \\ &\approx e^{-z_0 S_0} [1 + S_1^2/4 + S_1 \cos \Omega t + (S_2 + S_1^2/4) \cos 2\Omega t]. \end{aligned} \tag{6.29}$$

The first order component of modulated spectrum in (6.29) is proportional to the frequency deviation: $S_1 = 2\delta\omega(\omega_p - \omega_0)/(\Delta\omega)^2$; it is thus to be used as an error signal for feedback control. The odd number order components will equal to zero when the laser frequency ω_0 is scanned around the line peak ω_p; while the second order component is $S_2 \approx -m^2/2$, as shown in Fig. 6.3b. Figure 6.3c gives the first order and second order signals of a Lorentzian line.

When the gas pressure is high, Gaussian and Voigt lines may be the better description. The practical absorption lines may not be expressed as an analytic function. Tailor series or Fourier series are then used for a general line and even for overlapped lines. Taking dispersion into account, the transmission has a phase factor: $T(\omega) = \exp[j\varphi(\omega) - z(\omega)/2]$. The transmitted optical field is generally expressed as [13]:

$$E_{\text{out}} = E_0 e^{j\omega_0 t} \sum_{n=-\infty}^{\infty} J_n(\beta) \exp[jn\Omega t + j\varphi_n - z_n/2], \tag{6.30}$$

where $\varphi_n = \varphi(n\Omega)$ and $z_n = \alpha(n\Omega)l$ are their n-order components. The output intensity is

$$\begin{aligned} I_{\text{out}} &= |E_0|^2 \sum_{k,l=-\infty}^{\infty} J_k J_l \exp[j(k-l)\Omega t + j(\varphi_k - \varphi_l) - (z_k + z_l)/2] \\ &= I_0 \sum_{n=0}^{\infty} K_n \cos(n\Omega t). \end{aligned} \tag{6.31}$$

If the phase difference $(\varphi_k - \varphi_l)$ is small enough to be neglected, the coefficient is reduced to

$$K_n = 2 \sum_{l=-\infty}^{\infty} J_{l+n} J_l \exp[-(z_{l+n} + z_l)/2]. \tag{6.32}$$

Formulas (6.30)–(6.32) are general mathematical expressions of modulation spectroscopy, applicable both to FMS and WMS. The first order and second order components of transmitted intensity are mostly concerned for the sensitivity and dynamic range of frequency discrimination in laser frequency stabilization by modulation spectroscopy.

6.2.2 Frequency Stabilization with Modulation

The frequency stabilized laser referring to gas absorption lines is one of the key devices in many advanced technologies, such as high precision spectrum analyzers, high stability frequency standards, and detections of gas specimens. The frequency stabilized semiconductor laser based on the absorption spectroscopy is widely used in applications due to its advantages of low cost and easy operation. For laser frequency stabilization the current tuning and temperature tuning are often used to adjust and to control frequency of semiconductor lasers. For ECLD, tuning is also realized by translating the external mirror, or by rotating the planar grating. External optical phase modulators, such as electro-optic modulators (EOM) and acousto-optic modulators (AOM), can also be used to modulate the laser frequency.

(1) **Frequency Stabilization with Saturated Absorption Spectroscopy**

In the applications, such as the laser spectroscopy and the laser cooling of atoms, the laser frequency is required to be locked at one of the fine-structure absorption lines. It is understandable that the narrower the reference frequency the better the frequency stabilized laser works. However, the elements, such as Rubidium (Rb,

780 nm), cesium (Cs, 852 nm), and iodine (I_2, 532 nm), are generally in gaseous state; their absorption lines are often broadened by Doppler Effect, which must be suppressed by the saturated absorption spectroscopy.

The typical structure of frequency stabilized semiconductor lasers with the saturated absorption spectroscopy is shown in Fig. 6.4a, which is very similar to that used by Schawlow in his original publications [1, 2]. In the system the pump beam and the probe beam pass the gas cell in directions counter to each other with a small angle; the signal of probe gives Doppler-free spectrum. The beam reflected from the rear face of thicker plate passes the cell directly, giving Doppler broadened spectrum. The differential amplifier gives their differences as the error signal with a low background; its first order component obtained by serve electronics is fed back to the laser driver to adjust LD current and/or mechanical tuning components of ECDL. Based on the same principle, different structures of frequency stabilized LD and ECDL have been developed [21–25]. Figure 6.4b shows another structure, where the laser frequency is modulated externally by an AOM, and the pump and probe are in the exact antiparallel directions.

In the structure of Fig. 6.5a, the exact antiparallel pump and probe beams are separated by polarization components. The intensity of pump beam is adjusted by rotating the $\lambda/2$ waveplate, and is converted to a circle polarized bean by the $\lambda/4$

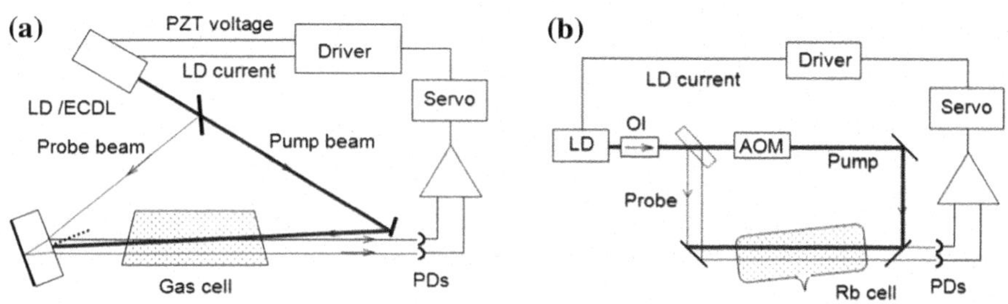

Fig. 6.4 Configurations of frequency stabilized laser modulated by driving current (**a**) and by AOM (**b**)

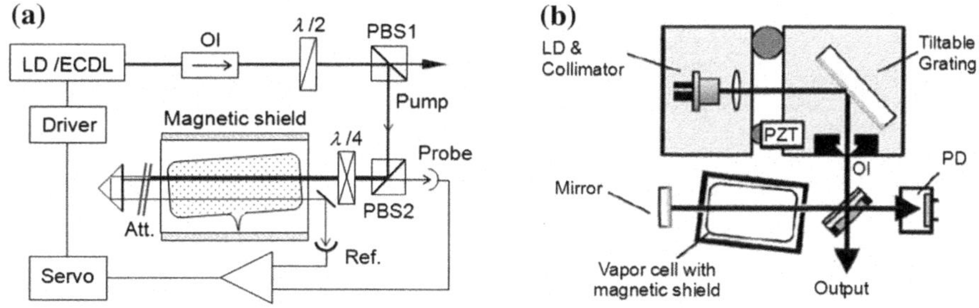

Fig. 6.5 **a** Frequency stabilized laser with a polarization beam splitter; **b** structure of a compact frequency stabilized LD. Reprinted with permission from [24]

waveplate. The beam reflected from the prism and attenuated twice is taken as the probe, which is converted back to a linearly polarized beam and output from PBS2 to a photodetector. Another beam gives a reference signal of Doppler broadened line. The gas cell is magnetically shielded to prevent from Zeeman Effect induced by external magnetic field. Figure 6.5b shows a compact frequency stabilized ECDL [24], where the grating can be tilted by the PZT to maximize the error signal.

(2) **Frequency Stabilization with Linewidth Broadening Retained**

In some applications the laser frequency must be locked on an absorption line, but its linewidth requirement is not critical. For example, detection and measurement of methane (CH_4) and carbon dioxide (CO_2) are important topics for environment protection. Lasers with frequency stabilized on absorption line of ethane (C_2H_2, ~1530 nm) and cyanic acid (HCN, ~1550 nm) are needed in wavelength division multiplex (WDM) optical communications. The line broadening by Doppler Effect and/or by collisions may be allowed in these cases, and the system can be simplified.

Figure 6.6a shows a typical configuration of optical path of a frequency stabilized LD [26, 27]. A portion of LD output is split by a beam splitter or a fiber coupler (BS), and is modulated by a phase modulator; and then passes through a gas cell. The signal of modulated absorption is detected by the photodetector (PD) and mixed with a RF source. The mixed signal is amplified in servo electronics and fed back to adjust LD driving current. An optical isolator (IO) is inserted in LD output beam to prevent the LD from influence of optical feedback. The modulation can also be realized by modulating LD's pump current, and its DC bias and temperature of its heat-sink are used to adjust and lock the frequency, as shown in Fig. 6.6b, where TEC stands for the thermo-electrical cooler based on Peltier effect.

The gas cell is one of the key components. It is usually made of a glass tube fused with two optical plates with anti-reflective coating. The two plates are intentionally designed to be deviated from parallel with each other to prevent from Fabry–Perot effect. The cell may be heated for some metal gasses for higher vapor density. For gasses with low absorption coefficient, long length cells are needed. Multiple back-and-forth paths are established in a cell with mirrors, so-called White type cavity or other modified types [28]. In recent years the hollow-core photonic

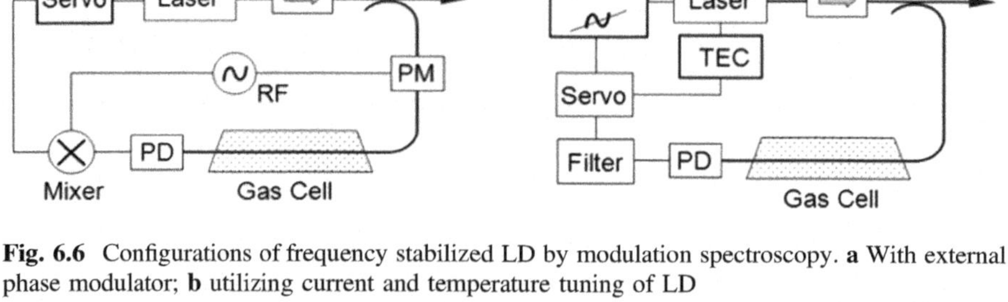

Fig. 6.6 Configurations of frequency stabilized LD by modulation spectroscopy. **a** With external phase modulator; **b** utilizing current and temperature tuning of LD

crystal fiber (HCPCF) shows attractive features. In HCPCF, a hollow core with diameter of tens micrometers is surrounded by the cladding, where multiple holes are arranged periodically in two dimensional space. Due to the photonic crystal effect, the cladding will reflect optical waves with certain wavelengths, playing a role of total internal reflection (TIE). The absorption gas can be filled into the hollow core. The outer diameter of HCPCF is similar to that of ordinary optical fiber, so that it can be bent and wound with negligible losses. It is therefore suitable to make a gas cell with a long optical path, and with merits of compact structure and better robustness. Its characteristics and applicability in spectroscopy were studied and reported in Refs. [29, 30]. A semiconductor with frequency stabilized at R24 absorption line of CO_2 ($\sim$1572 nm) by utilizing HCPCF as the gas cell was reported in Ref. [31]. Of cause a series technical issues have to be solved for practical applications, such as mitigation of harmful reflections at the two ends of HCPCF, and its hermitic seal.

Generally the following performances are required in the frequency stabilization.

(1) A higher sensitivity of frequency discrimination, i.e., a larger slope of the first order components of signal spectrum. It is the guarantee of lower frequency deviation and narrower linewidth of the frequency stabilized laser.
(2) A wider range of frequency locking, which is important for finding the aimed reference frequency as quick as possible and to reduce the possibility of out-of-lock. It requires a larger linear range of the first order component and a smaller second order component. However, the large linearity is contradictory to the high sensitivity, as seen from formula (6.28); therefore a trade-off design is needed.
(3) A fast response, which is necessary for dealing with the external disturbances, especially of the high frequency disturbances. It is also necessary for laser frequency sweeping, which will be discussed in the next chapter.
(4) A lower intensity noise and frequency noise induced by the optics and electronics, especially in low frequency bands. It is important for lasers working in long terms and in complicated environment.

The stability of absorption lines of atoms and molecules is generally good enough for most applications. But the line may be affected by temperature and gas pressure in some situations. Related technical issues should be taken into consideration.

6.2.3 Residual Amplitude Modulation and Its Removal

In TDLS, LD is usually modulated by its driving current. As stated in Chap. 2, its output power will be modulated at the same time as frequency modulation, so that the signal of frequency discrimination will be distorted by adding a parasitic signal. Such an effect is called the residual amplitude modulation (RAM) . It may also occur when lasers other than LD are used, and when the laser frequency is modulated externally by electro-optic modulators (EOM) or acousto-optic modulators (AOM) . This harmful effect has to be removed or mitigated in modulation spectroscopy and in laser frequency stabilization.

In case the L-I curve of semiconductor laser is approximated as a linear line above the threshold, and the first order component of output power modulation is considered, the transmission of modulated laser beam from absorption cell should be written as [18–20]

$$\begin{aligned} T(\omega) &= \frac{P_{\text{out}}}{P_0} = (1 + m_P \cos \Omega t) \exp \frac{-z_0}{1 + (x + m \cos \Omega t)^2} \\ &\approx \mathrm{e}^{-z_0 S_0} (1 + m_P \cos \Omega t)(1 - S_1 \cos \Omega t), \end{aligned} \tag{6.33}$$

where $m_P = \eta_d V_J \Delta I / P_0$ is the normalized power modulation coefficient with external differential efficiency of η_d and current modulation amplitude of ΔI. The frequency modulation amplitude is proportional to ΔI, so that $m_P \propto m$. The modulation amplitude of first order component is modified to $m_P - S_1 \approx m_P - 2mx$, which does not equal to zero at absorption peak. The frequency given by zero point will deviate from the peak by $\omega_0 - \omega_p \approx m_P \Delta\omega / 2m = m_P (\Delta\omega)^2 / 2\delta\omega$. Since both of m_P and $\delta\omega$, are proportional to ΔI, the deviation is independent of modulation amplitude, but is inversely proportional to the bias power P_0.

Several schemes of RAM compensation were reported in journals. References [32, 33] analyzed the RAM of LD in detail, taking the nonlinearity of L-I curve and the phase shift between amplitude modulation and frequency modulation into consideration. The method given by Refs. [34, 35] is shown in Fig. 6.7a, where the modulated laser beam is divided into three, one of them is used to measure the absorption of gas cell, the second is used as the power monitor, and the third passes a known resonator, used as a mark of optical frequency. The measured absorption spectrum can then be corrected in data processing.

The phase shift between the power modulation and the frequency modulation can be utilized to compensate RAM, as shown in Fig. 6.7b [36, 37]. Besides the power monitor and frequency marking, another reference path with a certain delay is mixed with the absorbed beam; two outputs of 50:50 fiber coupler will give the

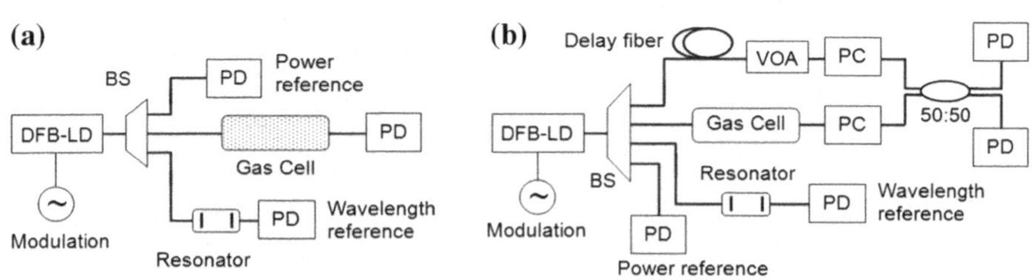

Fig. 6.7 Schemes of RAM compensation: **a** with a power monitor and a frequency mark; **b** with beating a delayed beam

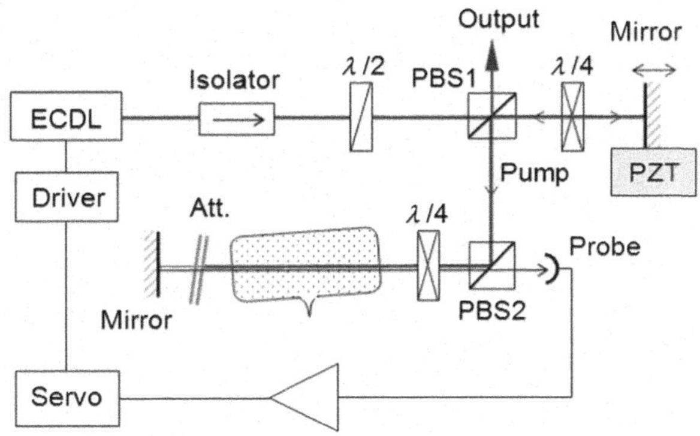

Fig. 6.8 RAM-free scheme by using external phase modulation

sum and the difference between the two beams; and the RAM will be removed when a proper delay is adjusted and the powers of two beams are matched with each other by adjusting the valuable optical attenuator (VOA).

The parasitic power modulation can be avoided by external modulations, instead of LD current modulation. Figure 6.8 shows a method by modulating the external optical path by a PZT [38], which is equivalent to external phase modulation and resultant frequency modulation, while the power keeps unchanged. One of its shortcomings is probably that the modulation frequency is limited.

EOM and AOM are the widely used external modulators. Although the accompanied power modulation by LD current is avoided, RAM may not be removed thoroughly. For example, it may arise from the F-P effect introduced by residue reflections at the facets of EOM and AOM crystals; the interference noises may also generated by residue reflections at the facets of other components in the path, and by scatterings of internal defects in the materials as well. Reference [39] studied the temporal characteristics of RAM, and found that the photorefractive effect is one of the mechanisms. Reference [40] found that uniformity of optical intensity transverse distribution is one of factor related to RAM, and proposed a flat top beam to replace the Gaussian beam. Reference [41] analyzed the effect of overlapping between the pump beam and the probe beam on RAM.

RAM has to be characterized quantitatively. The transmission of discriminator is expressed as $T_n = T(\omega_0 + n\Omega), \quad n = 0, \pm 1$ for the base band and the first order bands; the transmitted optical field and power are then written as [39, 40]:

$$E_R(t) = E_0 e^{j\omega_0 t}\left[J_0(\beta)\sqrt{T_0} + J_1(\beta)\sqrt{T_1}e^{j\Omega t} - J_1(\beta)\sqrt{T_{-1}}e^{-j\Omega t}\right], \tag{6.34a}$$

$$I_R = I_0\left[J_0^2(\beta)T_0 + 2J_0(\beta)J_1(\beta)\sqrt{T_0}\left(\sqrt{T_1} - \sqrt{T_{-1}}\right)\cos\Omega t\right]. \tag{6.34b}$$

The amount of RAM is characterized by the ratio of amplitudes of base band and the difference of two first order sideband components:

$$R_{\mathrm{AM}} = \left|\sqrt{T(\omega_0+\Omega)} - \sqrt{T(\omega_0-\Omega)}\right| = 2J_1\left(\sqrt{T_1} - \sqrt{T_{-1}}\right)/J_0\sqrt{T_0}. \quad (6.35)$$

It can be measured by heterodyne detected signals, which gives the in-phase component P and the quadrature component Q, and $R_{\mathrm{AM}} = \sqrt{P^2+Q^2}$; the phase difference between P and Q can be utilized for further analysis.

References [42, 43] analyzed a system with two frequency modulation, Ω and $\Omega/2$, indicating that the phase shift between the amplitude and frequency modulations is a function of modulation frequency. When the difference between the phase shifts of Ω and $\Omega/2$ reaches $\phi_{\Omega/2} - \phi_{\Omega} = \pi/4$, and the related modulation amplitudes meet condition of $A_{\Omega} = A_{\Omega/2}^2/4$, the RAM of $\pm\Omega$ modulations will be canceled each other. The property can be utilized to compensate the RAM.

6.3 Modulation-Free Frequency Stabilization

The frequency stabilization by means of modulation spectroscopy has been successfully used in many applications, showing attractive advantages and effectiveness. However, some shortcomings have also been noticed. Besides the complicity related to RAM, the frequency modulation sets a limit to the laser optimal linewidth suppression. Modulation-free frequency stabilizations are therefore developed. It is seen from the universal Kramers–Kronig (K-K) relations that the absorption is accompanied by index dispersion, and the dispersion spectrum has a similar behavior to the first order differential of absorption spectrum. Based on the property, the polarization spectroscopy and the interference discrimination were proposed and developed. Another scheme called the magnetic dichroism was also developed successfully based on Zeeman Effect. Optical negative feedback is another modulation-free method. This section is engaged in these technologies.

6.3.1 Frequency Stabilization by Polarization Spectroscopy

The polarization spectroscopy was developed based on the index dispersion spectrum around absorption lines [44]. Its basic structure seems similar to that of the saturated absorption spectroscopy, as shown in Fig. 6.9. The main difference is that the pump of absorption medium is now a circularly polarized beam, generated by the quarter wavelength plate (QWP), instead of a linearly polarized beam.

Its mechanism relates to the basic quantum mechanics principles of transitions. Electrons possess both of spin angular momentum and orbital angular momentum; electron transitions obey not only the energy conservation, but also the angular momentum conservation. The photons possess also spin angular momentum with values of $\pm\hbar$, corresponding to left and right circularly polarized waves,

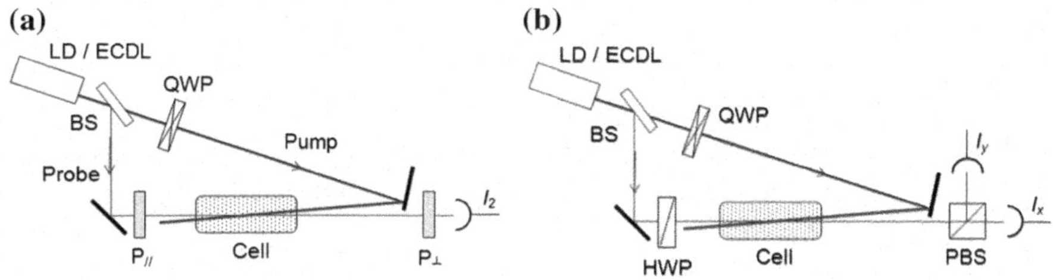

Fig. 6.9 Basic structure of polarization spectroscopy: **a** with perpendicular polarizers; **b** with polarization beam splitter

respectively [45]. The linearly polarized wave contains equally photons with the opposite angular momentum, whereas the circularly polarized wave contains photons with the same angular momentum. The left or right circularly polarized pump photons will be absorbed by electrons with correspondent spin state σ^+ or σ^-, leaving electrons with the other spin state at the ground. When the pump is strong enough, absorption saturation will occur for one of the two states. Different absorption coefficients $\alpha_\pm$ and different index $n_\pm$ will be measured by two circularly polarized probes. The Doppler broadening can thus be removed in the same way as stated in the saturated absorption spectroscopy.

The linearly polarized probe beam can be written as a sum of left and right circularly polarized beams [44, 46]:

$$\boldsymbol{E}_0 = E_0\begin{pmatrix}\cos\varphi\\ \sin\varphi\end{pmatrix} = \frac{E_0}{2}\left[\begin{pmatrix}1\\ j\end{pmatrix}\mathrm{e}^{-j\varphi} + \begin{pmatrix}1\\ -j\end{pmatrix}\mathrm{e}^{j\varphi}\right]. \tag{6.36}$$

The transmitted wave from the absorption cell is

$$\begin{aligned}\boldsymbol{E}_1 &= \frac{E_0}{2}\left[\begin{pmatrix}1\\ j\end{pmatrix}\mathrm{e}^{-j\varphi}\mathrm{e}^{j(n_+k+j\alpha_+/2)L} + \begin{pmatrix}1\\ -j\end{pmatrix}\mathrm{e}^{j\varphi}\mathrm{e}^{j(n_-k+j\alpha_-/2)L}\right]\\ &= \frac{E_0}{2}\mathrm{e}^{j(nk+j\alpha/2)L}\left[\begin{pmatrix}1\\ j\end{pmatrix}\mathrm{e}^{-j(\varphi-\Phi)} + \begin{pmatrix}1\\ -j\end{pmatrix}\mathrm{e}^{j(\varphi-\Phi)}\right].\end{aligned} \tag{6.37}$$

where
$\Phi = (\Delta nk + j\Delta\alpha/2)L, \Delta n = (n_+ - n_-)/2, \Delta\alpha = (\alpha_+ - \alpha_-)/2, n = (n_+ + n_-)/2,$ $\alpha = (\alpha_+ + \alpha_-)/2$. $\Delta nk = x\Delta\alpha_0/[2(1+x^2)]$ and $\Phi = (x+j)\Delta\alpha_0 L/[2(1+x^2)]$ are obtained from K-K relation, where $x = 2(\omega_0 - \omega)/\gamma$. With $\Phi = \Phi_r + j\Phi_i$, (6.37) is then rewritten as

$$\boldsymbol{E}_1 = E_0\begin{pmatrix}\cosh\Phi_i\cos(\varphi-\Phi_r) + j\sinh\Phi_i\sin(\varphi-\Phi_r)\\ \cosh\Phi_i\sin(\varphi-\Phi_r) - j\sinh\Phi_i\cos(\varphi-\Phi_r)\end{pmatrix}\mathrm{e}^{j(nk+j\alpha/2)L}. \tag{6.38}$$

The power transmitted from the polarizer is obtained [44, 46]:

$$\begin{aligned} I_2 &= E_0^2[\cos^2(\varphi - \Phi_r) + \sinh^2 \Phi_i] \\ &= I_0[\sin^2(\theta + \Phi_r) + \Phi_i^2 + \sinh^2 \Phi_i] \\ &\approx I_0(\theta^2 + 2\Phi_r\theta + \Phi_r^2 + \Phi_i^2) \propto I_0\theta x/(1+x^2), \end{aligned} \tag{6.39}$$

where the perpendicular principal axis is set at $\varphi + \pi/2 + \theta$, and θ is an adjustable angle deviation, The approximation above is for the case of $\theta + \Phi_r \ll 1$, which is usually valid. It is seen that the signal behaves similar to the derivative of a Lorenzian absorption line to be used as a frequency discriminating signal, while the laser is not modulated.

The scheme of Fig. 6.9b uses a polarization beam splitter (PBS) to decompose the transmitted wave, and then to differentiate the detected signals [46]:

$$\begin{aligned} I &= I_x - I_y = I_0 e^{-\alpha L}\cos(2\varphi + 2\Phi_r) \\ &\rightarrow I_0 e^{-\alpha L}\sin 2\Phi_r \approx I_0 e^{-\alpha L}\Delta\alpha_0 Lx/(1+x^2), \end{aligned} \tag{6.40}$$

where the last two expressions are for the polarization adjusted at $\varphi = \pi/4$; and a frequency discriminating signal is thus provided.

It is seen from the optical path configurations in Fig. 6.9 that the Doppler broadening has been removed. References [46–49] reported experimental results of frequency stabilizations of semiconductor lasers at ^{85}Rb and ^{87}Rb lines by means of polarization spectroscopy successfully. The laser in the technology is with high precision and suitable for analyses with fine spectral structure involved, very useful in researches of cooled atoms and other applications.

6.3.2 *Frequency Stabilization with Sagnac Interferometer*

Sagnac interferometer is well known as a basis of optical gyroscope. It is composed of a common loop and a beam splitter, giving interference output of optical beams propagating in clockwise (CW) and counter clockwise (CCW) directions. If an absorption cell and a neutral optical attenuator are inserted in the loop, the CW and CCW beams in the cell will have different intensities, and can be used as a pump and a probe, respectively, in the saturated absorption spectroscopy, as shown in Fig. 6.10a, b [50, 51].

The pump beam and the probe beam experience not only different absorptions in the cell, but also different phase shifts due to the different index. A phase difference of $\pi/2$ exists between the reflection and the transmission of the beam splitter (BS); the output of Sagnac interferometer is expressed as:

$$E_1 = E_0(t^2 e^{j\phi_1} - r^2 e^{j\phi_2})e^{jkL}, \tag{6.41}$$

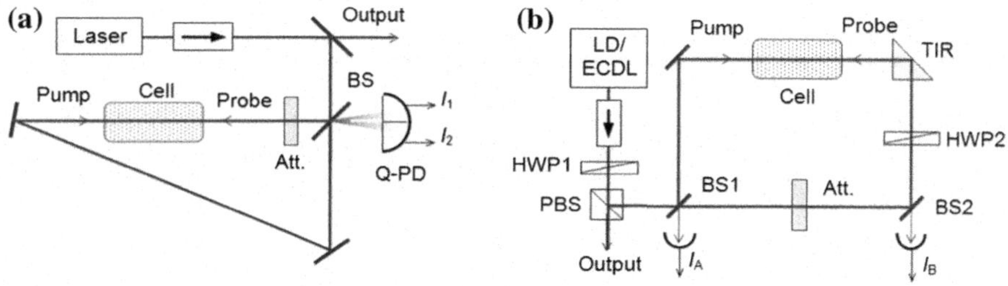

Fig. 6.10 Sagnac interferometers: **a** with fundamental mode and first order mode detected separately; **b** by utilizing phase difference between polarizations in total internal reflection

where t^2: r^2 is the split ratio of BS, L is the length of loop, ϕ_1 and ϕ_2 are the phase shifts of pump and probe in the cell, respectively. The output power is

$$I_1 = |E_1|^2 = I_0[(t^2 - r^2)^2 + 4t^2r^2\sin^2(\Delta\phi/2)], \tag{6.42}$$

where $\Delta\phi = \phi_2 - \phi_1$; $\phi_{1,2} = (n_{1,2} - n)kl \propto \alpha_{1,2}x_{1,2}l/2(1 + x_{1,2}^2)$ with $x_1 = 2(\omega_0 - \omega)/\Delta\omega_G, x_2 = 2(\omega_0 - \omega)/\Delta\omega_L$, and the cell length l. $\Delta\omega_G$ is the width of Gaussian line for pump beam, and $\Delta\omega_L$ is the width of Doppler-free line for probe beam. Because of $\Delta\omega_G \gg \Delta\omega_L$, approximation of $\Delta\phi = \phi_2 - \phi_1 \approx \alpha_2x_2l/2(1 + x_2^2)$ is given, by the similar analysis above for polarization spectroscopy.

It is seen from (6.42) that the signal is now proportional to x_2^2, which appears to be a sensitivity dead region near the absorption peak, unable to be used as a frequency discriminating signal. Several schemes were proposed to overcome the difficulty. In Fig. 6.10a, Ref. [50] built an interferometer with the optical path deviated in a certain degree so that a first order transverse mode is generated in the loop, besides the fundamental mode. The interference between the fundamental mode and the first order mode results in two lobes for $E_{00} + |E_{01}|$ and $E_{00} - |E_{01}|$, and detected by a quadrant photodetector (Q-PD). The difference of detected signals gives a frequency discriminating signal of $I_1 - I_2 \propto x_2$.

Reference [51] proposed another approach to introduce a phase bias, as shown in Fig. 6.10b. The scheme utilizes the polarization property of Fresnel reflection, i.e., the phase difference between two incident waves with polarization perpendicular and parallel to the incident plane is generated by the total internal reflection (TIR): $\Delta\delta = 2\tan^{-1}\sqrt{1 - 2/n^2}$ for incident angle of 45° and refractive index n [3]. For ordinary glasses, $n \approx 1.51$, $\Delta\delta \approx 40°$ are obtained, higher phase difference can be obtained if glass with higher index is used. In the loop of Fig. 6.10b, the pump beam is polarized parallel or perpendicular to the incident plane, and the polarization of probe beam is rotated 90° by the half wavelength plate (HWP2). The phase factor in (6.42) is now replaced by $\Delta\phi = \phi_2 + \Delta\delta$; the working point of the interferometer is then from zero to $\Delta\delta$, and the dead region is avoided. The output I_A is proportional to I_1 of (6.42) with attenuations taken into account, used as the frequency discrimination signal; I_B gives the absorption spectrum with Doppler

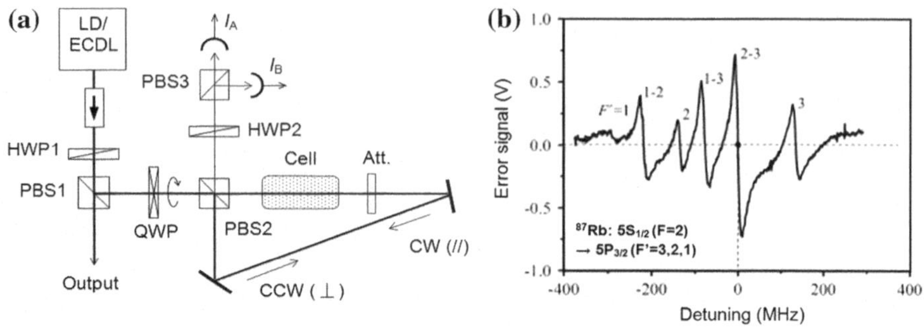

Fig. 6.11 **a** Discriminator with a polarization beam splitter Sagnac loop. **b** Error signal of ^{87}Rb saturated absorption line. Reprinted from Ref. [52] with permission

broadening, used to deduct the background in data processing. The HWP1 is to adjust the power amount entering the discriminator.

Reference [52] utilized the phase shift of $\pi/2$ between the two linearly polarized components of a circularly polarized wave to set the working point of interferometer at the quadrature. Figure 6.11a shows its structure, where the laser beam is converted by a quarter wavelength plate (QWP) to a circularly polarized beam before entering Sagnac loop: $\boldsymbol{E} = E_0[1 \quad j]^T/\sqrt{2}$. The Sagnac loop is composed of a polarization beam splitter (PBS2) and two reflectors, and with a cell and an attenuator inserted, to make the two counter direction propagating beams as pump and probe. It is necessary to notice that it is different from the conventional Sagnac interferometer, since the CW and CCW beams do not interfere with each other at the PBS; they pass the PBS separately and interfere at PBS3 with polarizations rotated by HWP2.

The beam transmitted from PBS2 is an elliptically polarized wave:

$$E_1 = \frac{E_0}{\sqrt{2}} \begin{pmatrix} e^{j\phi_1 - \alpha_1 L/2} \\ je^{j\phi_2 - \alpha_2 L/2} \end{pmatrix} e^{jkL} = \frac{E_0}{\sqrt{2}} \begin{pmatrix} e^{-\Delta\alpha L/2} \\ je^{j\Delta\phi} \end{pmatrix} e^{j\phi_1 - \alpha_2 L/2} e^{jkL}. \tag{6.43}$$

Its principal axis is rotated θ by HWP2; the optical field changes to

$$E_2 = \frac{E_0}{\sqrt{2}} \begin{pmatrix} e^{-\Delta\alpha L/2} \cos 2\theta + je^{j\Delta\phi} \sin 2\theta \\ e^{-\Delta\alpha L/2} \sin 2\theta - je^{j\Delta\phi} \cos 2\theta \end{pmatrix} e^{j\phi_1 - \alpha_2 L/2} e^{jkL}. \tag{6.44}$$

After beam split at PBS3, two detected signals are given:

$$\begin{aligned} I_A &= \frac{I_0}{2} \left[e^{-\Delta\alpha L} \cos^2 2\theta + \sin^2 2\theta + e^{-\Delta\alpha L/2} \sin \Delta\phi \, \sin 4\theta \right] e^{-\alpha_2 L} \\ I_B &= \frac{I_0}{2} \left[e^{-\Delta\alpha L} \sin^2 2\theta + \cos^2 2\theta - e^{-\Delta\alpha L/2} \sin \Delta\phi \, \sin 4\theta \right] e^{-\alpha_2 L}. \end{aligned} \tag{6.45}$$

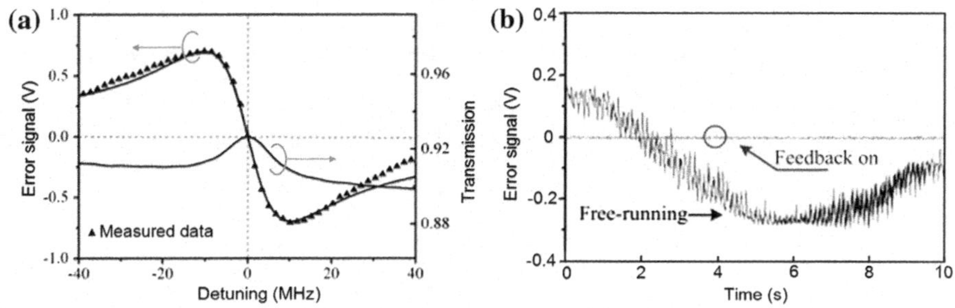

Fig. 6.12 **a** Transmission signal of Rb absorption cell and its derivative as error signal. **b** Error signals before and after frequency stabilization. Reprinted from Ref. [52]

$I_A + I_B = I_0(1 + e^{-\Delta\alpha L})/2$ and $I_A - I_B = I_0 e^{-\Delta\alpha L/2} \sin\Delta\phi \sin 4\theta$ are then obtained. The latter is just the frequency discriminating signal of saturated absorption; when the angle of HWP2 is adjusted to $\theta = \pi/8$, the error signal is maximized.

Figure 6.11b gives the measured error signal for ^{87}Rb saturated absorption when the ECDL is tuned, showing a series of transitions from level $5S_{1/2}$ (F = 2) to level $5P_{3/2}$ (F' = 3, 2, 1) and to crossover (CO) resonance level $5P_{3/2}$ (F' = 1–2, 1–3, 2–3). Figure 6.12a shows the transmission signal of at transition of ^{87}Rb $5S_{1/2}$ (F = 2) to $5P_{3/2}$ (CO F' = 2–3), and its derivative as error signal. Figure 6.12b is the error signals in free-running state and in locked state, demonstrating the effectiveness of frequency stabilization.

6.3.3 *Frequency Stabilization by Magnetic Dichroism*

It is well known that under the external magnetic field $\boldsymbol{B}$, the absorption line of atoms will split as $\hbar\omega = \hbar\omega_0 - m_J g \mu_B B$, where $m_J = -J, -J+1, \ldots J$, and J stands for the total angular momentum of atom, $\mu_B = \hbar e/(2m_e)$ is the Bohr

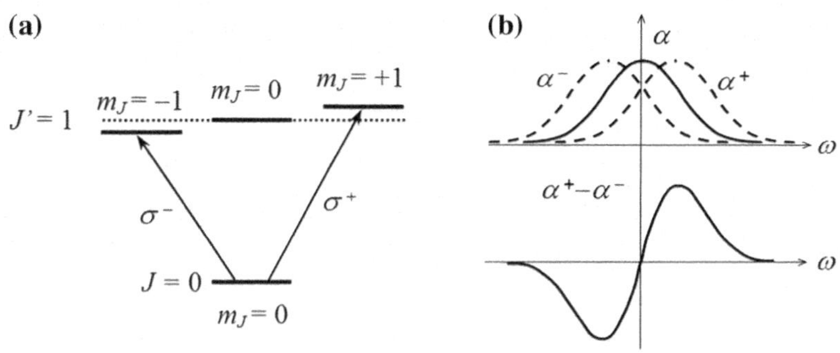

Fig. 6.13 **a** Levels and transitions of Zeeman Effect. **b** Absorption spectra for m_J = 0, ±1, and the difference of m_J = ±1 spectra

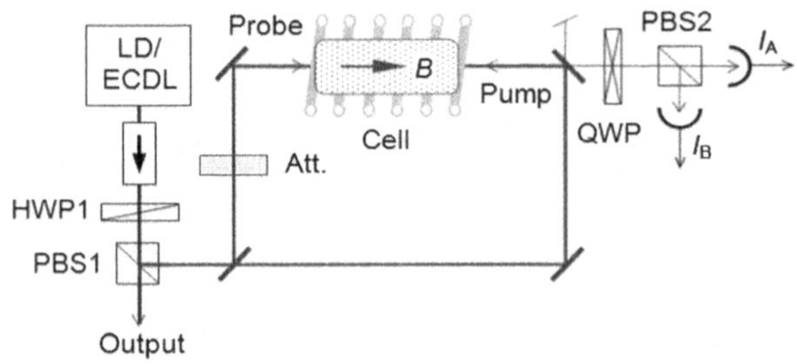

Fig. 6.14 Structure of a frequency stabilized laser based on magneto dichroic effect

Magneton, and g is the Lande factor. It is so-called Zeeman Effect. For most of the first order excited states, $J = 1$, and correspondent three lines have the same line shape, as shown in Fig. 6.13a, b.

The photons absorbed by the three transitions must have different angular momentums to obey the momentum conservation law. In other words, the transitions of $m_J = \pm 1$ states require photons with the left or right circular polarizations, respectively, marked by $\sigma^{\pm}$ in the figure. Such a phenomenon is called the magneto-induced dichroism. It is seen from Fig. 6.13b that the difference between spectra of $m_J = \pm 1$ passes zero at the absorption peak, so that it can be used as a discrimination signal. Such a function is valid both for Doppler broadened lines [53] and for Doppler-free lines [54–57]. Figure 6.14 shows a typical structure of a frequency stabilized laser based on the magnetic dichroic effect, where the absorption cell is placed inside an artificial magnetic field parallel to the beam direction.

The linearly polarized probe beam is regarded as a sum of left and right circularly polarized waves; they suffers absorptions with different peak positions, giving transmitted beam of

$$\boldsymbol{E}_1 = E_0\left[\begin{pmatrix}1\\ j\end{pmatrix}\mathrm{e}^{-\alpha_+ l/2} + \begin{pmatrix}1\\ -j\end{pmatrix}\mathrm{e}^{-\alpha_- l/2}\right] = E_0\begin{pmatrix}\mathrm{e}^{-\alpha_+ l/2} + \mathrm{e}^{-\alpha_- l/2}\\ j(\mathrm{e}^{-\alpha_+ l/2} - \mathrm{e}^{-\alpha_- l/2})\end{pmatrix}. \tag{6.46}$$

Passing through a quarter wavelength plate (QWP) with Jones matrix of $\frac{1}{2}\begin{pmatrix}1+j & 1-j\\ 1-j & 1+j\end{pmatrix}$, it is converted to an elliptically polarized beam $\boldsymbol{E}_2 = E_0\begin{pmatrix}(1+j)\mathrm{e}^{-\alpha_+ l/2}\\ (1-j)\mathrm{e}^{-\alpha_- l/2}\end{pmatrix}$, and then split by PBS2 to give signals of optical intensity to be detected:

$$\begin{aligned} I_A &= I_0 \exp[-\alpha_+(\omega)l] \approx I_0[1 - \alpha_+(\omega)l] \\ I_B &= I_0 \exp[-\alpha_-(\omega)l] \approx I_0[1 - \alpha_-(\omega)l]. \end{aligned} \tag{6.47}$$

Their difference gives an error signal for frequency discrimination.

The frequency stabilization by magneto-dichroism has attractive merits. In the scheme of Ref. [53] the magnetic field is modulated sinusoidal with noise and drift out of the frequency band suppressed in data processing, while the laser keeps modulation-free with the related RAM avoided.

Inspired by the idea, Refs. [58, 59] used an AOM to generate two side bands of $\omega \pm \omega_{AO}$. The differential of measured absorption spectra corresponding to the two frequencies gave an error signal similar to that of dichroism. Reference [60] used a pump beam inclined to the cell to generate different absorption spectra with displaced saturated dips, giving also a similar error signal. These frequency discrimination schemes do not need to modulate the laser to be stabilized. It is believed that novel techniques will be developed continuously to meet the requirement of applications.

6.3.4 Frequency Stabilization by Optical Negative Feedback

Similar to RIN suppression of semiconductor lasers by APC, as stated in Sect. 3.3.3, the frequency noise and drift can also be suppressed by electrical feedback. The frequency discriminator for the feedback signal is usually the edge of absorption band of a certain medium or of a filter, which may provide a locked band in the order of hundreds MHz, and meet the requirement of many applications, such as DWDM optical fiber communications.

By transplanting the electrical feedback to optical domain, Refs. [61–63] proposed and demonstrated a scheme of optical negative feedback to stabilize laser's frequency at the edge of absorption line. The optical path looks similar to an external cavity laser, as shown in Fig. 6.15a [64, 65], where a gas cell and an attenuator are inserted in the external path with the roundtrip beams as the pump and probe for saturated absorption. The difference from ECDL is that a quarter wavelength plate (QWP) is inserted in the path, so that the polarization of returned wave is perpendicular to that of laser mode; it does not interfere with the laser mode. The method is thus called also the non-coherent feedback. On the contrary, the feedback enhances the intensity of non-lasing mode, and consumes more

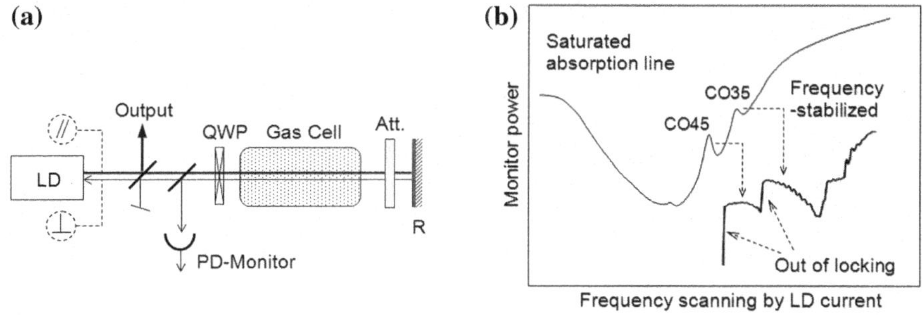

Fig. 6.15 **a** Structure of a frequency stabilized laser with optical negative feedback. **b** Result of frequency stabilization on cesium lines

populations. Therefore it is a negative feedback for the lasing mode. The variation of carrier concentration will change the frequency of laser oscillation, which can then be used for frequency stabilization.

The feedback process can be analyzed by using laser's rate equations. The power of feedback into the cavity is denoted as $\Delta P = \eta_{\perp}(\omega)P$, where the coupling efficiency $\eta_{\perp}$ is a function of frequency due to the gas cell absorption, its subscript stands for perpendicular polarization. The feedback induces carrier concentration change, determined by

$$\frac{\mathrm{d}N}{\mathrm{d}t} = \frac{J}{ed} - \frac{N + \Delta N}{\tau_{\mathrm{sp}}} - v_g g_N (N + \Delta N - N_t)(P + \Delta P) = 0. \tag{6.48}$$

Neglecting the product $\Delta N \cdot \Delta P$, the carrier concentration change is obtained:

$$\Delta N = -\frac{v_g \tau_{\mathrm{sp}} g \Delta P}{1 + v_g \tau_{\mathrm{sp}} g_N P} = -\frac{v \tau_{\mathrm{sp}} g \eta_{\perp}(\omega)}{1 + v_g \tau_{\mathrm{sp}} g_N P} P. \tag{6.49}$$

Taking the frequency fluctuation $\Delta\omega = \omega - \bar{\omega}$ into account, the mode frequency is written as $\omega = M\pi c/(nL) + \Delta\omega$ for F-P cavity LD, and $\omega = 4\pi c/(n\Lambda) + \Delta\omega$ for the DFB-LD. The frequency fluctuation is now rewritten as $\Delta\omega = \delta\omega_{\Delta N} + \delta\omega$, the sum of that induced by feedback and that without feedback, which should be zeroed in frequency stabilization. From $\delta\omega_{\Delta N} = (\partial\omega/\partial n)(\partial n/\partial N)\ \Delta N = -(\bar{\omega}/n)(\partial n/\partial N)\Delta N = -\delta\omega$, it is required that

$$\frac{\partial N}{\partial \omega} = \frac{n}{\bar{\omega}} \left(\frac{\partial n}{\partial N} \right)^{-1} = -\frac{v \tau_{\mathrm{sp}} g P}{1 + v_g \tau_{\mathrm{sp}} g_N P} \frac{\partial \eta_{\perp}(\omega)}{\partial \omega} \tag{6.50}$$

With the coupling efficiency written as $\eta_{\perp} = \eta_c \exp[-\alpha(\omega)l]$, its derivative over frequency is $\partial\eta_{\perp}/\partial\omega = -\eta_{\perp} l \alpha'(\omega)$. Since $\partial n/\partial N < 0$, a negative $\alpha'(\omega)$ is thus required. Therefore, the frequency to be stabilized should be located at the decline edge of absorption line in frequency domain.

Figure 6.15b shows the result of an LD with frequency stabilized on absorption line of cesium [65], where the abscissa is for the LD pump current, corresponding to optical frequency scan. The saturated absorption curve was measured out of locking. When the laser was stabilized, the frequency is locked at crossover resonance peak of CO35 and CO45, where the power detected by PD monitor shows a little change, whereas it drops suddenly when the frequency scanned too far from the decline edges of the absorption line. With the optical negative feedback, the linewidth of DFB-LD was reduced by 40, and the frequency drift was reduced by 14.

The optical negative feedback does not rely on electrical circuit. Its response is determined by the roundtrip time in the external optical path. The response of electrical feedback is limited not only by electronics, but also by the packaging of LD. Especially, the response of high power lasers, working on CW or QCW, is very low, so that the result of electrical feedback stabilization is often not satisfactory. The merit of optical negative feedback is then attractive.

Compared with stabilizations with frequency locked at the line peak, the discrimination at line edge has a shortcoming, i.e., a wider uncertainty range of frequency exists. Besides, the carrier concentration change due to the optical feedback will cause output power fluctuations.

6.4 Pound–Drever–Hall (PDH) Frequency Stabilization

In the research and development of microwave technology during 1940, R.V. Pound proposed a frequency stabilization method with a high Q resonator [66]. In 1980, R.W.P. Drever and J.L. Hall transplanted the idea to the stabilization of laser frequency [67], established an effective frequency stabilization method termed Pound–Drever–Hall (PDH) method. It is widely used in related applications, especially for extremely precise frequency standards [68–72].

6.4.1 *Basic Principle of PDH Method*

The resonator used as frequency discriminator in PDH is a Fabry–Perot (F-P) cavity composed of two high reflective mirrors. Its characteristics are introduced in Sect. 5.4.1. Its resonant peaks have very narrow lines and Lorenzian-like shape near the top, similar to absorption lines of atoms. Compared with the fixed absorption lines of atoms, it has an important advantage: the resonant frequencies are determined by the cavity length and can thus be designed and fabricated arbitrarily. But F-P resonances may be affected by external conditions, such as temperature and mechanical vibrations, not as stable as atom's energy levels. Fortunately, due to technological progresses, highly stable F-P cavities with extremely high finesse have been successfully made; its linewidth is even smaller than that of most atom absorption lines.

The transmission and the reflection have the same resonance linewidth, which decreases with increasing of the finesse. However, high finesse means a longer photon lifetime in cavity and longer group delay. If the transmission of F-P is used to provide frequency discrimination signal, the response will be very slow. Differently, the reflected beam responses to the frequency fluctuation instantly. It is because that the F-P reflection can be regarded as sum of the direct reflection from the front mirror and the leakage of multiple roundtrip oscillations inside the cavity: $r_{FP} = -r + rt^2 e^{j\Delta}/(1 - r^2 e^{j\Delta})$ with $\Delta = 4\pi nL/\lambda$, as given in Sect. 5.4.1. The first term gives a direct and transient response to the input wave, whereas the second term shows high spectral resolution. At the resonances, the two terms have the same amplitude and opposite phases, resulting in zero reflection. When the frequency of input wave deviates from the resonance, reflection will increase immediately, giving error signals.

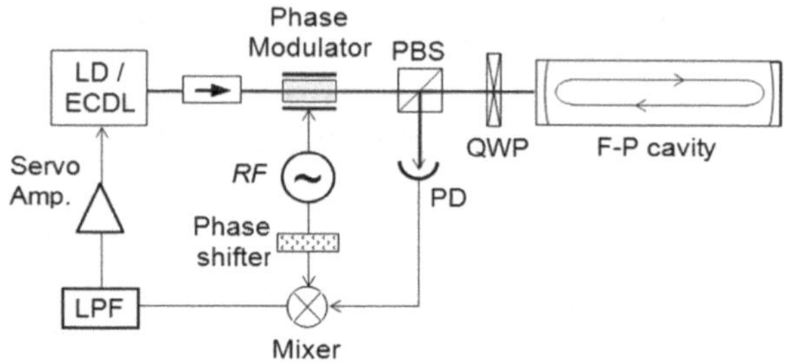

Fig. 6.16 Structure of a PDH-stabilized laser

When the F-P cavity is used as a frequency discriminator, similar to the frequency stabilization by spectral modulation, the laser beam to be stabilized has to be modulated. Figure 6.16 is a schematic diagram of PDH-stabilized semiconductor laser [68], where an external phase modulator is used to modulate the laser frequency. A polarization beam splitter and a quarter wavelength plate are used as an optical isolator to the laser, and to give discriminating signal as well. The PD detected signal mixes with the RF phase modulation signal, which is maximized by adjusting the RF phase shifter; and the sine component and the cosine component of mixed signal can also be obtained by adjusting the phase shifter, respectively. The low pass filter (LPF) bulks high frequency components; and the amplified servo signal is used to control the LD driver and/or the tuning mechanics of ECDL and to lock the laser frequency on one of the F-P resonance peaks.

The process of error signal generation is described quantitatively as follows. The phase modulation is written as $\beta \sin(\Omega t)$. The modulated optical field contains the base band ω_0 and a series of side bands, as shown in (6.23). With the side bands above the first order neglected due to their much smaller amplitudes, the field is written as:

$$\begin{aligned} E_{\text{in}} &= E_0\left[J_0(\beta)e^{j\omega_0 t} + J_1(\beta)e^{j(\omega_0+\Omega)t} - J_1(\beta)e^{j(\omega_0-\Omega)t}\right] \\ &= \left(\sqrt{P_c} + 2j\sqrt{P_s}\sin\Omega t\right)e^{j\omega_0 t}, \end{aligned} \tag{6.51}$$

where $P_c = E_0^2 J_0^2(\beta)$ and $P_s = E_0^2 J_1^2(\beta)$ are powers of the carrier and the first order band. By referring basic formulas given in Sect. 5.4.1, with $x = (\omega - \omega_m)/(2\pi\nu_{\text{FSR}}) = (\omega - \omega_m)\tau_c$, the F-P reflectance, i.e., its transfer function, is written as:

$$F(\omega) = r_{\text{FP}} = \frac{r[(1+R)(\cos x - 1) + j(1-R)\sin x]}{1 + R^2 - 2R\cos x}. \tag{6.52}$$

The amplitude and power of reflected wave are then obtained to be [69]

$$E_{\text{ref}} = \left[F(\omega_0)\sqrt{P_c} + F(\omega_0+\Omega)\sqrt{P_s}e^{j\Omega t} - F(\omega_0-\Omega)\sqrt{P_s}e^{-j\Omega t}\right]e^{j\omega_0 t}, \tag{6.53}$$

$$P_{\text{ref}} = P_{\text{dc}} + P_{\text{ac1}}\cos\Omega t + P_{\text{ac2}}\sin\Omega t + P_{2\Omega}, \tag{6.54}$$

where $P_{2\Omega}$ stands for the components of interference between the two side bands and between the base band and the second order bands. It is much smaller than the former terms, and can be filtered out in data processing, and omitted below in the analysis. The DC and AC components generated by interference of base band and first order band are deduced as

$$\begin{aligned} P_{\mathrm{dc}} &= P_c|F(\omega_0)|^2 + P_s\left(|F(\omega_0+\Omega)|^2 + |F(\omega_0-\Omega)|^2\right) \\ P_{\mathrm{ac1}} &= 2\sqrt{P_cP_s}\mathrm{Re}[F(\omega_0)F^*(\omega_0+\Omega) - F^*(\omega_0)F(\omega_0-\Omega)] \\ P_{\mathrm{ac2}} &= 2\sqrt{P_cP_s}\mathrm{Im}[F(\omega_0)F^*(\omega_0+\Omega) - F^*(\omega_0)F(\omega_0-\Omega)]. \end{aligned} \tag{6.55}$$

Expression of $F(\omega_0)F^*(\omega_0+\Omega) - F^*(\omega_0)F(\omega_0-\Omega)$ in (6.55) is denoted by ρ; its characteristics are discussed below for cases of low ($\Omega \ll 2\pi\delta_{\mathrm{FP}}$) and high ($\Omega \gg 2\pi\delta_{\mathrm{FP}}$) modulation frequencies, respectively.

In the low modulation frequency, the transfer function of F-P is expanded as a Tailor series in the region of $\Omega \ll \omega_0 - \omega_m$ and $F(\omega_0 \pm \Omega) = F(\omega_0) \pm F'_\Omega(\omega_0)\Omega$ with the higher order terms neglected. It is deduced that

$$\begin{aligned} \rho &= F(\omega_0)\left[F^*(\omega_0) + F'^*_\Omega(\omega_0)\Omega\right] - F^*(\omega_0)\left[F(\omega_0) - F'_\Omega(\omega_0)\Omega\right] \\ &= \left[F(\omega_0)F'^*_\Omega(\omega_0) + F^*(\omega_0)F'_\Omega(\omega_0)\right]\Omega = 2\,\mathrm{Re}\left[F(\omega_0)F'^*_\Omega(\omega_0)\right]\Omega, \end{aligned} \tag{6.56}$$

and $P_{\mathrm{ac2}} = 0$. Differentiating (6.52) gives

$$P_{\mathrm{ac1}} \approx 4\sqrt{P_cP_s}\frac{\Omega}{\nu_{\mathrm{FSR}}}\frac{R\sin x_0}{\left[1+s^2\sin^2(x_0/2)\right]^2}, \tag{6.57}$$

where $x_0 = (\omega_0 - \omega_m)\tau_c, s = 2\sqrt{R}/(1-R)$. A calculated P_{ac1} as a function of $(\omega_0 - \omega_m)$ is shown in Fig. 6.17a. It is an odd function of frequency deviation, good to be used as an error signal. It is seen from (6.57) that the error signal increases with the modulation frequency.

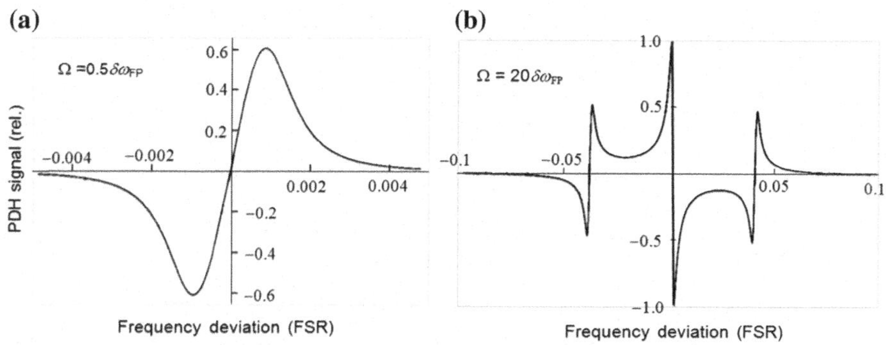

Fig. 6.17 PDH signal spectra for low (**a**) and high (**b**) modulation frequencies. Reproduced from [69] with permission of the American Association of Physics Teachers

In case of $\Omega \gg 2\pi\delta_{FP}$, the reflectance at the resonance is near that of the front mirror, $r_{FP} \approx -r$, for a high finesse F-P. It is deduced that $\rho \approx -j2r\mathrm{Im}F(\omega_0)$ and $P_{ac1} = 0$. The sine component of detected reflection signal (6.54) can be used as error signal for frequency discrimination:

$$P_{ac2} = 2\sqrt{P_c P_s}\mathrm{Im}\rho \approx -4\sqrt{P_c P_s}s^{-1}\cot(x/2). \tag{6.58}$$

The form of (6.58) resembles the phase factor of F-P reflection, formula (5.77), which has a negative slope at the resonance frequency and is towards infinite when $R \to 1$. Figure 6.17b shows P_{ac2} as a function of $(\omega_0 - \omega_m)$, where the upper and down first order side bands are located at ± 0.04 *FSR* just as an example. The steep slope at resonances provides a very high sensitivity to the frequency deviation. The error signal can also be expressed as [69]:

$$S \approx -8\sqrt{P_c P_s}(\omega_0 - \omega_m)/\delta\nu_{FP} \equiv -D(\omega_0 - \omega_m), \tag{6.59}$$

where $D = 8\sqrt{P_c P_s}/\delta\nu_{FP}$ is termed the frequency discriminate (FD). It is deduced from (6.51) that $P_c + 2P_s \approx P_0$, and $D \propto J_0(\beta)J_1(\beta)P_0$. The FD is proportional to the input laser power to the cavity, and dependent on the phase modulation amplitude. It is deduced that the maximum FD is obtained when $\beta = 1.08$ [69]. Although such a large modulation amplitude is used, the second order modulation component is still much smaller than that of the first order component: $J_2^2(1)/J_1^2(1) \approx 0.07$. It is noticed that the above analysis of PDH under high modulation frequency is also suitable for the frequency modulation spectroscopy (FMS) , but their optimized parameters are different, because the purpose of PDH is to pursuit FD as high as possible, rather than to measure the profile of resonance line in FMS. Apart from the high error signal, higher modulation frequency is beneficial for suppressing the 1/*f* noise.

The above analyses are for the ideal case without any loss and with equal reflectance of the two mirrors. Taking into consideration the case of mirrors with different reflectance and with losses, denoting $r_1^2 + t_1^2 = a < 1$, the field and power reflections of F-P resonator are deduced as

$$r_{FP} = \frac{-r_1 + ar_2 e^{j\Delta}}{1 - r_1 r_2 e^{j\Delta}}, \tag{6.60a}$$

$$R_{FP} = \frac{a^2 r_2^2 + r_1^2 - 2ar_1 r_2 + 4ar_1 r_2 \sin^2(\Delta/2)}{(1 - r_1 r_2)^2 + 4r_1 r_2 \sin^2(\Delta/2)}, \tag{6.60b}$$

In such a case, the reflection at resonances is no longer zero, unless the following critical condition is satisfied:

$$r_1 = ar_2 = (r_1^2 + t_1^2)r_2. \tag{6.61}$$

The case of $ar_2 < r_1$ is called the sub-critical coupling; and the opposite case is called over-critical coupling. The use of asymmetric F-P cavity increases flexibility in design and fabrication of PDH system, and also provides better applicability for different cases.

6.4.2 Technical Issues of PDH Frequency Stabilized Laser

(1) **High Finesse F-P Cavity**

High finesse of F-P cavity is one of the keys of PDH technology, ensuring its narrow linewidth and high frequency discriminate and giving a large suppression ratio of laser's linewidth in frequency stabilization. High finesse requires the reflectance of cavity mirrors near unity as much as possible. Reference [70] used cavities with finesse as high as 250,000 and obtained results of frequency stabilized laser with linewidth of 2 Hz and stability of 1×10^{-15}. Reference [71] used a cavity with length of 77.5 mm and finesse as high as 300,000, and demonstrated a frequency stabilized Nd:YAG laser with linewidth of 1 Hz.

(2) **High Stability Resonator**

The high finesse cavity must be very sable. Any minor temperature fluctuation, or mechanical vibrations, including environmental sounds, will cause fluctuations of cavity length and index, resulting in fluctuations and drift of resonant frequencies. References [70–72] described the structure of high stability cavity, as shown in Fig. 6.18 schematically [71]. Features of the structure are as follows.

(a) All of the cavity components, including the supporter of two mirrors and their collimation mechanics, and the base of whole structure, must be made of materials with ultra-low thermal expansion (ULE). Among them invar and micro-crystalline glass are often used; and novel special materials for the purpose are being invented and developed.

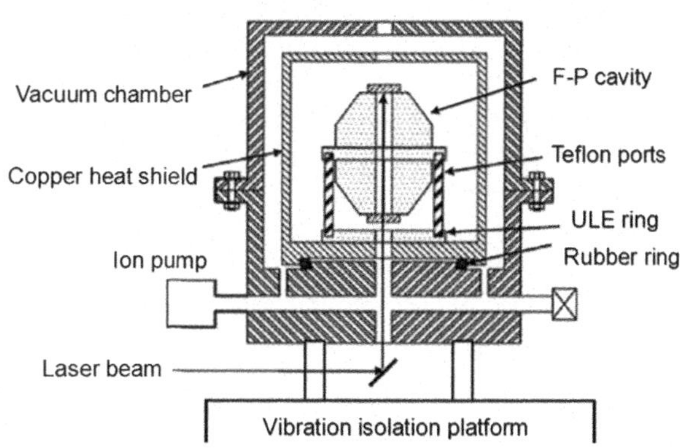

Fig. 6.18 Structure of high stability F-P cavity. Reprinted from Ref. [71] with permission of Springer

(b) The cavity should be placed inside a heat shield with high thermal conductivity, such as copper; and the temperature of the shield is stabilized precisely, so that the cavity is ensured with temperature fluctuations and spatial nonuniformity negligibly small.
(c) The heat shield and its enclosure should be packaged in a vacuum container to prevent the structure from external mechanical disturbances, and from the influences of thermal conduction and convection. The vacuum system must use pumps without movable components, such as the ion pump.
(d) The materials and mechanical parts, including connectors and supporters, should be optimized to ensure strict mechanical and thermal isolations. The cavity is set up vertically to prevent from structure deformation caused by the earth gravity.

By using such a careful design and fabrication, the stability of PDH-stabilized laser reaches an extremely high level of ±7 MHz/s [72].

(3) **Removal of Residue Amplitude Modulation**

The laser frequency modulation in PDH is mostly carried out by using external modulator, especially EOM, no matter the laser is semiconductor laser or not. Although the RAM induced by current modulation of LD is avoided, it still exists in case of external modulation. Reference [73] listed possible mechanisms of RAM in EOM.

(a) F-P interferences induced by the residue reflections of EO crystal facets, and by scatterings inside the material;
(b) The piezoelectric effect of EOM material occurring at the same time as the EO effect, which causes the optical beam deflection and drifting;
(c) The inhomogeneous distribution of modulating electric field;
(d) The temperature effect of optical length and the birefringence of EO crystal. As discussed in Sect. 6.2.3, RAM will result in instability of reference frequency and decrease of FD.

As one of the basic and main properties of EO crystals, birefringence causes the polarization rotation, and RAM correspondingly. Reference [74] measured the temperature effect of so-induced RAM, and found that it varied periodically with the temperature. By controlling the temperature on the point when RAM reaches zero, a good result with RAM less than 25 ppm was obtained. Reference [71] used an active feedback method, where the variation of transmitted optical power of EOM was measured, and taken as a feedback signal to control the transmission of the acoustic optical modulator (AOM), which was placed before the EOM in their system. Such a configuration brings about a merit that the AOM plays also a role of optical isolator, since the frequency is shifted by AOM.

References [39, 40] pointed that RAM depends not only on the material and environment conditions, but also on the arrangement and design of optical path.

Some of the factors have feature of fast response, whereas others act slowly. The photo-refraction effect may also induce RAM; it is usually a slow effect, giving minor influence to PDH with high frequency modulation; but for long term applications it is necessary to pay attention to its effect. References [39–43] proposed various methods of RAM reduction in frequency modulation spectroscopy, which are also useful in PDH technology.

6.4.3 *Different Schemes Based on Resonant Cavities*

(1) **Transfer Cavity and Multiple Stage Stabilization**

Oppositely to the above process of using F-P cavity to stabilize laser frequency, the F-P resonant frequency can be locked on a laser with frequency stabilized at an atom's absorption line. For the purpose, a piezoelectric transducer (PZT) is used to translate the cavity mirror position and adjust the resonance frequency. Furthermore, the frequency stabilized F-P cavity can be used as a frequency discriminator to stabilize the frequency of another laser. Their frequencies may differ from each other by using different resonances. The F-P cavity with such a function is called the transfer cavity. Reference [75] stabilized a 776 nm laser on a frequency stabilized 780 nm laser by using the transfer cavity method. In addition, the work used a laser locked to a current modulated master laser by injection locking, so that RAM was avoided.

One of the important applications of frequency stabilized lasers is to provide time service. More advanced technologies are thus developed for the purpose, such as two stage stabilization [70] and stabilization with combination of multiple cavities [76]. Up-to-recent, the laser noise has been reduced to below the limitation of thermal noise; the frequency uncertainty has been decreased to below 10^{-18} [77].

(2) **Stabilization with Miniature and Compact Cavity**

Practical applications of frequency stabilized laser require small and compact size, and robust structures. Many designs of miniature cavity are proposed and demonstrated experimentally [78]. Reference [79] used a spiral waveguide as a reference frequency element, which is fabricated on silicon substrate with size near the coin of quarter dollar. The length of spiral waveguide reaches 1.2 m, and the quality factor (Q) up-to 140 million is obtained. The miniature cavity was used in frequency stabilized fiber lasers based on the principle of PDH, demonstrating linewidth of 100 Hz, and Allan deviation of 3.9×10^{-13} at 400 μs averaging time.

It is found [80] that the monolithic crystal cavity has better stability than that composed of separated components. The crystal has very low loss in related wavelength band, with a high Q factor; and some sophisticated structures and crafts may be avoided, such as techniques of vacuum cavity. A single crystal silicon cavity is used in PDH system in Ref. [81], demonstrating a sub-40 mHz linewidth laser.

6.5 Noise Reduction by Self-mixing Interference

Self-mixing interference occurs in many cases. The external feedback into a ECDL induces interference with the internal field; such a process is termed the self-mixing interference, and is used to describe the characteristics of ECDL and to develop a sensor [82, 83]. The MI and MZI can also be regarded as self-mixing interferometers. Since the interference signal varies with the frequency noise of the source, they are used in the self-homodyne and self-heterodyne methods for laser noise measurement, as introduced in Chap. 3. The self-mixing interference signal can also play a role of frequency discrimination to be used in frequency stabilization [84–88]. The approach is sometime called the self-referenced frequency stabilization [84]; it is suitable especially in frequency swept lasers, as discussed in Chap. 7.

Figure 6.19 is a typical MI configuration of laser noise reduction system; a MZI configuration can also be used for the purpose. It is similar to the structure of self-heterodyne system shown in Fig. 3.4b, but the PD received signal is mixed with doubled AOM driving signal, since the frequency shift is doubled by the roundtrip propagation. If a coupler of split ratio 50:50 is used, and the fiber loss is neglected for simplicity, the PD detected optical power can be written as

$$I_{\mathrm{PD}} \propto E_0^2\{1 + \cos[\Omega t - \omega_0\tau_d - \Delta\phi(t, \tau_d)]\}, \tag{6.62}$$

where E_0 and ω_0 are the laser's amplitude and central frequency, $\Delta\phi(t, \tau_d) = \phi(t + \tau_d) - \phi(t)$ is the differential phase shift. The signal beat with the local oscillation is then obtained:

$$I_{\mathrm{beat}} \propto \cos[\varphi + \omega_0\tau_d + \Delta\phi(t, \tau_d)] \rightarrow \propto \sin(2\pi\tau_d\delta v). \tag{6.63}$$

where φ is the phase shift between signal and local oscillation, and $\Delta\phi(t, \tau_d) = 2\pi\tau_d\delta v$ with frequency noise of δv. The last expression is obtained when the interferometer works at quadrature, i.e., $(\varphi + \omega_0\tau_d)$ equals an integer multiple of $\pi/2$ by adjusting PZT in Fig. 6.19. The beat signal can then be used to control the laser's driver and its tuning mechanism. Its linewidth is then effectively reduced when a close loop is established.

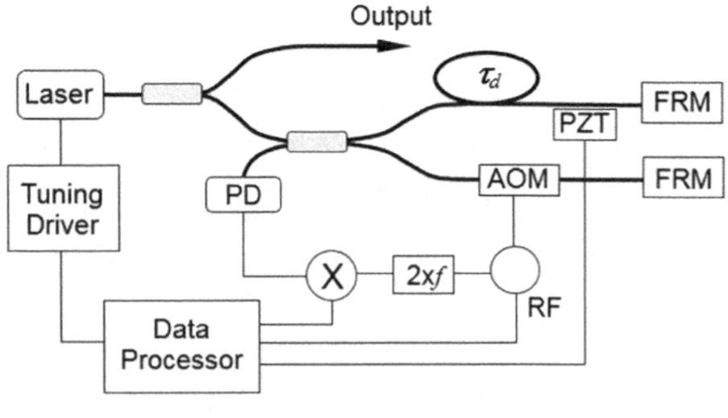

Fig. 6.19 Schematic diagram of frequency stabilization by self-mixing interference

It is seen that a longer delay is needed for a higher sensitivity of noise detection. Some technical problems have to be resolved in implement of this scheme in practice, as described in Refs. [84–88]. The interferometer has to be packaged in a box with temperature stabilized and vibration isolated to avoid frequency drifting. High performance low noise lasers are achieved by the self-mixing interference; e.g., the frequency noise is reduced by more than 40 dB for Fourier frequencies ranging from 1 Hz to 10 kHz, with PSD of 1 Hz^2/Hz, which is comparable to that by PDH method [88].

References

1. Schawlow AL (1978) Laser spectroscopy of atoms and molecules. Science 202(4364):141–147
2. Schawlow AL (1982) Spectroscopy in a new light. Rev Mod Phys 54(3):697–707
3. Born M, Wolf E (1999) Principles of optics, 7th edn. Cambridge University Press
4. Saleh BEA, Teich MC (2007) Fundamentals of photonics. Wiley, New York
5. Yariv A (1997) Optical electronics in modern communications, 5th edn. Oxford University Press, Inc, USA
6. Shen YR (1984) The principle of nonlinear optics. Wiley, New York
7. Maitland A, Dunn MH (1970) Laser physics. Elsevier
8. Ohtsubo J (2013) Semiconductor lasers—stability, instability and chaos, 3rd edn. Springer
9. Varghese PL, Hanson RK (1984) Collisional narrowing effects on spectral line shapes measured at high resolution. Appl Opt 23(14):2376–2385
10. Bennett WR Jr (1962) Hole burning effects in a He–Ne optical maser. Phys Rev 126(2):580–593
11. Lamb WE (1964) Theory of an optical maser. Phys Rev 134(6A):A1429–A1450
12. Bomse DS, Stanton AC, Silver JA (1992) Frequency modulation and wavelength modulation spectroscopies: comparison of experimental methods using a lead-salt diode laser. Appl Opt 31(6):718–731
13. Supplee JM, Whittaker EA, Lenth W (1994) Theoretical description of frequency modulation and wavelength modulation spectroscopy. Appl Opt 33(27):6294–6302
14. Arndt R (1965) Analytical line shape for Lorentzian signal broadened by modulation. J Appl Phys 36(8):2522–2524
15. Bjorklund GC, Levenson MD (1983) Frequency modulation (FM) spectroscopy theory of lineshapes and signal-to-noise analysis. Appl Phys B 32:145–152
16. Uehara K (1998) Dependence of harmonic signals on sample-gas parameters in wavelength-modulation spectroscopy for precise absorption measurements. Appl Phys B 67:517–523
17. Iseki T (2003) Calculation of the ratio between the second and first harmonic signals in wavelength-modulation spectroscopy for absorption measurement. Opt Rev 10(1):24–30
18. Zhu X, Cassidy DT (1997) Modulation spectroscopy with a semiconductor diode laser by injection-current modulation. J Opt Soc Am B 14(8):1945–1950
19. Wang Y, Cai H, Geng J et al (2007) Behaviors of harmonic signals in wavelength-modulated spectroscopy under high absorption strength. Chin Opt Lett 5(9):552–555
20. Wang Y, Cai H, Geng J et al (2009) Logarithmic conversion of absorption detection in wavelength modulation spectroscopy with a current-modulated diode laser. Appl Opt 48 (21):4068–4076
21. de Labachelerie M, Nakagawa K, Awaji Y et al (1995) High-frequency-stability laser at 1.5 μm using Doppler-free molecular lines. Opt Lett 20(6):572–574
22. Bruner A, Mahal V, Kiryuschev I et al (1998) Frequency stability at the kilohertz level of a rubidium-locked diode laser at 192.114 THz. Appl Opt 37(20):6410–6414

23. Lazar J, Cip O, Jedlicka P (2000) Tunable extended-cavity diode laser stabilized on iodine at $\lambda = 633$ nm. Appl Opt 39(18):3085–3088
24. Affolderbach C, Mileti G (2005) Tuneable, stabilised diode lasers for compact atomic frequency standards and precision wavelength references. Opt Lasers Eng 43:291–302
25. Fang H, Wang S, Shy J (2006) Frequency stabilization of an external cavity diode laser to molecular iodine at 657.483 nm. Appl Opt 45(13):3173–3176
26. Yanagawa T, Saito S, Machida S et al (1985) Frequency stabilization of an InGaAsP distributed feedback laser to an NH_3 absorption line at 15137 Å with an external frequency modulator. Appl Phys Lett 47(10):1036–1038
27. Numata K, Chen JR, Wu ST et al (2011) Frequency stabilization of distributed-feedback laser diodes at 1572 nm for lidar measurements of atmospheric carbon dioxide. Appl Opt 50 (7):1047–1056
28. Altmann J, Raumgart R, Weitkamp C (1981) Two-mirror multipass absorption cell. Appl Opt 20(6):995–999
29. Benabid F, Couny F, Knight JC et al (2005) Compact, stable and efficient all-fibre gas cells using hollow-core photonic crystal fibres. Nature 434:488–491
30. Lurie A, Baynes FN, Anstie JD et al (2011) High-performance iodine fiber frequency standard. Opt Lett 36(24):4776–4778
31. Huang C, Chen D, Cai H et al (2014) Transmission characteristics of photonic crystal fiber gas cell used in frequency stabilized laser. Chin Opt Lett 12(8):080602(1-5)
32. Kluczynski P, Axner O (1999) Theoretical description based on Fourier analysis of wavelength-modulation spectrometry in terms of analytical and background signals. Appl Opt 38(27):5803–5815
33. Li H, Rieker GB, Liu X et al (2006) Extension of wavelength-modulation spectroscopy to large modulation depth for diode laser absorption measurements in high-pressure gases. Appl Opt 45(5):1052–1061
34. McGettrick AJ, Duffin K, Johnstone W et al (2008) Tunable diode laser spectroscopy with wavelength modulation: a phasor decomposition method for calibration–free measurements of gas concentration and pressure. J Lightwave Technol 26(4):432–440
35. Johnstone W, McGettrick AJ, Duffin K et al (2008) Tunable diode laser spectroscopy for industrial process applications: system characterization in conventional and new approaches. IEEE Sens J 8(7):1079–1088
36. Chakraborty AL, Ruxton K, Johnstone W et al (2009) Elimination of residual amplitude modulation in tunable diode laser wavelength modulation spectroscopy using an optical fiber delay line. Opt Express 17(12):9602–9607
37. Ruxtona K, Chakraborty AL, Johnstone W et al (2010) Tunable diode laser spectroscopy with wavelength modulation: elimination of residual amplitude modulation in a phasor decomposition approach. Sens Actuators B 150:367–375
38. Debs JE, Robins NP, Lance A et al (2008) Piezo-locking a diode laser with saturated absorption spectroscopy. Appl Opt 47(28):5163–5166
39. Sathian J, Jaatinen E (2013) Reducing residual amplitude modulation in electro-optic phase modulators by erasing photorefractive scatter. Opt Express 21(10):12309–12317
40. Sathian J, Jaatinen E (2013) Dependence of residual amplitude noise in electro-optic phase modulators on the intensity distribution of the incident field. J Opt 15:125713
41. Jaatinen E, Hopper DJ, Back J (2009) Residual amplitude modulation mechanisms in modulation transfer spectroscopy that use electro-optic modulators. Measur Sci Technol 2:025302(1-8)
42. Whittaker EA, Shum CM, Grebel H et al (1988) Reduction of residual amplitude modulation in frequency-modulation spectroscopy by using harmonic frequency modulation. J Opt Soc Am B 5(6):1253–1256
43. Burck Fd, Lopez O, Basri AE (2003) Narrow-band correction of the residual amplitude modulation in frequency-modulation spectroscopy. IEEE Trans Instrum Measur 52(2):288–291
44. Wieman C, Hänsch TW (1976) Doppler-free laser polarizations spectroscopy. Phys Rev Lett 36(20):1170–1173

45. Simpson NB, Dholakia K, Allen L et al (1997) Mechanical equivalence of spin and orbital angular momentum of light: an optical spanner. Opt Lett 22(1):52–54
46. Pearman CP, Adams CS, Cox SG et al (2002) Polarization spectroscopy of a closed atomic transition: applications to laser frequency locking. J Phys B: At Mol Opt Phys 35(24): 5141–5151
47. Lancaster GPT, Conroy RS, Clifford MA et al (1999) A polarization spectrometer locked diode laser for trapping cold atoms. Opt Commun 170:79–84
48. Yoshikawa Y, Umeki T, Mukae T et al (2003) Frequency stabilization of a laser diode with use of light-induced birefringence in an atomic vapor. Appl Opt 42(33):6645–6649
49. Jiang K, Wang J, Tu X et al (2003) Polarization spectra of Rb atoms and their application in laser frequency stabilization. Chin Opt Lett 1(7):377–379
50. Robins NP, Slagmolen BJJ, Shaddock DA et al (2002) Interferometric, modulation-free laser stabilization. Opt Lett 27(21):1905–1907
51. Wei F, Chen D, Fang Z et al (2010) Modulation-free frequency stabilization of external-cavity diode laser based on a phase-difference biased Sagnac interferometer. Opt Lett 35(22): 3853–3855
52. Wei F, Chen D, Sun Y et al (2013) Modulation-free frequency stabilization based on polarization-split Sagnac loop. IEEE Photonics Technol Lett 25(11):1031–1034
53. Corwin KL, Lu ZT, Hand CF et al (1998) Frequency-stabilized diode laser with the Zeeman shift in an atomic vapor. Appl Opt 37(15):3295–3298
54. Lu W, Milic D, Hoogerland MD et al (1996) A practical direct current discharge helium absorption cell for laser frequency locking at 1083 nm. Rev Sci Instrum 67(9):3003–3004
55. Gertsvolf M, Rosenbluh M (1999) Injection locking of a diode laser locked to a Zeeman frequency stabilized laser oscillator. Opt Commun 170:269–274
56. Wasik G, Gawlik W, Zachorowski Z et al (2002) Laser frequency stabilization by Doppler-free magnetic dichroism. Appl Phys B: Lasers Opt 75(6):613–619
57. Wang J, Yan S, Wang Y et al (2004) Modulation-free frequency stabilization of a grating-external-cavity diode laser by magnetically induced sub-Doppler dichroism in cesium vapor cell. Jpn J Appl Phys 43(3):1168–1171
58. Sukenik CL, Busch HC, Shiddiq M (2002) Modulation-free laser frequency stabilization and detuning. Opt Commun 203:133–137
59. Van Ooijen ED, Katgert G, Van der Straten P (2004) Laser frequency stabilization using Doppler-free bichromatic spectroscopy. Appl Phys B 79:57–59
60. Park SE, Lee HS, Kwon TY (2001) Dispersion-like signal velocity-selective saturated-absorption spectroscopy. Opt Commun 192:49–55
61. Yasaka H, Kawaguchi H (1988) Linewidth reduction and optical frequency stabilization of a distributed feedback laser by incoherent optical negative feedback. Appl Phys Lett 53 (15):1360–1362
62. Yasaka H, Yoshikuni Y, Kawaguchi H (1991) FM noise and spectral linewidth reduction by incoherent optical negative feedback. IEEE J Quantum Electron 27(2):193–204
63. da Rocha AFA, Segundo PCS, Chevrollier M (2004) Diode laser coupled to an atomic line by incoherent optical negative feedback. Appl Phys Lett 84(2):179–181
64. Ying K, Chen D, Cai H et.al. (2012) Frequency stabilization of a DFB laser to molecular Cesium at 852 nm by polarization-rotated optical feedback. In: Proceedings of CLEO/QELS JW2A, pp 87–88
65. Sun Y, Wei F, Dong Z et al (2014) All-optical frequency stabilization and linewidth reduction of distributed feedback diode lasers by polarization rotated optical feedback. Opt Express 22 (13):15757–15762
66. Pound RV (1946) Electronic frequency stabilization of microwave oscillators. Rev Sci Instrum 17(11):490–505
67. Drever RWP, Hall JL, Kowalski FV et al (1983) Laser phase and frequency stabilization using an optical resonator. Appl Phys B 31(2):97–105
68. Dahmani B, Hollberg L, Drullinger R (1987) Frequency stabilization of semiconductor lasers by resonant optical feedback. Opt Lett 12(11):876–878

69. Black ED (2001) An introduction to Pound–Drever–Hall laser frequency stabilization. Am J Phys 69:79–87
70. Ludlow AD, Huang X, Notcutt M et al (2007) Compact, thermal-noise-limited optical cavity for diode laser stabilization at 1×10^{-15}. Opt Lett 32(6):641–643
71. Jiang Y, Fang S, Bi Z et al (2010) Nd:YAG lasers at 1064 nm with 1-Hz linewidth. Appl Phys B 98:61–67
72. Dubé P, Madej AA, Bernard JE et al (2009) A narrow linewidth and frequency-stable probe laser source for the $^{88}Sr+$ single ion optical frequency standard. Appl Phys B 95(1):43–54
73. Ishibashi C, Ye J, Hall JL (2002) Analysis/reduction of residual amplitude modulation on phase/frequency modulation by an EOM. In: Proceedings of QELS QTuF27, pp 91–92
74. Li L, Liu F, Wang C et al (2012) Measurement and control of residual amplitude modulation in optical phase modulation. Rev Sci Instrum 83:043111(1-10)
75. Liekhus-Schmaltz CE, Mantifel R, Torabifard M et al (2012) Injection-locked diode laser current modulation for Pound-Drever-Hall frequency stabilization using transfer cavities. J Opt Soc Am B 29(6):1394–1398
76. Jiang YY, Ludlow AD, Lemke ND et al (2011) Making optical atomic clocks more stable with 10^{-16}-level laser stabilization. Nat Photonics 5(3):158–161
77. Nicholson TL, Campbell SL, Hutson RB et al (2015) Systematic evaluation of an atomic clock at 2×10^{-18} total uncertainty. Nat Commun 6:6896(1-8)
78. Vahala KJ (2003) Optical microcavities. Nature 424:839–846
79. Lee H, Suh M, Chen T et al (2013) Spiral resonators for on-chip laser frequency stabilization. Nat Commun 3468:1–6
80. Savchenkov AA, Ilchenko VS, Matsko AB et al (2004) Kilohertz optical resonances in dielectric crystal cavities. Phys Rev A 70:051804(R)1-4
81. Kessler T, Hagemann C, Grebing C et al (2012) A sub-40-mHz-linewidth laser based on a silicon single-crystal optical cavity. Nat Photonics 687–692
82. Wang WM, Grattan KTV, Palmer AW et al (1994) Self-mixing interference inside a single-mode diode laser for optical sensing applications. J Lightwave Technol 12(9):1577–1586
83. Randone EM, Donati S (2006) Self-mixing interferometer: analysis of the output signals. Opt Express 14(20):9188–9196
84. Greiner C, Boggs B, Wang T et al (1998) Laser frequency stabilization by means of optical self-heterodyne beat-frequency control. Opt Lett 23(16):1280–1282
85. Cranch GA (2002) Frequency noise reduction in erbium-doped fiber distributed-feedback lasers by electronic feedback. Opt Lett 27(13):1114–1116
86. Cliché JF, Allard M, Têtu M (2006) High-power and ultra-narrow DFB laser: the effect of linewidth reduction systems on coherence length and interferometer noise. In: Proceedings of SPIE 6216: 62160C(1-11)
87. Sheard BS, Gray MB, McClelland DE (2006) High-bandwidth laser frequency stabilization to a fiber-optic delay line. Appl Opt 45(33):8491–8499
88. Kéfélian F, Jiang H, Lemonde P et al (2009) Ultralow-frequency-noise stabilization of a laser by locking to an optical fiber-delay line. Opt Lett 34(7):914–916

Chapter 7
Frequency Sweeping of Semiconductor Lasers

7.1 Applications of Frequency-Swept Laser

The single-frequency semiconductor lasers and their tunability are introduced in Chaps. 2–5. Their characteristics are discussed, but limited mainly in the static state. As applications develop continuously, lasers are required not only with narrow linewidth in continuous wave (cw) operations, but also in dynamic tuning operations. The laser's central frequency is swept in a high speed with a certain temporal function, mostly in a linear function of time (a serrated form). The frequency sweeping laser is also called the frequency agile laser or the chirped laser, in some publications. Three methods are usually used for frequency sweeping: by the direct current tuning of LD, by intra-cavity modulators of ECDL, and by extra-cavity modulations.

The frequency sweeping semiconductor lasers have been used in many advanced and important technologies. This section introduces briefly some of the examples.

(1) **High-Resolution, High-Precision Optical Spectrum Analyzer**.

The narrow linewidth of lasers has to be measured by self-heterodyne or self-homodyne, as described in Chap. 3. However these methods cannot tell an absolute spectral position of the line peak. The scanning Fabry–Perot etalon can give a calibrated spectral position, but its resolution is usually not high enough. If a frequency-swept narrow line laser is available with its peak being precisely calibrated, the line shape of the laser under test can be measured by heterodyne, as shown in Fig. 7.1, and the exact peak frequency and linewidth can be obtained from the spectrum of beat signal [1, 2]. Such an optical spectrum analyzer has emerged commercially now; and its performance is being improved and developed.

Z. Fang et al., *Single Frequency Semiconductor Lasers*, Optical and Fiber Communications Reports 9, DOI 10.1007/978-981-10-5257-6_7

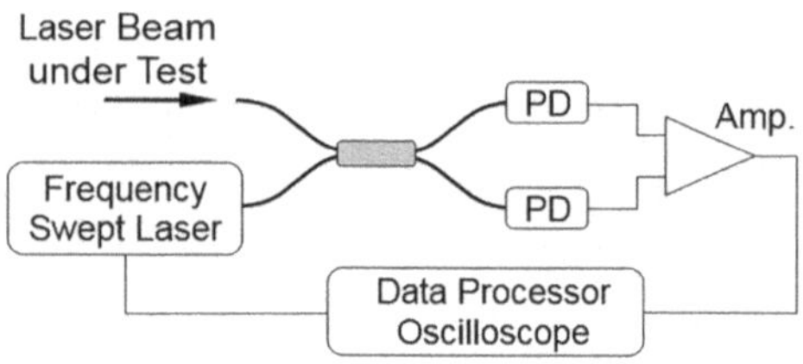

Fig. 7.1 Optical spectrum analyzer based on a frequency-swept laser

(2) **High-Resolution Laser Spectroscopy**.

The frequency-swept laser is a powerful tool for investigating properties of materials, e.g., measuring precisely absorption spectra, photoluminescence spectra, gain spectra of active medium, Raman and Brillouin spectra. These spectra will give abundant information of material structures and physical mechanisms. A periodically high speed swept source will greatly enhance signal-to-noise ratio (SNR), and suitable for measuring dynamical properties of the objects. The swept laser is also useful for high-precision sensing and metrology applications. The interferometer with a single frequency source is usually suitable to measure dynamic changes of optical path difference (OPD); the measurement of static OPD may need two or more sources to eliminate uncertainty due to multiple interference fringes. The frequency-swept laser can then fulfill both of static and dynamic measurements [3]. It is also beneficial to real-time measurements of other parameters, such as displacement, velocity and acceleration of objects, information of vibration and sound. Furthermore, the precision and sensitivity will be greatly improved by using a swept source [4].

(3) **FMCW Lidar**.

Frequency-modulated continuous wave (FMCW) technology is developed earlier for radars in microwave band [5]. The same principle implanted to the optical band shows many advantages, which uses a frequency modulated laser to detect target position precisely by mixing the local oscillation with the returned wave. When the laser frequency is linearly and periodically swept, the beat frequency will be proportional to the distance between target and the range finder [6]. One of the applications is the optical range finder, which plays important roles both in military fields and civil uses. The dependence of FMCW performance on the source characteristics of swept is analyzed in [7].

(4) **Distributed Optical Fiber Sensors**.

By utilizing Rayleigh scattering in optical fibers the OTDR technology has been developed and used widely in optical fiber communication systems. Rayleigh scattering, Raman scattering, and Brillouin scattering in fibers are also utilized to build distributed optical fiber sensors. Based on the narrow linewidth single-frequency lasers, the coherent OTDR and the phase-sensitive OTDR (ϕ-OTDR) show superior advantages to those with low coherent lasers. Especially, the ϕ-OTDR with a temporally sequenced multi-frequency source showed ultra-broadband responses for vibration detections [8]; the ϕ-OTDR with frequency-swept laser pulses showed high spatial resolution [9]. Obviously, lasers with precise tunability and high speed frequency sweeping are the precondition for such high performance ϕ-OTDR sensors. The FMCW technology is also implanted to distributed optical fiber sensors to build OFDR, which has much higher spatial resolutions [10–13].

(5) **Optical Frequency Domain Imaging**.

The principle of FMCW technology is also used in the optical coherence tomography (OCT), which is to give a three-dimensional image of object covered by diffusive media. Generally a low coherence source is used in OCT with its low coherence distance utilized to remove the influence of diffusive scattering. Compared with such an ordinary OCT, a frequency-swept source will greatly improve the OCT performances on imaging speed, resolution, sensitivity, etc. The technology is termed the optical frequency domain imaging (OFDI) [14], and a new kind laser called Fourier domain mode locked laser (FDML) is developed [15], as shown in Fig. 7.2. The OFDI is also used to dynamically monitor the laser ablation of biological tissue [16]. The application of OFDI is expanded to different areas. For example, it is used to measure the transverse mode properties of large mode area fibers and mode content of hollow-core photon crystal optical fiber (PCF), both of the spectral and the spatial distributions [17, 18].

(6) **Optical Communications and Microwave Photonics**.

The laser with high speed and wide range tunability is used both in the optical transmitter and in coherent detection as a local oscillation optical source. In the free space optical (FSO) communications between moving terminals, Doppler frequency shift is an important effect for the optical coherent communication, which is to be compensated in correlation detection by using a frequency-swept local oscillator. In the communication networks, it is necessary to combine the optical fiber communication and the wireless microwave communication. The radio over fiber (ROF) technology and, extensively, the microwave photonics are therefore developed rapidly [19, 20]. The frequency-swept laser plays indispensable roles in the mutual conversions between microwave band and optical band of the data to be transported, and of the controlling signals of the devices and systems.

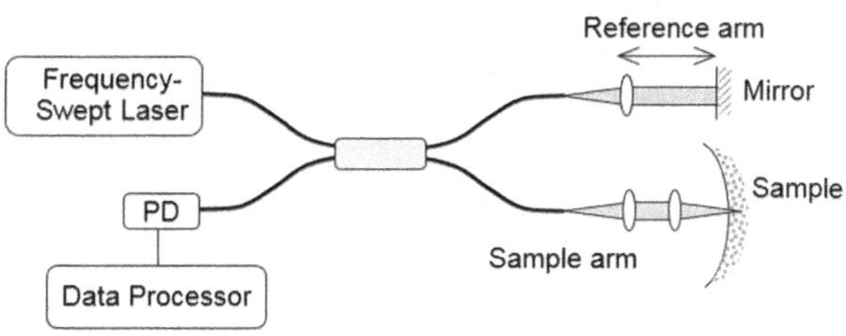

Fig. 7.2 Schematic diagram of OCT with swept source

(7) **Fundamental Scientific Researches**.

The measurement and analysis of optical properties and nonlinear optical effects of materials, especially those of newly emerged materials, are always hot spots in the front of scientific researches, in which the laser technology, including the frequency-swept semiconductor laser, plays important roles. Especially, the frequency-stabilized and tunable high coherence laser is one of the dispensable tools in the research of laser cooling of atoms and cooled atomic clock [21, 22]. The laser beam is also used as optical tweezers to manipulate micro particles, such as biological cells. Technical requirements of the laser used in researches on physics of atomic systems are analyzed in [23].

The frequency sweeping can be realized by various methods. Basically three schemes are usually used, i.e., by using LD current tuning, by using tuning components inside laser cavity, and by using phase/frequency modulator inserted in the beam path outside cavity. These methods are introduced in the following sections, respectively.

7.2 Frequency Sweeping by LD Current Modulation

7.2.1 LD Current Tuning and Its Linearization

The capability of current tuning is one of the important features of semiconductor laser. It is the simplest way of frequency sweeping. However, several issues have to be solved for a frequency sweeping laser. One is the residue amplitude modulation (RAM), as analyzed in Chap. 6. The second is the control of sweeping rate, especially its linearity. In general, LD tuning rate $\Delta\nu/\Delta i$ is not a constant, as discussed in Chap. 2 and in Ref. [11]. It is because the rate is determined by index variation with carrier concentration and temperature, whereas the rates of carrier concentration and temperature variations with the injected current are dependent on the modulation frequency and modulation amplitude. In case of repeated sweeping in a serrated waveform, the modulation frequencies contain infinitive components from the base band to high order bands, expressed by Fourier series. Therefore it is with difficulty to sweet the laser frequency linearly, especially for cases of high modulation frequency and high modulation amplitude. The third issue is the coherence, i.e., whether its instantaneous linewidth can keep narrow enough in the process of frequency sweeping.

Some methods of dealing with the nonlinearity of frequency modulation (FM) have already been proposed and demonstrated, such as pre-chirping of current variation and data processing post-acquisitions. However, the function of pre-chirping has to be designed beforehand for particular situations, and thus lack of flexibility and wide applicability. The data processing post-acquisitions is a time-consuming method, and can hardly meet the requirement of real-time measurement. It is necessary to develop linearly scanned FM technologies with real-time feedback and controlling.

One of the key steps in most methods of real-time controlling is to pick up the error signal by an interferometer. The linear frequency variation of rt corresponds to a phase factor of $\phi(t) = \pi rt^2$; the optical field with linearly swept frequency is written as

$$E(t) = E_0 \exp\left[j2\pi(\nu_0 t + rt^2/2)\right]. \tag{7.1}$$

The sweeping rate can be measured by using self-homodyne or self-heterodyne, the same as that used for laser linewidth and frequency noise measurement in Chap. 3. In the homodyne setup, the optical wave to be analyzed is split into two beams of an interferometer with delay of τ, and two waves are summed up at the beam splitter, giving

$$\begin{aligned} E_T(t) &= E_0\left\{p \exp[j2\pi(\nu_0 t + rt^2/2)] + q \exp[j2\pi\nu_0(t+\tau) + j2\pi r(t+\tau)^2/2]\right\} \\ &= E_0 e^{j2\pi(\nu_0 t + rt^2/2)}\{p + q \exp[j2\pi(\nu_0\tau + r\tau t + r\tau^2/2)]\}, \end{aligned} \tag{7.2}$$

where p and q stand for the field split ratio, and $p^2 + q^2 = 1$ in lossless cases. The linewidth is supposed narrow enough with visibility of unity here. The intensity of interference signal is

$$\begin{aligned} I(t) &= E_0^2\{p^2 + q^2 + 2pq \cos[2\pi(\nu_0\tau + r\tau t + r\tau^2/2)]\} \\ &\rightarrow I_0 \sum_n \cos[2\pi r\tau(t - nT) + \phi_0]. \end{aligned} \tag{7.3}$$

It is an oscillating signal with beat frequency of $r\tau$ and phase factor of $\phi_0 = 2\pi(\nu_0\tau + r^2/2)$. The second expression of (7.3) takes repeated modulations into account. The spectrum of beat signal is given by Fourier transform with a sinc function multiplied, expressed as [24]

$$\tilde{I}(f) \propto I_0 \frac{\sin^2[\pi(f - r\tau)(T - \tau)/2]}{[\pi(f - r\tau)(T - \tau)/2]^2}. \tag{7.4}$$

The deviation of sweeping rate from linear or other required waveform can then be obtained by a spectrum analyzer in data processing; and used as a feedback error signal to adjust and control the modulation current.

In practical applications such a measurement and control should be completed in real time. Some schemes were then proposed and demonstrated [25, 26]. Figure 7.3a shows a self-homodyne system composed of a Michelson interferometer (MI). The beat signal is mixed with the required waveform (Ramp) and feedback to control the LD driver. The homodyne is also used for spectral analysis

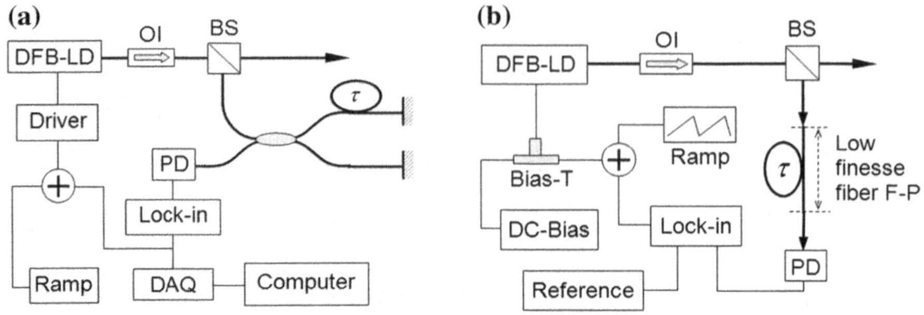

Fig. 7.3 Linear frequency sweeping lasers, feedback controlled with homodyne (**a**) and fiber F-P (**b**). *OI*: optical isolator; *DAQ*: data acquisition card

in the computer. The MI can be replaced by an F-P interferometer, as shown in Fig. 7.3b [11], where a section of fiber is connected with the fiber system, and the low reflectivity at connections makes it as a low finesse F-P cavity. Its output contains interference between the direct transmitted wave and the wave with a round trip delay. The Bias-T is used to add the DC bias current and the AC feedback current, which is widely used in LD driver. In the two schemes, the single frequency operation is ensured by using DFB-LD, instead of an ordinary LD chip.

It is necessary to characterize quantitatively the linearity of frequency sweeping, i.e., the deviation of swept frequency waveform from the required temporal function. Reference [26] proposed a method based on Hilbert transform. The beat signal Exp. (7.3) is normalized as

$$p(t) = I(t)/I_0 \propto \cos[2\pi r(t)\tau t + \phi_0] = \cos[\Phi(t)], \tag{7.5}$$

where the sweeping rate is expressed as a function of time. The composite phase $\Phi(t)$ can be solved from $p(t)$ by using Hilbert transform. It is known that Hilbert transform of function $u(t)$ is its convolution with function $h(t) = 1/(\pi t)$, expressed as

$$H[u(t)] = \hat{u}(t) = P\int_{-\infty}^{\infty} u(x)h(t-x)\mathrm{d}x = \frac{1}{\pi}P\int_{-\infty}^{\infty} \frac{u(x)}{t-x}\mathrm{d}x, \tag{7.6}$$

where P stands for the principal value of Cauchy integral. Hilbert transforms of sine function and cosine function are $H[\sin(t)] = -\cos(t)$ and $H[\cos(t)] = \sin(t)$. The phase factor is thus solved by using Hilbert transform:

$$\Phi(t) = \tan^{-1}[\hat{p}(t)/p(t)], \tag{7.7}$$

the transient frequency sweeping rate is obtained as

$$r(t) = \frac{1}{2\pi\tau}\frac{\mathrm{d}}{\mathrm{d}t}\tan^{-1}[\hat{p}(t)/p(t)]. \tag{7.8}$$

Since the algorithm of Hilbert transform can be found directly from ready-made software, it is a convenient and widely applicable method.

In addition of using a ramp electrical signal to tune the laser, Ref. [27] uses a confocal F-P cavity to read the swept frequency of ECDL periodically, and to check and feedback control its linearity. The sweeping period covers several F-P resonances; the laser gives serrated waveform outputs with frequency varied precisely; the laser is used in high speed phase shift interferometry.

The residue amplitude modulations (RAM), accompanied with LD frequency modulation must be removed in most applications. The related methods are discussed in Chap. 6. Reference [28] proposed a scheme with an amplitude controller, as shown in Fig. 7.4. The self-homodyne signal is obtained by an MZI with delay of τ. In electronics domain, a reference frequency is used to convert the beat signal to a middle frequency band, which is beneficial for reducing $1/f$ noise and for more agility. A sweeping rate as high as 100 GHz/ms was reported by such a scheme. By using the similar technology, Ref. [29] demonstrated a linear frequency sweeping laser with a range up to 5 THz and frequency errors deviated from linearity as low as 170 kHz.

The effectiveness of sweeping linearization is obviously related to the response speed of feedback control, caused by optical delay and electrical delay. The sweeping rate and the repetition frequency are limited by the response. Especially, a certain delay is needed in turning on of the feedback control at the starting point of every repetition period. Obviously, the ratio of response time over sweeping period is required as smaller as possible; the optimization of system design, including optics and electronics is surely needed. It is conjectured that a higher beat signal frequency is beneficial for a short response time of the control, whereas the beat frequency can be designed by the delay of interferometer. Related mechanism and characteristics were analyzed in detail in Ref. [5].

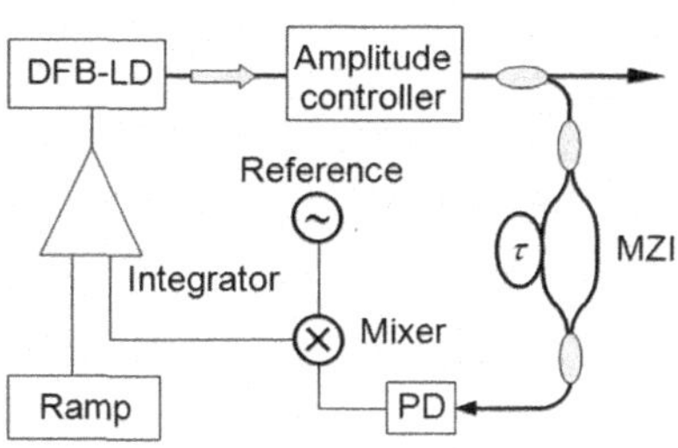

Fig. 7.4 Agile frequency-swept laser with MZI self-homodyne

7.2.2 Linewidth Reduction of Frequency-Swept Lasers

As discussed in Chap. 3, LD frequency is very susceptible to the state of operation. Its linewidth will be broadened by current modulation. It is required for most applications that the laser's coherence is maintained high enough during the frequency sweeping. The interferometer for linearity characterization and controlling can actually be used as a system of noise reduction by self-mixing interference, which is introduced in Sect. 6.5 for a fixed single-frequency laser. Taking frequency noise into account, the PD detected signal of a self-heterodyne system by using an AOM with frequency shift f is written as

$$I(t) \propto I_0 \cos[(\pi r\tau - 2\pi f)t + \phi_0 + \Delta\phi(t,\tau)], \tag{7.9}$$

where $\Delta\phi(t,\tau) = \phi(t+\tau) - \phi(t) = 2\pi\delta v(t)\tau$ is the differential phase shift caused by the OPD of interferometer, taking the frequency noise δv into account. The signal is then mixed with the local oscillator in an electronic mixer, giving the in-phase and out-of-phase beat signals of

$$\begin{aligned} V_I &\propto \cos[\pi r\tau t + \Delta\phi(t,\tau) + \phi_0] \\ V_Q &\propto \sin[\pi r\tau t + \Delta\phi(t,\tau) + \phi_0], \end{aligned} \tag{7.10}$$

which can be used to characterize and control the sweeping linearity, and also to reduce the linewidth, as done in frequency stabilization by self-mixing interference.

Several schemes of linewidth reduction of frequency-swept laser were proposed and demonstrated. In the structure shown in Fig. 7.5a [30], the MZI is used as a self-heterodyne system, where AOM plays roles of both beam splitter and frequency shifter. Compared with self-homodyne, the beat signal is shifted to a middle frequency band for suppression of $1/f$ noise. The phase-locked loop (PLL) is used to demodulate the FM signal, providing error detection with higher sensitivity for linewidth reduction. The self-mixing interference method provides compatibility for both of frequency sweeping and linewidth reduction.

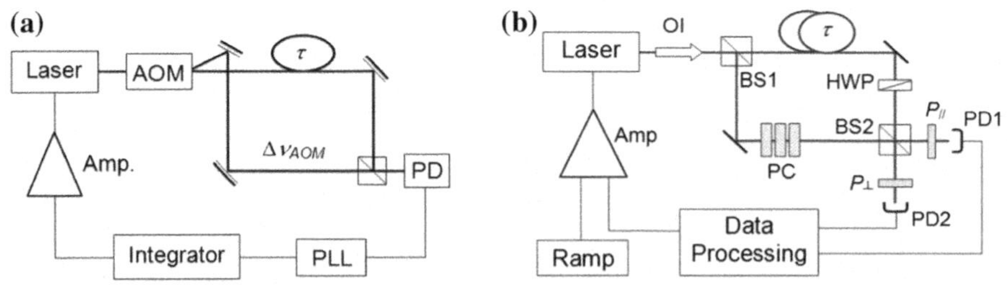

Fig. 7.5 Frequency-swept laser both with sweeping rate control and linewidth reduction: **a** Feedback with self-heterodyne; **b** Based on polarization split interferometer

Reference [31] proposed another method to pick up the signal related to $\Delta\phi$, which is based on the polarization split interferometry, as shown in Fig. 7.5b. A polarization controller (PC) is inserted in one arm of the MZI, and a half-wave plate (HWP) is inserted in the other arm to rotate its polarization direction. The two outputs from BS2 pass through perpendicular polarizers ($P_{//}$,$P_{\perp}$), and detected by photodetectors respectively:

$$\begin{aligned} I_1 &\propto \cos[2\pi r\tau t + \phi_0 + \Delta\phi(t,\tau)]\cos^2\theta \\ I_2 &\propto \cos[2\pi r\tau t + \phi_0 + \Delta\phi(t,\tau) + \varphi]\sin^2\theta, \end{aligned} \tag{7.11}$$

where φ is the phase shift induced by PC, and θ is the rotation angle of HWP. In the data processing, the detected signals are mixed with the digital signal of sin $(2\pi rt + \phi_0)$ and $\cos(2\pi rt + \phi_0)$, which are averaged over the linewidth and stored beforehand in the field programmable gate array (FPGA) memory. In case of $\varphi = \pi/2$ and $\theta = \pi/4$, a signal proportional to laser's frequency noise is obtained:

$$V(t) \propto I_1(t)\sin(2\pi r\tau t + \phi_0) - I_2(t)\cos(2\pi r\tau t + \phi_0) \propto \sin[\Delta\phi(t)] \tag{7.12}$$

which can be used to feedback control the LD's driving current and to reduce its linewidth in real time.

It is necessary to characterize the coherence of frequency-swept laser in real time. The methods introduced in Chap. 3 are basically for cw-working lasers, i.e., the spectrum is obtained by integration in a certain duration, which may cover several sweeping periods for high speed swept lasers. The measured linewidth should be its high limit, but it cannot tell how the linewidth changes during sweeping. The instantaneous linewidth is not only one of the important specifications in applications but also of interest for understanding physical effect of the frequency-swept lasers. For example, in the swept source optical coherence tomography (SS-OCT) and the optical frequency domain reflectometry (OFDR), the instantaneous linewidth determines the maximum detected depth or distance, whereas the sweeping range determines the spatial resolution.

Reference [32] proposed a method for direct measurement of the instantaneous linewidth of swept source in OCT, which is based on a snapshot of instantaneous spectrum chopped by an EOM switch in pulse width of 1.6 ns, and scanning synchronized with the swept waveform. The broadening of spectrum by such a short gate exists inevitable. The wavelength range of sweeping in SS-OCT is as large as near 100 nm in about 20 μs period, the instantaneous linewidth is in the order of 0.1 nm, the precision of such a measurement is satisfactory.

The ordinary delayed self-homodyne method can still be used for the instantaneous linewidth measurement, so long as the measured spectrum can be analyzed quickly enough. Reference [33] uses an MZI composed of a 3×3 fiber coupler with delay of τ to get time-resolved measurement of instantaneous linewidth in a single shot of frequency-swept laser, as shown in Fig. 7.6. By the 120° phase shift 3×3 coupler, the adjustment of quadrature working point is avoided in

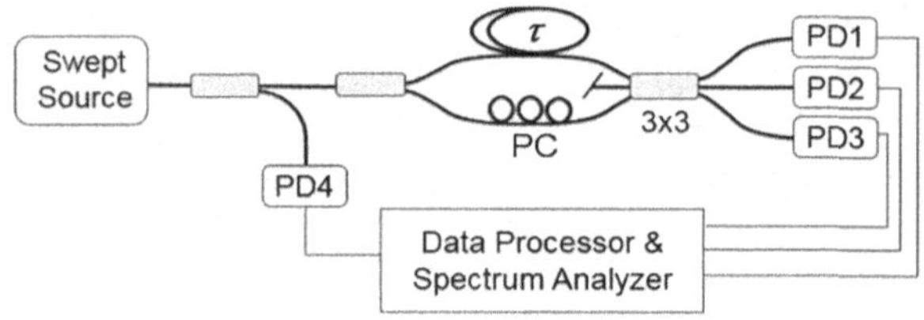

Fig. 7.6 Measurement setup of instantaneous linewidth of frequency-swept laser

demodulating the deferential phase of MZI, as discussed in Sect. 3.3; thus the instantaneous linewidth can be interrogated quickly and continuously. In addition, high-speed photodetectors up to 12 GHz are used in the setup, ensuring the measurement reliable.

7.3 Frequency Sweeping with Intra-cavity Tuning Devices

Quite many approaches are available for frequency sweeping of external cavity diode lasers, including methods of rotating and/or translating the external cavity mirror or planar grating, and methods by using phase modulators inserted inside the cavity. Besides, a beam deflector can serve the similar function to the rotation of grating. Phase modulators based on electro-optic (EO) effect are widely used for tuning and frequency stabilization, as stated in Chaps. 5 and 6. The properties of EO materials and devices are introduced in details here.

7.3.1 Electro-optic Materials and Devices

Basically three kinds of materials are used to make EO modulators, i.e., the electro-optic crystals, the liquid crystals, and the EO ceramics.

(1) **EO Crystals**

The electro-optic effect is the change in refractive index induced by a DC voltage or a low frequency modulated voltage applied on the medium. The effect includes two basic kinds: the linear EO effect and the quadratic EO effect. In the former the index change is proportional to the applied voltage, known as Pockels effect; in the latter it is proportional to the square of voltage in the latter, known as Kerr effect. From relation of $n^2 = \varepsilon/\varepsilon_0$, the effect can be expressed by the change of dielectric constant:

$$\Delta\varepsilon = \alpha E_{\rm ex} + \beta E_{\rm ex}^2 + \cdots, \tag{7.13}$$

where two terms correspond to Pockels effect and Kerr effect, respectively. The above description is valid for the isotropic media. The materials used for EO device are mostly anisotropic crystals due to its high EO coefficients. The dielectric constant of an anisotropic crystal is not a scalar quantity but a tensor, which relates the electric displacement with the electric field vector as [34]:

$$\begin{pmatrix} D_x \\ D_y \\ D_z \end{pmatrix} = \begin{pmatrix} \varepsilon_{xx} & \varepsilon_{xy} & \varepsilon_{xz} \\ \varepsilon_{yx} & \varepsilon_{yy} & \varepsilon_{yz} \\ \varepsilon_{zx} & \varepsilon_{zy} & \varepsilon_{zz} \end{pmatrix} \begin{pmatrix} E_x \\ E_y \\ E_z \end{pmatrix}. \tag{7.14}$$

The nine elements of ε_{ij} must satisfy the symmetry of $\varepsilon_{ij} = \varepsilon_{ji}$, based on the physical principle. By converting the coordinator axes to coincide with the principal axes of crystal lattice the tensor can be reduced to a diagonal tensor, i.e., $\varepsilon_{ij} = \varepsilon_{ii}\delta_{ij}$, where δ_{ij} is Kronecker delta. It is indicated by (7.14) that the electric displacement is no longer parallel to the electric vector, if ε_{ii} are not equal with each other for the three dimensions x, y, and z. Since the electromagnetic energy flow, i.e., Poynting vector, is perpendicular to the electric vector, whereas the wavevector is perpendicular to the electric displacement, the refraction direction of optical beam at the crystal surface will depend on its polarization and the orientation to the crystal, resulting in different refractive indexes for the ordinary wave and for the extraordinary wave, termed the birefringence (double refraction). The crystal with $\varepsilon_x = \varepsilon_y \neq \varepsilon_z$ is termed the uniaxial crystal, and the crystal with $\varepsilon_x \neq \varepsilon_y \neq \varepsilon_z$ is termed the biaxial crystal. If $\varepsilon_x = \varepsilon_y = \varepsilon_z$, the material is isotropic.

The EO effect of an anisotropic crystal is dependent on its lattice structure. Due to the symmetric reason of structure, Pockels effect may disappear for some crystals. In addition, the EO-induced index change is dependent on direction of incident beam, and on its polarization. Furthermore, the external field will induce certain anisotropy in isotropic materials, called the artificial birefringence, and change the amount of anisotropy of the anisotropic material. The birefringence is usually described by the index ellipse with the inverse tensor of index dielectric tensor, termed impermeability, $\boldsymbol{B} = \varepsilon^{-1}$. Its increment induced by the external field is expressed as:

$$\Delta B_{ij} = \Delta[\boldsymbol{\varepsilon}^{-1}]_{ij} = \sum_{k=1}^{3} \gamma_{ijk} E_k + \sum_{k,l=1}^{3} R_{ijkl} E_k E_l \quad (i, j, k, l = x, y, z) \tag{7.15}$$

The coefficients above will be reduced by the symmetry of crystal structure. Because of $\Delta B_{ij} = \Delta B_{ji}$, the 27 coefficients of Pockels effect above can be reduced to 18; and some of them must be equal to zero for certain symmetries. The widely used linear EO crystals include the lithium niobate ($LiNbO_3$), lithium tantalate ($LiTaO_3$), KDP (KH_2PO_4), etc. $LiNbO_3$ is an important crystals used mostly in laser technology area. It has symmetry of trigonal 3m with EO impermeability matrix as

$$\begin{pmatrix} \Delta B_1 \\ \Delta B_2 \\ \Delta B_3 \\ \Delta B_4 \\ \Delta B_5 \\ \Delta B_6 \end{pmatrix} = \begin{pmatrix} 0 & -\gamma_{22} & \gamma_{13} \\ 0 & \gamma_{22} & \gamma_{13} \\ 0 & 0 & \gamma_{33} \\ 0 & \gamma_{51} & 0 \\ \gamma_{51} & 0 & 0 \\ -\gamma_{22} & 0 & 0 \end{pmatrix} \begin{pmatrix} E_1 \\ E_2 \\ E_3 \end{pmatrix}, \tag{7.16}$$

where $\Delta B_i = \Delta B_{ii}$ $(i = 1,\ 2,\ 3)$, $\Delta B_4 = \Delta B_{23}$, $\Delta B_5 = \Delta B_{13}$, and $\Delta B_6 = \Delta B_{12}$. The EO coefficients of $LiNbO_3$ of $\gamma_{13} = 8.6 \times 10^{-12}$ m/V and $\gamma_{33} = 30.8 \times 10^{-12}$ m/V have been measured experimentally; the data for other crystals can also be found in literature [34–36].

The crystals of EO devices are mostly cut to be a cuboid with one or more edges parallel to its principal axis, and the driving voltage is applied on two opposite faces, as shown in Fig. 7.7. The modulating voltage is applied in z-direction of $LiNbO_3$ crystal, as an example, in the configuration of Fig. 7.7a; the indexes for three directions of optical field take different values:

$$n_x = n_y = n_o - n_o^3\gamma_{13}V/(2d), \tag{7.17a}$$

$$n_x = n_e - n_e^3\gamma_{33}V/(2d), \tag{7.17b}$$

where n_o and n_e are the indexes of ordinary wave and extraordinary wave, respectively. The beam is incident along y-direction and perpendicular to the applied field. Formula (7.17a) is for the beam polarized in x-direction, marked as s-component; formula (7.17b) is for the p-component. The effect occurring in such a configuration is called the transverse EO effect.

If the beam direction is parallel to the applied field, as shown in Fig. 7.7b, the effect is called the longitudinal EO effect. For such a device, transparent electrodes have to be used. N.B. the directions marked in Fig. 7.7 are only for crystals with 3m symmetry; if other kind crystals are used the cut direction has to be designed in different way. The phase variation with the applied voltage for Fig. 7.7a structure is obtained as:

$$\Delta\phi = \Delta nkL = \pi\gamma n_0^3 VL/(\lambda d). \tag{7.18}$$

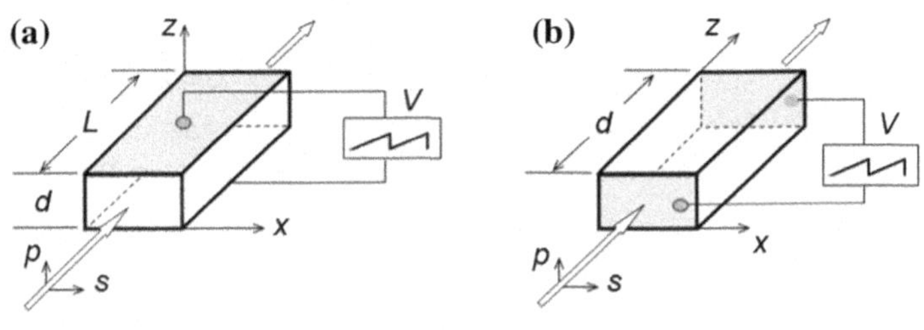

Fig. 7.7 **a** Transverse EO device; **b** Longitudinal EO device

The performance of phase modulator is usually characterized by the half-wave voltage, which is the voltage for the phase shift of π:

$$V_\pi = \lambda d/(\gamma n_0^3 L). \tag{7.19}$$

The lower the half-wave voltage is, the lower is the working voltage, and the lower is the device dissipation power. Therefore a smaller d and a longer L are preferable. In longitudinal EO modulators, the optical path is just the distance between two electrodes, so that its half-wave voltage is not dependent on the crystal length; it is usually used in binary optics systems.

It is seen also if the polarization of incident beam is not parallel to x, or to z in Fig. 7.7a, it will be rotating along with propagating in the crystal. The output polarization direction is then changed, and the rotated angle depends on the applied voltage, and on the crystal length. Such a device is used as a polarization controller.

The applied voltage can either be a DC voltage, or an AC voltage with arbitrary waveforms. If the modulation frequency is low enough, the above formulas are applicable. When the response of modulator is concerned, the propagation delay of the applied AC electric field in the crystal has to be taken into consideration, especially in high frequency modulations.

(2) **Liquid Crystals**

Liquid crystal (LC) is a special liquid, in which anisotropic molecules dissolve uniformly; these molecules move freely and line up some way in the liquid. Such an arrangement can be varied by the externally applied electric field, resulting in changes of its refractive index. It has been used widely in variety of areas, especially in household electric appliances as displayers and indicators, such as screens of TV, computer, and mobile phone. A simplified structure of LC displayer (LCD) is shown in Fig. 7.8a, where "*P*" is a thin film polarizer, "*A*" is a thin film analyzer, and LED is a light-emitting diode illuminator. Liquid crystal is a special EO crystal used in scientific apparatus and in optoelectronics field. If the two electrode plates are parallel with each other, the structure will become a phase modulator. The structure of Fig. 7.8a can also be utilized as an optical switch, or as a variable optical attenuator. If the electrodes are set up in a certain angle, as shown

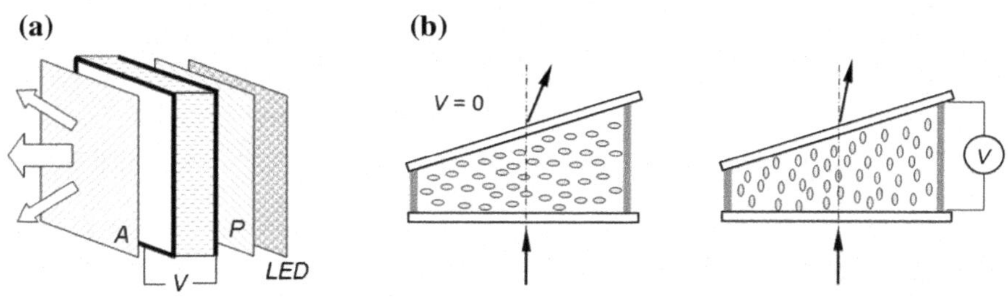

Fig. 7.8 Schematic diagram of **a** LC displayer and **b** LC deflector

in Fig. 7.8b, the spatial variation of LC index can be changed by the applied voltage, resulting in change of refraction angle, used as a controllable light beam deflector.

The LC device has important advantages. It needs very low working voltage; its materials and fabrication processing are cheap. For more functions and improved performances, various LC materials have been developed with different features, such as nematic, chiral nematics, cholesteric, and smectics. Interested readers may find detailed introductions in literature [35].

(3) **EO Ceramics**

As stated above, the crystals may have the second-order EO effect, i.e., Kerr Effect, but it is usually much weaker than the linear EO effect. It was found that ferroelectric materials have higher electric susceptibility. Among them, the lead lanthanum zirconate titanate (PLZT) ceramic has strong EO effect, and is a transparent material, suitable to be developed as optical devices [37–39]. Since the ferroelectric domains exist in ceramic randomly, its linear EO effect is smoothened, and only the second-order EO effect is presented. The index response to the external electric field is

$$\Delta n = n_0^3 R_{\mathrm{ij}} E_{\mathrm{ex}}^2 / 2 \tag{7.20}$$

The optimized composition of PLZT is $(La_{0.09}Pb_{0.91})(Zr_{0.65}Ti_{0.35})O_3$, marked as (9/65/35). Its index measured at 633 nm is n_0 = 2.5 [37] and nearly the same value in near-infrared band. The ceramic is isotropic macroscopically. Its Kerr coefficient is a tensor, expressed as

$$\hat{R} = \begin{pmatrix} R_{33} & R_{13} & R_{13} & 0 & 0 & 0 \\ R_{13} & R_{33} & R_{13} & 0 & 0 & 0 \\ R_{13} & R_{13} & R_{33} & 0 & 0 & 0 \\ 0 & 0 & 0 & R_{44} & 0 & 0 \\ 0 & 0 & 0 & 0 & R_{44} & 0 \\ 0 & 0 & 0 & 0 & 0 & R_{44} \end{pmatrix}, \tag{7.21}$$

where $R_{44} = (R_{33} - R_{13})/2$. The index increment of Kerr effect is thus dependent on the polarization of incident wave. If the polarization is parallel to the applied field $\boldsymbol{E}_{\mathrm{ex}}$, the EO coefficient is R_{33} in (7.20); it is R_{13} for the case of polarization perpendicular to $\boldsymbol{E}_{\mathrm{ex}}$. Different values of coefficients R_{ij} of PLZT were reported in journals, due to diversity of material compositions and processing; as examples, $R_{13} = -0.37 \times 10^{-16}\ \mathrm{m}^2/\mathrm{V}^2$ and $R_{33} = 2.1 \times 10^{-16}\ \mathrm{m}^2/\mathrm{V}^2$ was measured in Ref. [39]. It is worthy of noticing that the coefficients are not only different in values, but also in signs.

The ceramic EO materials are beneficial for fabricating larger size component, and have the advantage of low cost. New kinds of EO ceramics other than PLZT are being developed, such as the lead magnesium niobate titanate (PMNT) [40]. Besides the devices with block ceramics, the ceramic waveguide devices are also developed.

7.3.2 Frequency Sweeping with Intra-cavity Modulator

The frequency of ECDL can be swept by modulating the EO phase modulator inserted inside the external cavity. Figure 7.9 shows two lasers with intra-cavity phase modulator. Structure (a) is a Littrow ECDL [41] with an EO crystal (EOC) inserted in the cavity as a phase modulator. Structure (b) is a Littman ECDL with a LC phase modulator inserted [42]. The index changes with the applied voltage, and the laser's frequency is swept.

Reference [43] reported an external cavity laser with a volume Bragg grating (VBG) as the external mirror, working at 810 nm band, and with a PLZT phase modulator inserted, as shown in Fig. 7.10. The frequency sweeping range of 2.5 GHz was obtained with mode hop free under voltage of 540–860 V applied on the PLZT phase modulator. The continuous frequency sweeping range of these schemes is limited within one mode spacing; beyond the range, mode hopping will occur. Continuous sweeping over several modes was realized by LD current assisted with precise control [29].

Fabry-Perot filter is often used in ECDL as a tuning component, as introduced in Sect. 5.4.1. The laser frequency can be swept by tuning the F-P, which is available commercially now. Figure 7.11 shows a ring cavity ECDL configured with a fiber coupled SOA and fiber coupled F-P, which can be tuned by modulating the voltage

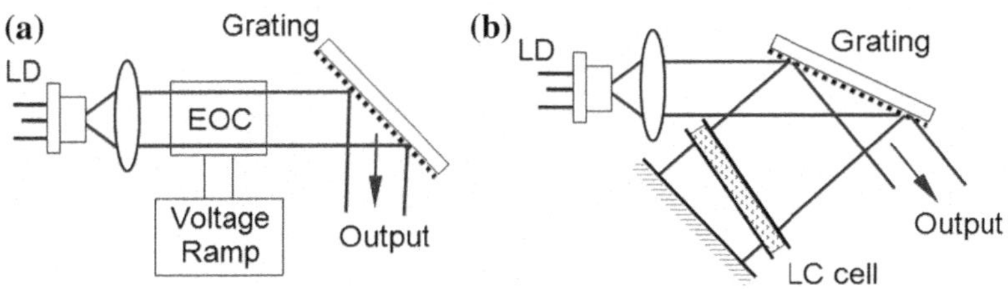

Fig. 7.9 **a** Littrow ECDL with an EO phase modulator; **b** Littman ECDL with a LC phase modulator

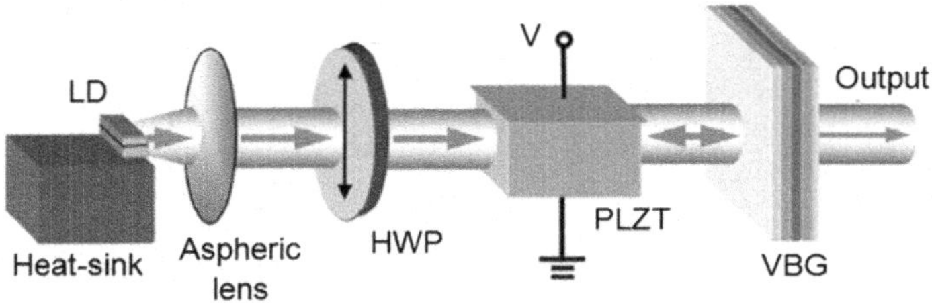

Fig. 7.10 VBG ECDL with PLZT phase modulator

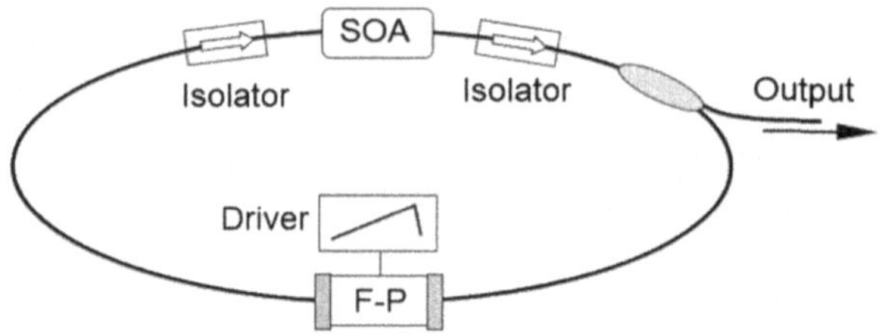

Fig. 7.11 An SOA ring laser swept by a tunable F-P filter

applied to piezoelectric transducer (PZT). By such a configuration, the laser was used in OCT satisfactorily with sweeping rate of 20 kHz and tuning range of 120 nm [44]. The length of ring cavity is usually quite long, thus the longitudinal mode spacing is so small that the F-P resonance peak may cover several modes, not exactly a single-frequency laser. However, it is still useful so long the envelop linewidth is narrow enough to meet the requirement of application, such as OCT. The phase shift fiber Bragg grating can play the same role in the ring cavity laser as F-P filter, since it has a narrow transmission peak and can be tuned precisely, as introduced in Sect. 5.3.2.

7.3.3 *Frequency Sweeping by Intra-cavity Beam Deflection*

As stated in Chap. 5, the frequency of planar grating ECDL can be tuned by rotating the grating. The grating may be moved by two PZT both in rotation and in straight line direction, as shown in Fig. 5.7 of Chap. 5; continuous sweeping without mode hopping may be realized under some conditions as discussed in Ref. [45]. The scheme proposed by Ref. [46] uses a similar structure with two PZT, where the linearity and continuity of frequency sweeping are actively controlled by polarization spectroscopy, and a mode-hop free tuning range up to 130 GHz was achieved experimentally.

By using a beam deflector inserted in the cavity, the frequency of ECDL can also be tuned with the grating fixed. A typical scheme is shown in Fig. 7.12 [47], where an EO crystal prism is inserted in the Littrow cavity. The optical length of its cavity is $L = AB + CD + n * BC$; the index of prism is varied with the applied voltage,

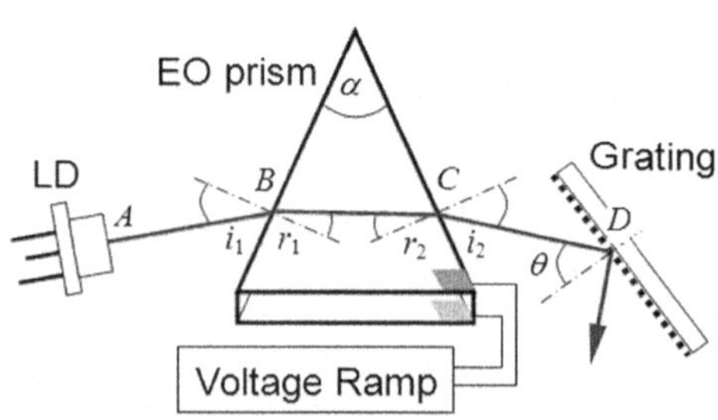

Fig. 7.12 ECDL with EO prism beam deflector

$\Delta n = \gamma n_0^3 V/(2d)$, where d is the thickness between two electrodes. The frequency change is then introduced: $\Delta \nu = \nu\, l\Delta n/L$, where $l = BC$. The index change causes also the beam deflection change $\Delta i_2 = \Delta\theta$, which is determined by the prism angle α, the incident angle i_1, and the index increment. According to Snell law, the angles in the figure satisfy relation of $r_1 + r_2 = \alpha$; it is deduced that $di_2/dn = \sin\alpha/(\cos r_1 \cos i_2)$. The angular dispersion of grating is $d\lambda_g/d\theta = \lambda_g/\tan\theta$, as discussed in Sect. 5.2.1. The continuous tuning without mode hopping requires the angular dispersion coincident with the tuning rate on cavity length, that is

$$\frac{di_2}{dn} = \frac{\sin\alpha}{\cos r_1 \cos i_2} = \frac{l \tan\theta}{L}, \tag{7.22}$$

which gives the condition of geometrical design of components. Reference [48] reported the experimental results of such a Littrow ECDL with sweeping linearity controlled by feedback from a M–Z interferometer, demonstrating sweeping rate of 3 GHz/5 μs, and frequency deviation less than 100 kHz.

Reference [49] presented a beam deflector made of an EO wedge, as shown in Fig. 7.13. The electric field inside the wedge is inversely proportional to the local thickness, resulting in a spatial distribution of OE index with linear change in y-direction. According to the principle of ray optics, the incident beam will propagate in a curved way toward the higher index region. It is deduced that the resultant optical path change is proportional to the applied voltage $\Delta L(y) \sim \Delta nl = -n^3\gamma_{33}Vl/2h(y)$, where l is the crystal length, $h(y)$ is its thickness. The beam output from the wedge is then deflected, and thus the incident angle to the grating is changed. According to the lasing condition, the optical path change should be proportional to the effective cavity length for continuous tuning without mode hopping, i.e., $\Delta L(y) \propto L(y) = L(0) + y \tan\theta$. It is then deduced that the wedge thickness should meet condition of

$$h(y) = h(0)[1 - y \tan\theta/L(0)]. \tag{7.23}$$

In Ref. [49], a $LiTaO_3$ wedge with $\gamma_{33} = 30.4 \times 10^{-12}$ m/V and sizes of 40 mm × 15 mm × (1–1.51)mm was used as a deflector in a 793 nm ECDL; and continuous sweeping range of 50 GHz, sweeping speed of 1.5 GHz/μs, and sweeping rate of 13 MHz/V were demonstrated.

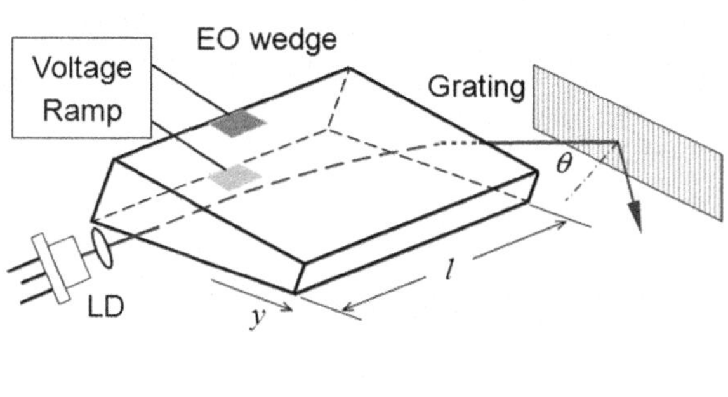

Fig. 7.13 ECDL with EO wedge beam deflector

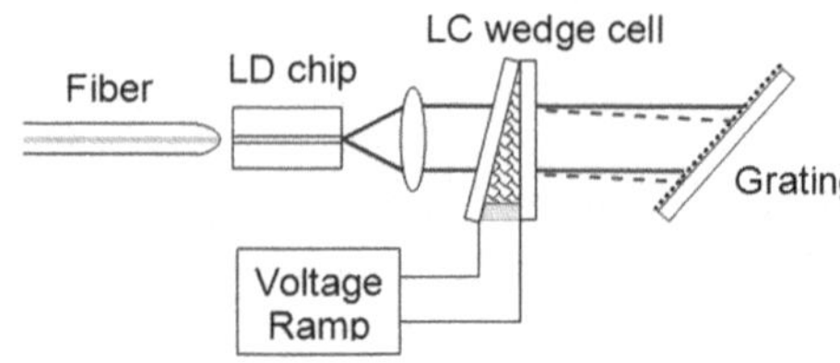

Fig. 7.14 Frequency sweeping with an LC deflector

A liquid crystal wedge, as shown in Fig. 7.8b, is a good beam deflector with low driving voltage. Figure 7.14 depicts an ECDL with an LC deflector inserted in the cavity [50]. It was reported that the tuning range of 12 nm (with mode hops) at 1550 nm band was obtained, and the side mode suppression ratio of single longitudinal mode of 30 dB was measured. The collimated beam of LD output can be deflected by translating the collimation lens transversely. Reference [51] utilized this plain geometric optics to build a beam deflector in a Littrow ECDL, as shown in Fig. 7.15. Continuous tuning range of 1 nm at 780 nm band was obtained under lens translation of 1 μm.

If the electrodes on cuboid EO crystal or EO ceramic are with a right triangle shape, refraction at the interface of two regions will occur, where the index difference Δn between them is induced by applied voltage. Reference [43, 52] presented a Littrow ECDL with such a beam deflector, as shown in Fig. 7.16. From Snell law, $(n + \Delta n)\sin\alpha = n\sin(\alpha + \Delta\alpha)$; usually $\Delta n << n$, then $\Delta\alpha \approx \Delta n\ \tan\alpha/n$. The refraction at the right surface changes the incident angle to the grating from θ to $\theta + \alpha_1$, where $\alpha_1 \sim \sin\ \alpha_1 = n\ \sin(\Delta n\ \tan\alpha/n) \sim \Delta n\ \tan\alpha$, meaning that the deflection angle is proportional to the applied voltage. From grating's angular dispersion, the refraction peak wavelength changes as $\Delta\lambda_g = \alpha_1\ \lambda_g/\tan\theta$.

The cavity length changes accordingly. The increment of optical path inside the EO material is $\Delta l = (n+\Delta n)l/2 + nl/(2\cos\Delta\alpha) - nl \approx l\Delta n/2$. The increment of distance between the EO material and the grating is $\Delta l_1 \approx \Delta n(l_1 + l/2n)\tan\alpha\ \tan\theta$, as depicted in the figure. The total cavity length change is

$$\Delta L = \Delta l + \Delta l_1 \approx (\Delta n)[l/2 + (l_1 + l/2n)\tan\alpha\ \tan\theta] \tag{7.24}$$

It is required for continuous tuning without mode hopping that the wavelength change of longitudinal mode equals the change of grating peak wavelength, i.e., $\Delta L\lambda_m/L_0 = \Delta\lambda_g$, where L_0 is the cavity length at zero voltage. The condition for continuous tuning is thus deduced as:

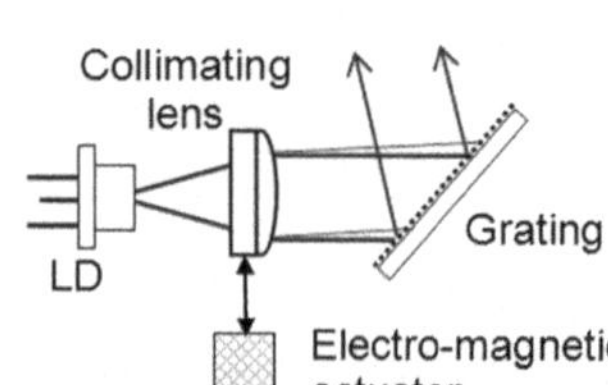

Fig. 7.15 Tuning by collimation lens translation

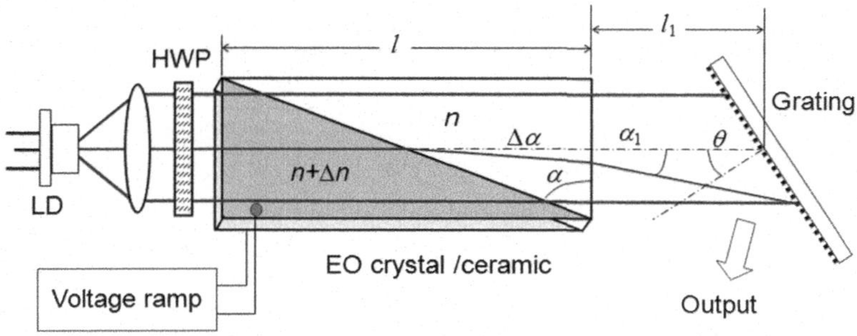

Fig. 7.16 A frequency sweeping laser based on deflection of refractive angle inside EO crystal

$$\frac{\tan \alpha}{\tan \theta} = \frac{l/2 + (l_1 + l/2n) \tan \alpha \tan \theta}{L_0}. \tag{7.25}$$

Experimental studies of such a structure were implemented in Refs. [43, 52] with PLZT EO ceramic deflector. The measured spectra of frequency sweeping laser are shown in Fig. 7.17a. The curve of frequency variation with the applied voltage in Fig. 7.17b shows the feature of quadratic EO effect. It is seen that the tuning range of 500 GHz ($\sim$1 nm) at 780 nm band is obtained under 0–800 V voltage. The high working voltage is hopefully lowed by two ways: one is to develop new EO ceramic material, such as PMNT [40]; the other is to develop optical waveguide devices, as reported by Ref. [53] for optical switches made of PLZT waveguide.

The acousto-optic modulator (AOM) is a dynamic grating; its diffraction angle can be changed by the RF frequency of its driver. However, the AOM cannot be used for the frequency sweeping by directly inserting in the ECDL, because the frequency of incident beam will also be tuned at same time of deflection, as described in the next section. To avoid such an obstacle, Refs. [54, 55] proposed a scheme with two AOMs driven for opposite frequency shifts, as shown in Fig. 7.18,

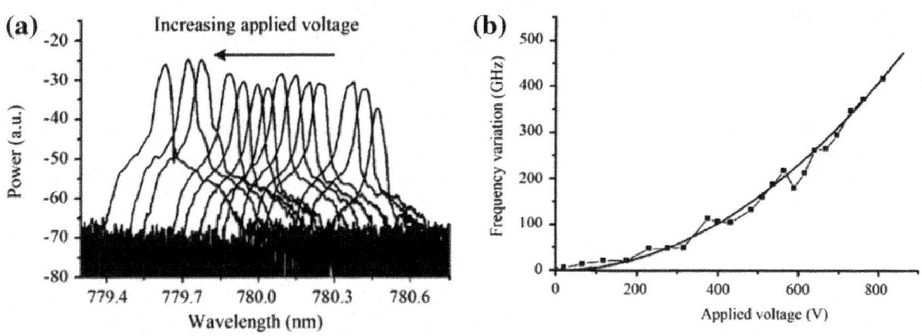

Fig. 7.17 **a** Spectra of frequency sweeping laser with PLZT deflector. **b** Curve of frequency vs. applied voltage. Reprinted from Ref. [52]

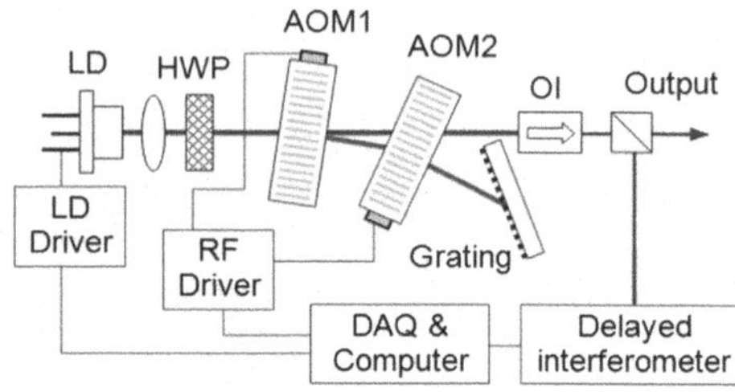

Fig. 7.18 A Littrow ECDL with paired AOM deflector

where their frequency shifts are canceled out with each other in the round trip way, while the deflection is remained to tune the laser frequency. The zeroth-order diffraction of AOM is used as the laser output. The sweeping can be linearized by feedback controlling and digital synthesis in the computer.

In addition, the waveguide grating made of EO crystal can be used to build a DBR laser, which can be tuned by applied voltage. Experimental results of sweeping rate of 55.5 MHz/V, and linewidth of 18 kHz were demonstrated by such an ECDL [56].

7.4 Frequency Sweeping with Extra-Cavity Modulators

The schemes of frequency sweeping introduced above have compact structures. It is welcome for some applications due to the easiness of operation. However, their performances are not satisfactory for some important applications; and it lacks flexibilities. The external modulation can provide possibilities of flexible combination of lasers and modulators optimized separately. High quality EO and AO modulators have been available commercially and widely used in laser technologies.

7.4.1 Frequency Shifting and Sweeping with Acousto-Optic Modulators

The acousto-optic effect is attributed to the photoelastic effect; that is, when a material is deformed, its index will be changed. The strain state of elastic solid is described by a tensor, which has six components, three normal strain e_{ii}, and three shearing strains e_{ij}, where $i, j = x, y, z$. The photoelastic effect is characterized by a 6×6 matrix. Determined by the structure of material, not all the elements have independent significances, some of them have the same value, some equal zero. The increment of impermeability is proportional to the strain tensor; for isotropic materials it is expressed as

$$\Delta\left(\frac{1}{\varepsilon}\right) = \frac{-1}{\varepsilon^2}\begin{pmatrix} \Delta\varepsilon_x \\ \Delta\varepsilon_y \\ \Delta\varepsilon_z \\ \Delta\varepsilon_{yz} \\ \Delta\varepsilon_{zx} \\ \Delta\varepsilon_{xy} \end{pmatrix} = \begin{pmatrix} p_{11} & p_{12} & p_{12} & 0 & 0 & 0 \\ p_{12} & p_{11} & p_{12} & 0 & 0 & 0 \\ p_{12} & p_{12} & p_{11} & 0 & 0 & 0 \\ 0 & 0 & 0 & p_{44} & 0 & 0 \\ 0 & 0 & 0 & 0 & p_{44} & 0 \\ 0 & 0 & 0 & 0 & 0 & p_{44} \end{pmatrix}\begin{pmatrix} e_{xx} \\ e_{yy} \\ e_{zz} \\ e_{yz} \\ e_{zx} \\ e_{xy} \end{pmatrix}. \quad (7.26)$$

A planar acoustic wave in an elastic medium is actually a propagating periodic strain wave, and thus a propagating index wave, which plays a role of dynamic grating. The wavelength of acoustic wave is determined by the frequency of acoustic wave f_A and the acoustic velocity V_A, $\lambda_A = V_A/f_A$. The acoustic velocity is determined by Young's modulus and density of the material. The acoustic wave in AOM made of AO crystal is generated with an RF transducer attached on one of the facets of cuboic crystal. Based on the piezoelectric effect, the elastic deformation can be stimulated by RF electric field. The incident optical beam to the dynamic grating is thus diffracted. Two typical acoustic waves are considered and used in AO devices. One is a standing acoustic wave; if the facet opposite to the RF electrode is a reflector of acoustic wave, the wave generated by RF transducer will propagate in roundtrip way between the two facets to form a standing wave. The other is a traveling wave; if the opposite facet is connected with some acoustic absorbing material so that the reflection there is fully eliminated.

The interaction between the optical wave and the acoustic wave can be regarded as collision of photons and phonons. Two cases will occur then. One is the elastic collision, where the energy of photons does not change. The other is the inelastic collision, when the photon absorbs or releases (emits) a phonon. The energy conservation law and the momentum conservation law must be satisfied in the process. In the quantum mechanics, the momentum is proportional to the wavevector, i.e. $\hbar k$. Therefore, the two conservations are reduced to

$$h\nu_d = h\nu_i \pm hf_A, \quad (7.27)$$

$$\boldsymbol{k}_d = \boldsymbol{k}_i \pm \boldsymbol{k}_A, \quad (7.28)$$

where the subscripts *d*, *i*, and *A* stand for valuables of diffracted photon, incident photon, and phonon. Figure 7.19 depicts the relationship of wavevectors and frequencies, where the traveling acoustic wave is considered. The standing acoustic wave composed of two waves with opposite directions builts a fixed grating, which will change the optical beam direction, but not change the photon's energy.

The frequency of acoustic wave is much lower than that of the optical waves, so that the wavevector length of diffracted wave is nearly equal to that of incident optical wave: $k_d = k_i = 2\pi n/\lambda$. From the phonon's wavevector $k_A = 2\pi f_A/V_A$, the diffraction angle is deduced as

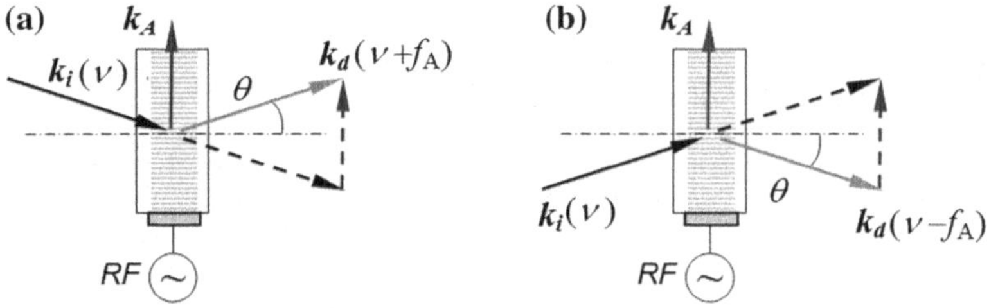

Fig. 7.19 Schematic diagram of AO frequency shifter: **a** Blueshift; **b** Redshift

$$\sin\theta = k_A/(2k_i) = \lambda f_A/(2nV_A). \tag{7.29}$$

It is noticed that the direction of output beam will deflect more than that given by (7.29) due to diffraction at the crystal surface. The acousto-optic effect can be utilized to make either an optical switch, if the transmitted beam is used, or a frequency shifter, if the diffracted beam is used. For practical devices, collimation lenses or fiber coupling have to be implemented for the input and output ports. Instead of blocky crystals, the AO device can also be made of waveguide structure, which is based on the surface acoustic wave effect. The performances of AO devices are dependent on the material parameter and the device design. Detailed analyses on related scientific and technical issues can be read in Refs. [34–36].

The scheme of frequency-swept laser based on AO modulator is plain, so long as a frequency sweeping RF driver is used. It is seen from (7.29) that the diffraction angle will vary with the frequency shift f_A, so that the output coupling efficient will be changed. The resultant residue amplitude modulation (RAM) has to be overcome. Reference [57] designed and demonstrated a scheme with a roundtrip path traversing the AOM to make the modulated beam reflect back antiparallel with the incident beam, as shown in Fig. 7.20. The polarization of reflected beam is rotated 90° by the quarter wavelength plate (QWP), so that the beam can be output from the polarization beam splitter (PBS). The RAM due to beam deflection induced by AOM is thus avoided. In the setup, the polarization maintained fiber (PMF) is used as an output port, and its power is adjusted by the half wavelength plate (HWP).

Fig. 7.20 Frequncy sweeping scheme with external AOM frequency shifter

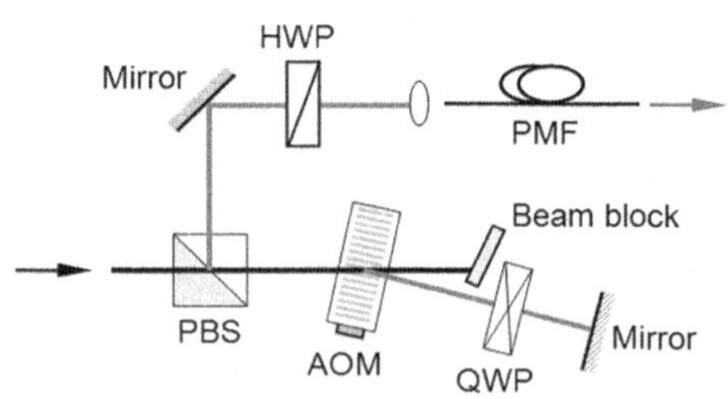

7.4.2 Frequency Sweeping with Electro-optic Modulators

Based on the property of EO effect introduced in Sect. 7.2, various EO modulators are developed. For the applications of frequency-swept LD, especially those with fiber pigtail input and output, the optical waveguide OE modulator is used widely. The small size of waveguide brings about obvious advantages: greatly decreased working voltage and enhanced efficiency of fiber coupling. The basic principle, performances, and fabrication processing are expounded in many monographs and journal papers [34–36, 58]. This book gives only a brief introduction, mainly to the issues concerned in applications of frequency sweeping.

The material mostly used for EO waveguide devices is the lithium niobate crystal ($LiNbO_3$). The waveguide region has a index higher than $LiNbO_3$ substrate, which can be fabricated by using techniques of diffusion or ion-exchanging, e.g., Ti-doping in $LiNbO_3$. The metal electrodes are coated on the top surface; the transverse field will generate under the applied voltage, and the optical beam propagating in the waveguide will experience the transverse EO effect. Typical structures of EO phase modulator are shown in Fig. 7.21, where structure (a) uses *X*-cut crystal, and (b) uses *Z*-cut one. The waveguide of structure (b) has to be placed under one of the electrodes to experience *z*-direction voltage. A buffer layer underneath the electrodes is needed to reduce propagation loss. The modulating voltage is now in *z*-direction, the EO index change is generated as $\Delta n_x = \Delta n_x \propto \gamma_{13}$ E_z in both of the structures. The dependence of modulation efficiency on input beam polarization can thus be well removed.

When an AC or RF voltage is applied, the optical phase of propagating beam will be modulated by $\Delta\Phi = \beta \sin \Omega t$. The modulation amplitude is proportional to the voltage and to the waveguide length. By using Fourier expansions of exponential function,

$$\begin{aligned} e^{j\beta \sin \Omega t} &= \sum_{n=-\infty}^{\infty} J_n(\beta) e^{jn\Omega t} \\ e^{j\beta \cos \Omega t} &= \sum_{n=-\infty}^{\infty} j^n J_n(\beta) e^{jn\Omega t}. \end{aligned} \tag{7.30}$$

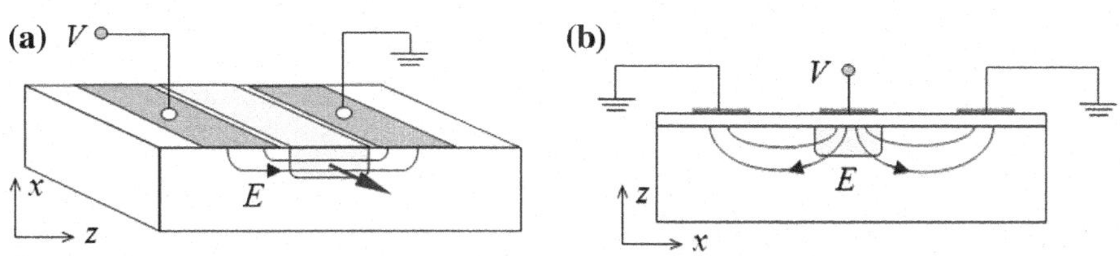

Fig. 7.21 EO waveguide phase modulators with *X*-cut (**a**) and Z-cut (**b**) crystals

The modulated optical field contains a series of harmonic components, expressed as:

$$E(t) = E_0 e^{j(\omega_0 t + \beta \sin \Omega t)} = E_0 e^{j\omega_0 t} \sum_{n=-\infty}^{\infty} J_n(\beta) e^{jn\Omega t}, \tag{7.31}$$

where J_n is the nth order Bessel function. It is seen that a series side bands generates, distributed symmetrically in two sides of input beam frequency with $\omega = \omega_0 \pm n\Omega$. That is the same effect as discussed in the modulation spectroscopy and PDH. The amplitude of side bands is determined by modulation amplitude β. Figure 7.22 shows the four lowest order Bessel functions. It is seen that when the argument $x = 1.84$, the first-order Bessel function reaches its maximum of 0.58, and the zeroth-order and the second-order functions have the same value of 0.3. At $x = 2.405$, the base band reaches zero and the first order and the second order take similar values near 0.5. This is the working point for clear frequency shifts with the base band suppressed.

The optical phase modulator of Fig. 7.21 can be directly used as a frequency shifter, used outside of laser cavity and driven by a serrated electric voltage. Such a scheme is termed the serrodyne modulation in literature [59–62]. In the ascending and descending period of the sawtooth wave, the modulated phase can be written as $\Delta\phi = 2\pi f_m t$, and the optical frequency is converted to $v = v_0 + f_m$. If the applied voltage is modulated in a quadratic waveform, a linear frequency sweeping can be obtained. However, technical issues related to electronics have to be solved. References [60, 61] analyzed influences of the interval between sawtooth waves by using Fourier transform.

More often used EO devices are waveguide interferometers. Figure 7.23 is a schematic diagram of optical waveguide Mach–Zehnder interferometer; (a) is its overlook, and (b) is the profile of A–A cross section of the X-cut crystal. The RF signal is applied on the middle electrode from the left end, with its right end and the two side electrodes connected to the ground. Electric fields with opposite directions are thus applied on the two waveguides located between the electrodes. The MZI is

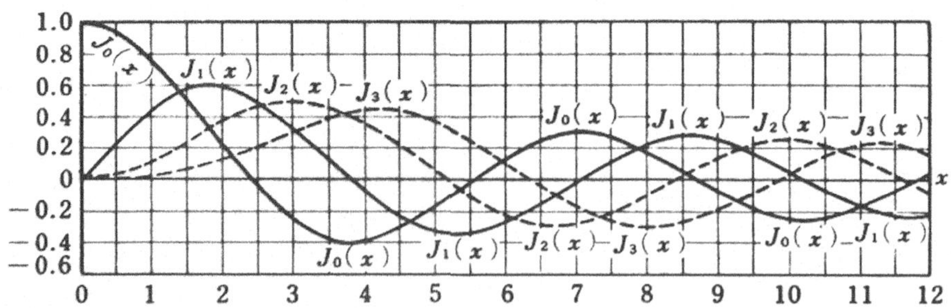

Fig. 7.22 Bessel functions

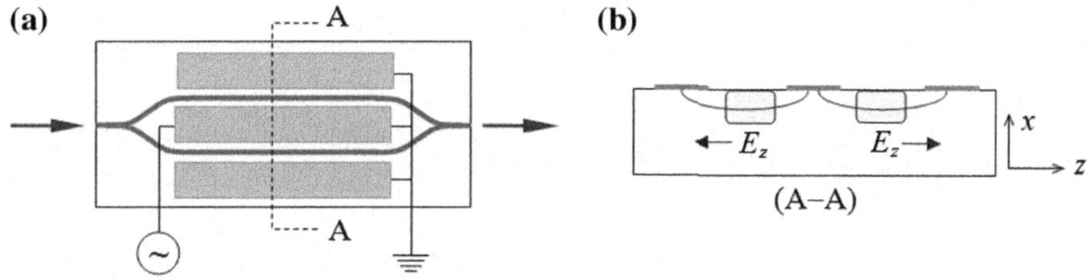

Fig. 7.23 EO waveguide MZI modulator, overlook (**a**) and A–A profile (**b**)

composed of the two waveguide arms and two *Y*-type waveguide couplers. The optical field is modulated by the MZI, expressed as

$$E_{\text{out}} = E_{\text{in}} \cos(\Delta\phi + \phi_{\text{dc}}) = E_{\text{in}} \cos\left[n_o^3 \gamma_{13}(V_{\text{ac}} + V_{\text{dc}})l/d_{\text{eff}}\right], \tag{7.32}$$

where l is the waveguide length, d_{eff} is the effective width between the waveguides, and V_{ac} and V_{dc} are the applied AC and DC voltages. The output amplitude of MZI is modulated by V_{ac}, and the modulation depth can be controlled by the DC phase bias, with the maximum at the quadrature point of $\phi_{\text{dc}} = \pi/2$. If the driving voltage is with a square waveform, square waveform intensity pulses will be obtained, which is just the signal used in pulse code modulation (PCM) optical fiber communication system.

Similar to the phase modulation, EO amplitude modulation contains a series of side band components if a sinusoidal voltage is applied to the EO modulator, since $E_{\text{out}} \propto \cos(\phi_{\text{dc}} + \phi_{\text{dc}} \sin \Omega t)$ can be expanded by formula (7.30). It is usually required to have a single side band component. The method proposed by Ref. [63] is to apply a pair of RF voltage with quadrature phases. The higher order side bands can thus be filtered off with only the base band and the first-order band remained; and the former is further filtered by an FBG filter. The first-order band frequency can then be swept by tuning the RF frequency.

The tuning efficiency of EO modulation is one of the concerned performances. It is lowered by using too many filters in such schemes as used in Ref. [63]. A high efficiency and convenient single side band (SSB) modulator is needed for frequency sweeping, and attracts attention of researchers and industries [64–68]. It has now been available commercially. A typical SSB modulator combines waveguide phase modulators and MZIs, called the dual parallel Mach–Zehnder modulator (DPMZM), as shown in Fig. 7.24a, which is developed based on the mature $LiNbO_3$ waveguide technology. An *X*-cut crystal is used in the structure; the modulating field is in *z*-direction. The RF signals with a quadrature phase difference, V_{ac1} and V_{ac2}, and two DC biases are applied to the two MZM, whose structure is the same as that in Fig. 7.23. The two outputs of the first step MZM, with the phases adjusted by DC voltage of V_{dc3}, are combined at *Y*-type coupler of the second step MZM.

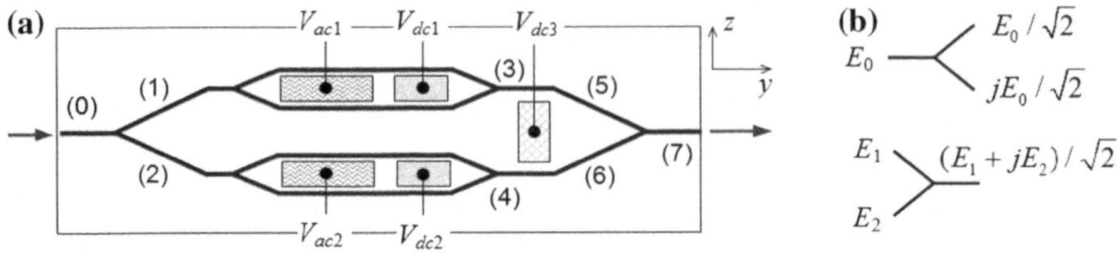

Fig. 7.24 **a** Structure of DPMZM SSB modulator. **b** Output–input relations of a directional coupler

Characteristics of DPMZM are discussed as follows, in which the split ratios of all 6 couplers are supposed to be 50:50 for simplicity. According to the basic principle of directional coupler, $\pi/2$ phase shift exists between the two output fields, as shown in Fig. 7.24b. The modulated phases of the two first step MZM are written as $\beta \sin \Omega t$ and $\beta \cos \Omega t$, respectively. Their output fields are

$$\begin{aligned} E_3 &= \frac{j}{2} E_1 e^{j\omega t}\left[e^{j(\beta \sin \Omega t + \phi_{dc1})} + e^{-j(\beta \sin \Omega t + \phi_{dc1})}\right] = -\frac{1}{2} E_1 e^{j\omega t}\left(e^{j\beta \sin \Omega t} - e^{-j\beta \sin \Omega t}\right) \\ E_4 &= \frac{j}{2} E_2 e^{j\omega t}\left[e^{j(\beta \cos \Omega t + \phi_{dc2})} + e^{-j(\beta \cos \Omega t + \phi_{dc2})}\right] = -\frac{1}{2} E_2 e^{j\omega t}\left(e^{j\beta \cos \Omega t} - e^{-j\beta \cos \Omega t}\right). \end{aligned} \tag{7.33}$$

The last expressions of (7.33) are for the case of $\phi_{dc1} = \phi_{dc2} = \pi/2$. By using Fourier expansions, (7.30), with components higher than the second omitted (7.31) are reduced to

$$\begin{aligned} E_3 &= E_1 J_1(\beta)\left[e^{j(\omega+\Omega)t} - e^{j(\omega-\Omega)t}\right] = \frac{1}{\sqrt{2}} E_0 J_1(\beta)\left[e^{j(\omega+\Omega)t} - e^{j(\omega-\Omega)t}\right] \\ E_4 &= jE_2 J_1(\beta)\left[e^{j(\omega+\Omega)t} + e^{j(\omega-\Omega)t}\right] = -\frac{1}{\sqrt{2}} E_0 J_1(\beta)\left[e^{j(\omega+\Omega)t} + e^{j(\omega-\Omega)t}\right]. \end{aligned} \tag{7.34}$$

With DC phase biases adjusted by V_{dc3}, a single side band output is obtained at the Y-type beam combiner:

$$E_7 = \frac{1}{\sqrt{2}} E_0 J_1(\beta) e^{j(\omega+\Omega)t} \quad \text{or} \quad E_7 = \frac{1}{\sqrt{2}} E_0 J_1(\beta) e^{j(\omega-\Omega)t}. \tag{7.35}$$

A real device may deviate from the ideal structure; the split ratio may not be 3 dB exactly, and the doubled waveguide may not be symmetric perfectly. The device is then needed to be optimized by adjusting the DC voltages.

The operation of DPMZM must use a double-channel RF generator, which has two output signals with equal amplitudes and quadrature phases. Its frequency can be tuned quickly and linearly, or in an arbitrary waveform controlled by a computer.

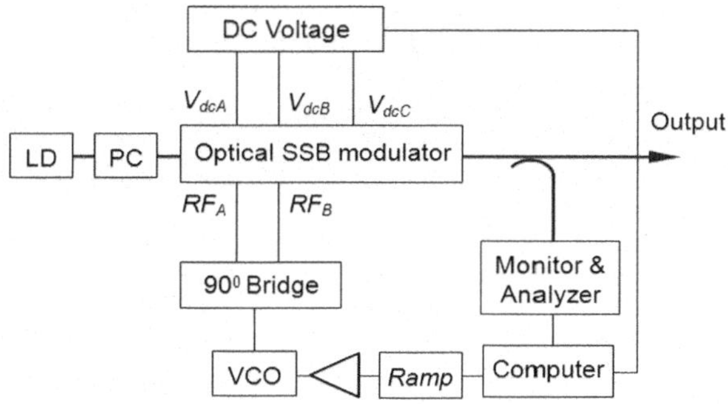

Fig. 7.25 Schematic diagram of frequency sweeping LD with SSB modulator. *PC*: polarization controller; *VCO*: voltage-controlled oscillator

Figure 7.25 shows an electric circuit diagram used to drive the SSB modulator. VCO in the figure stands for the voltage-controlled oscillator, whose oscillation frequency is proportional to the controlling voltage. When it is driven by a linearly serrated voltage (Ramp), a linearly sweeping RF signal will generate. The 90° bridge is used to convert the input to a pair of RF signals with in-phase and quadrature (*I*/*Q*) phases. The monitor and analyzer in the figure give characterization of sweeping performance, used to feedback control the VCO and DC voltages. Reference [67] used a frequency-swept source in 300–500 MHz band, which was doubled several times by electronics to be an RF signal in 9.6–16 GHz band. The SSB modulator was then driven by the RF signal, and a laser frequency sweeping rate of 12.8 GHz/μs was obtained. The advanced optical waveguide modulator has very high modulation response speed; incorporated with high quality electronic devices and technologies, it is promising to be used widely in laser frequency sweeping.

Experimental results of fast optical frequency seeping were described in Ref. [69] by using a wideband VCO and a high speed EO DPMZM, working in carrier suppressed single sideband (CS–SSB) modulation. Sweeping range of 3.85 GHz and rate of 80 GHz/μs were measured.

Many physical mechanisms and technologies are involved in AO and EO modulators, related to their materials, device designs, and device fabrications. Interested readers can find those in monographs and journals.

References

1. Baney DM, Szafraniec B, Motamedi A (2002) Coherent optical spectrum analyzer. IEEE Photon Technol Lett 14(3):355–357
2. Szafraniec B, Law JY, Baney DM (2002) Frequency resolution and amplitude accuracy of the coherent optical spectrum analyzer with a swept local oscillator. Opt Lett 27(21):1896–1898
3. Ohba R, Uehira I, Kakuma S (1990) Interferometric determination of a static optical path difference using a frequency swept laser diode. Meas Sci Technol 1:500–504
4. Roos PA, Reibel RR, Berg T et al (2009) Ultrabroadband optical chirp linearization for precision metrology applications. Opt Lett 34(23):3692–3694

5. Hymans AJ, Lait J (1960) Analysis of a frequency modulated continuous wave ranging system. Proceedings IEE 107-B, pp 365–372
6. Karlsson CJ, Olsson FAA (1999) Linearization of the frequency sweep of a frequency-modulated continuous-wave semiconductor laser radar and the resulting ranging performance. Appl Opt 38(15):3376–3386
7. Zheng J (2004) Analysis of optical frequency-modulated continuous-wave interference. Appl Opt 43(21):4189–4198
8. Wang Z, Pan Z, Fang Z et al (2015) Ultra-broadband phase-sensitive optical time-domain reflectometry with a temporally sequenced multi-frequency source. Opt Lett 40(22):5192–5195
9. Lu B, Pan Z, Wang Z et al (2017) High spatial resolution phase sensitive optical time domain reflectometer with frequency-swept pulse. Opt Lett 42(3):391–394
10. Passy R, Gisin N, von der Weid JP et al (1994) Experimental and theoretical investigations of coherent OFDR with semiconductor laser sources. J Lightwave Technol 12(9):1622–1630
11. Iiyama K, Wang LT, Hayashi K (1996) Linearizing optical frequency-sweep of a laser diode for FMCW reflectometry. J Lightwave Technol 14(2):173–178
12. Koshikiya Y, Fan X, Ito F (2008) Long range and cm-level spatial resolution measurement using coherent optical frequency domain reflectometry with SSB-SC modulator and narrow linewidth fiber laser. J Lightwave Technol 26(18):3287–3294
13. Li W, Bao X, Chen L (2014) High resolution high sensitivity and truly distributed optical frequency domain reflectometry for structural crack detection. Proceedings SPIE 9157: 91579T(1–4)
14. Yun SH, Tearney GJ, de Boer JF et al (2003) High-speed optical frequency-domain imaging. Opt Express 11(22):2953–2963
15. Biedermann BR, Wieser W, Eigenwillig CM et al (2009) Dispersion, coherence and noise of Fourier domain mode locked lasers. Opt Express 17(12):9947–9961
16. Oh WY, Yun SH, Vakoc BJ et al (2006) Ultrahigh-speed optical frequency domain imaging and application to laser ablation monitoring. Appl Phys Lett 88:103902(1–3)
17. Nicholson JW, Yablon AD, Ramachandran S et al (2008) Spatially and spectrally resolved imaging of modal content in large-mode-area fibers. Opt Express 16(10):7233–7243
18. Nicholson JW, Meng L, Fini JM et al (2012) Measuring higher-order modes in a low-loss, hollow-core, photonic-bandgap fiber. Opt Express 20(18):20494–20505
19. Seeds AJ (2002) Microwave photonics. IEEE Trans Microw Theory Tech 50(3):877–887
20. Yao J (2009) Microw Photonics J Lightwave Tech 27(3):314–335
21. Chu S (1998) The manipulation of neutral particles. Rev Mod Phys 70(3):685–703
22. Cohen-Tannoudji CN (1998) Manipulating atoms with photons. Rev Mod Phys 70(3):707–719
23. Thom J, Wilpers G, Riis E et al (2013) Accurate and agile digital control of optical phase, amplitude and frequency for coherent atomic manipulation of atomic systems. Opt Express 21 (16):18712–18723
24. Uttam D, Culshaw B (1985) Precision time domain reflectometry in optical fiber systems using a frequency modulated continuous wave ranging technique. J Lightwave Technol 3 (5):971–977
25. Kakuma S, Ohmura K, Ohba R (2003) Improved uncertainty of optical frequency domain reflectometry based length measurement by linearizing the frequency chirping of a laser diode. Opt Rev 10(4):182–184
26. Ahn TJ, Kim DY (2007) Analysis of nonlinear frequency sweep in high-speed tunable laser sources using a self-homodyne measurement and Hilbert transformation. Appl Opt 46 (13):2394–2400
27. Kang CS, Kim JA, Eom TB et al (2009) High speed phase shifting interferometry using injection locking of the laser frequency to the resonant modes of a confocal Fabry-Perot cavity. Opt Express 17(3):1442–1446
28. Satyan N, Vasilyev A, Rakuljic G et al (2009) Precise control of broadband frequency chirps using optoelectronic feedback. Opt Express 17(18):15991–15999
29. Führer T, Walther T (2008) Extension of the mode-hop-free tuning range of an external cavity diode laser based on a model of the mode-hop dynamics. Opt Lett 33(4):372–374

30. Greiner C, Boggs B, Wang T et al (1998) Laser frequency stabilization by means of optical self-heterodyne beat-frequency control. Opt Lett 23(16):1280–1282
31. Gorju G, Jucha A, Jain A et al (2007) Active stabilization of a rapidly chirped laser by an optoelect digital servo-loop control. Opt Lett 32(5):484–486
32. Biedermann BR, Wieser W, Eigenwillig CM et al (2010) Direct measurement of the instantaneous linewidth of rapidly wavelength-swept lasers. Opt Lett 35(22):3733–3735
33. Butler T, Slepneva S, OShaughnessy B et al (2015) Single shot, time-resolved measurement of the coherence properties of OCT swept source lasers. Opt Lett 40(10):2277–2280
34. Born M, Wolf E (1999) Principles of optics. Seventh Edition, University Press, Cambridge
35. Saleh BEA, Teich MC (2007) Fundamentals of photonics. Wiley, London
36. Yariv A (1997) Optical electronics in modern communications, 5th edn. Oxford University Press, Oxford
37. Song QW, Wang XM, Bussjager R et al (1996) Electro-optic beam-steering device based on a lanthanum-modified lead zirconate titanate ceramic wafer. Appl Opt 35(17):3155–3162
38. Ye Q, Dong Z, Fang Z et al (2007) Experimental investigation of optical beam deflection based on PLZT electro-optic ceramic. Opt Express 15(25):16933–16944
39. Dong Z, Ye Q, Qu R et al (2007) Characteristics of a PLZT electro-optical deflector. Chin Opt Lett 5(9):540–542
40. Ye Q, Zhang X, Qiao L et al (2012) Analysis and implementation of reflection-type electro-optic phase diffraction grating. J Lightwave Technol 30(17):2796–2802
41. Boggs B, Greiner C, Wang T et al (1998) Simple high-coherence rapidly tunable external-cavity diode laser. Opt Lett 23(24):1906–1908
42. Lan YP, Pan RP, Pan CL (2004) Mode-hop-free fine tuning of an external-cavity diode laser with an intracavity liquid crystal cell. Opt Lett 29(5):510–512
43. Shen L, Ye Q, Cai H et al (2011) Mode-hop-free electro-optically tuned external cavity diode laser using volume Bragg grating and PLZT ceramic. Opt Express 19(18):17244–17249
44. Huber R, Wojtkowski M, Taira K et al (2005) Amplified, frequency swept lasers for frequency domain reflectometry and OCT imaging: design and scaling principles. Opt Express 13(9):3513–3528
45. Hult J, Burns IS, Kaminski CF (2005) Wide-bandwidth mode-hop-free tuning of extended-cavity GaN diode lasers. Appl Opt 44(18):3675–3685
46. Fuhrer T, Stang D, Walther T (2009) Actively controlled tuning of an external cavity diode laser by polarization spectroscopy. Opt Express 17(7):4991–4996
47. Ménager L, Cabaret L, Lorgeré I et al (2000) Diode laser extended cavity for broad-range fast ramping. Opt Lett 25(17):1246–1248
48. Crozatier V, Gorju G, Bretenaker F et al (2006) Phase locking of a frequency agile laser. Appl Phys Lett 89:261115(1–3)
49. Levin L (2002) Mode-hop-free electro-optically tuned diode laser. Opt Lett 27(4):237–239
50. Wang P, Seah LK, Murukeshan VM et al (2006) External-cavity wavelength tunable laser with an electro-optic deflector. Appl Opt 45(34):8772–8776
51. Okamura H (2010) Shift lens external-cavity diode laser for broad wavelength tuning and switching. Opt Lett 35(8):1175–1177
52. Wei F, Sun Y, Chen D et al (2011) Tunable external cavity diode laser with a PLZT electro-optic ceramic deflector. IEEE Photon Technol Lett 23(5):296–298
53. Thapliya R, Okano Y, Nakamura S et al (2003) Electrooptic characteristics of thin-film PLZT waveguide using ridge-type Mach-Zehnder modulator. J Lightwave Technol 21(8):1820–1827
54. Bösel A, Salewski KD, Kinder T (2007) Fast mode-hop-free acousto-optically tuned laser with a simple laser diode. Opt Lett 32(13):1956–1958
55. Bösel A, Salewski KD (2009) Fast mode-hop-free acousto-optically tuned laser: theoretical and experimental investigations. Appl Opt 48(5):818–826
56. Crozatier V, Das BK, Baili G et al (2006) Highly coherent electronically tunable waveguide extended cavity diode laser. IEEE Photon Technol Lett 18(14):1527–1529
57. Donley EA, Heavner TP, Levi F et al (2005) Double-pass acousto-optic modulator system. Rev Sci Instr 76:063112(1–6)

58. Kaminow IP, Li T (2002) Optical fiber communications iv (components). Elsevier Inc, Amsterdam
59. Wong KK, Delarue RM, Wright S (1982) Electro-optic-waveguide frequency translator in $LiNbO_3$ fabricated by proton-exchange. Opt Lett 7(11):546–548
60. Johnson LM, Cox CH (1988) Serrodyne optical frequency translation with high sideband suppression. J Lightwave Technol 6(1):109–112
61. Laskoskie C, Hung H, El-Wailly T et al (1989) Ti-$LiNbO_3$ waveguide serrodyne modulator with ultrahigh sideband suppression for fiber optic gyroscopes. J Lightwave Technol 7(4):600–606
62. Houtz R, Chan C, Muller H (2009) Wideband efficient optical serrodyne frequency shifting with a phase modulator and a nonlinear transmission line. Opt Express 17(21):19235–19240
63. Smith GH, Novak D, Ahmed Z (1997) Technique for optical SSB generation to overcome dispersion penalties in fiber-radio systems. Electron Lett 33(1):74–75
64. Izutsu M, Shikama S, Sueta T (1981) Integrated optical SSB modulator/frequency shifter. IEEE J Quant Electron 17(11):2225–2227
65. Higuma K, Oikawa S, Hashimoto Y et al (2001) X-cut lithium niobate optical single sideband modulator. Electron Lett 37(8):515–516
66. Shimotsu S, Oikawa S, Saitou T et al (2001) Single side-band modulation performance of a LiNbO3 integrated modulator consisting of four-phase modulator waveguides. IEEE Photon Technol Lett 13(4):364–366
67. Kawanishi T, Sakamoto T, Izutsu M (2006) Fast optical frequency sweep for ultra-fine real-time spectral domain measurement. Electron Lett 42(17):999–1000
68. Shioda T, Yamamoto T, Sugimoto T et al (2007) 1 MHz-resolution spectroscopy based on light frequency sweeping using a single-sideband optical modulator. Jpn J Appl Phys 46(6A):3626–3629
69. Wang J, Chen D, Cai H et al (2015) Fast optical frequency sweeping using voltage controlled oscillator driven single sideband modulation combined with injection locking. Opt Express 23(6):7038–7043

Chapter 8
Optical Phase Locked Loop and Frequency Transfer

8.1 Optical Phase Locked Loop

8.1.1 Principles of OPLL

Laser applications, especially in frontier science researches and advanced technology developments, require the laser frequency transferring precisely and locked at arbitrary frequencies, and with high coherence maintained during the transfer. That is, not only the peak frequency is stably locked onto a certain reference frequency, but also the linewidth is maintained unchanged. Chap. 6 introduces the laser frequency stabilizations with atomic absorption line and resonant peak of passive resonator as the reference frequency. This Chapter involves frequency stabilization with another high coherence, high stable laser as reference. In the electrical engineering and electronics the phase locked loop (PLL) technology has been developed to meet the requirements of stabilization in wireless frequency bands; OPLL can be regarded as the technology with the same function in optical frequency band.

The PLL in electronics is a mature technology, demonstrating attractive advantages in suppression of noise and drift. Its concept is shown in Fig. 8.1a, where the phase discriminator PD detects the difference between signal frequency f_{out} and reference frequency f_{ref}, which is the target frequency being locked on. LPF is the low pass filter; VCO is the voltage controlled oscillator, which gives the output signal with frequency locked on the reference frequency. If the available reference frequency is different from the target, a frequency converter should be inserted in the feedback loop, as denoted by f_{out}/N in the doted frame for a frequency divider. The signal of phase discriminator may be processed with other functions, besides low-pass filtering, then the LPF may be understood as a loop filter.

Figure 8.1b gives a schematic diagram of OPLL [1, 2]. It is required to lock the frequency of slave laser (SL) to the reference frequency provided by the master

Z. Fang et al., *Single Frequency Semiconductor Lasers*, Optical and Fiber Communications Reports 9, DOI 10.1007/978-981-10-5257-6_8

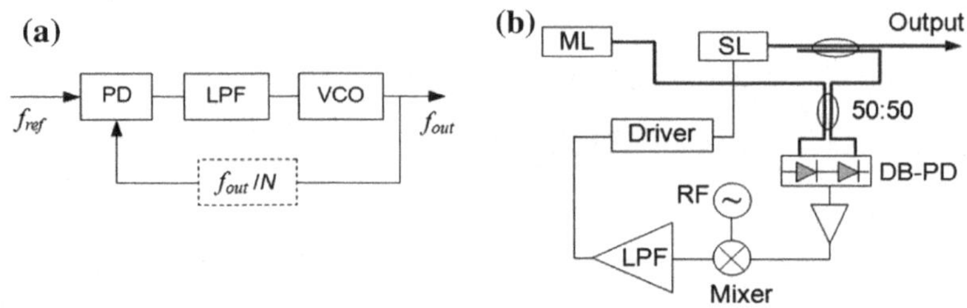

Fig. 8.1 **a** Concept of electrical PLL. **b** Schematic diagram of OPLL

laser (ML). The two laser beams beat at the 50:50 fiber coupler, and its outputs are detected by the double balanced photo-detector (DB-PD) with DC component removed in its output. The beat signal contains the difference between two frequencies, and the phase shift between them. The correlation detection here plays a role of optical frequency discriminator. The output of DB-PD is used as error signal to feedback control the slave laser. Such a feedback controlled slave laser is sometime called an optical voltage controlled oscillator (O-VCO) [3]. If the DB-PD is replaced by a single PD, the OPLL will work as well, but a band-pass filter is needed to remove the DC component.

Similar to the PLL, the central frequencies of SL and ML may be same or different. The two cases require different detection schemes, i.e. the homodyne or the heterodyne, respectively. In the latter case, a radio frequency source (RF) is needed to mix with the optical beat signal to convert the beat signal frequency to a low frequency band. The mixed signal is used as error signal after being filtered by LPF and being amplified. Once the phase locking is established, the slave laser will output beams with the same frequency stability and the same linewidth as those of the master laser. The RF source and the mixer can be put away for the homodyne OPLL. Some other optical components are needed for OPLL, such as polarization controllers to make the polarizations of SL and ML to be parallel.

The principle of OPLL can be described analytically as follows. The optical fields of ML and SL are written as:

$$\begin{aligned} E_{\mathrm{m}} &= A_{\mathrm{m}} \exp j[\omega_{\mathrm{m}} t + \phi_{\mathrm{m}}(t)] \\ E_{\mathrm{s}} &= A_{\mathrm{s}} \exp j[\omega_{\mathrm{s}} t + \phi_{\mathrm{s}}(t)]. \end{aligned} \tag{8.1}$$

They are combined at the fiber coupler. For an ideal 3 db coupler, the outputs are expressed as

$$\begin{aligned} E_1 &= (E_{\mathrm{m}} + jE_{\mathrm{s}})/\sqrt{2} \\ E_2 &= (jE_{\mathrm{m}} + E_{\mathrm{s}})/\sqrt{2}. \end{aligned} \tag{8.2}$$

The current detected by DB-PD with responsivity of ρ_{PD} is then obtained:

$$i_{PD} = \rho_{PD}(|E_1|^2 - |E_2|^2) = 2\rho_{PD}A_m A_s \sin[(\omega_m - \omega_s)t + \phi_m(t) - \phi_s(t)]. \quad (8.3)$$

Mixed with RF signal $v_{RF} = V_{RF}\cos[\omega_{RF}t + \phi_{RF}(t)]$, the signal is reduced to

$$\begin{aligned} v_{mix} &= \rho_{mix} V_{RF} V_{beat} \sin[(\omega_m - \omega_s)t + \phi_m(t) - \phi_s(t)]\cos[\omega_{RF}t + \phi_{RF}(t)] \\ &= V_{mix}\{\sin[(\omega_m - \omega_s - \omega_{RF})t + \phi_m - \phi_s - \phi_{RF}] \\ &\quad + \sin[(\omega_m - \omega_s + \omega_{RF})t + \phi_m - \phi_s + \phi_{RF}]\} \end{aligned} \quad (8.4)$$

where ρ_{mix} is the mixing efficiency, and V_{beat} is the amplified beat signal. The two frequency components of v_{mix} can be separated by band pass filters. By denoting $V_{err} \propto \rho_{PD}\rho_{mix} V_{RF} A_m A_s$, $\Delta\omega_\pm = \omega_m - \omega_s \pm \omega_{RF}$, and $\Delta\phi_\pm = \phi_m - \phi_s \pm \phi_{RF}$, the error signal is expressed as

$$v_{err} = V_{err}\sin(\Delta\omega_\pm t + \Delta\phi_\pm). \quad (8.5)$$

The driver of slave laser is then controlled by the error signal to establish a close loop feedback. The phase variation of slave laser is proportional to the error signal, expressed as [1]:

$$\frac{\mathrm{d}}{\mathrm{d}t}\Delta\Phi = \kappa v_{err} + \Delta\omega^{[fr]} = -K\sin\Delta\Phi + \Delta\omega^{[fr]}, \quad (8.6)$$

where $\Delta\Phi = \Delta\omega_\pm t + \Delta\phi_\pm$, κ is the tuning response of laser driver, and $\Delta\omega^{[fr]} = \omega_m - \omega_s^{[fr]}$ is the frequency difference between SL and ML in open loop, i.e. in free running case of SL. $K = -\kappa V_{err}$ is the composite gain of the loop, which is generally a complex to involve the loop delay; the minus sign indicates that a negative feedback is needed for the loop to be locked-in. Equation (8.6) can be analyzed by Laplace transform, as used in electrical PLL theory.

The phase lock-in is established under condition of $\mathrm{d}\Delta\Phi/\mathrm{d}t = 0$ and $\omega_s = \omega_m \pm \omega_{RF}$, with a constant phase remained: $\sin\varphi_0 = (\omega_m - \omega_s^{[fr]})/K$, ideally, which does not generate phase and frequency noise, but contributes a minimal error signal to maintain the loop working at locked-in state [2].

The linewidth of a free running SL is basically determined by the phase noise in low frequency band, which can be suppressed effectively by OPLL, so long as the loop response is fast enough and no extra noise from loop components is added in. The sensitivity to the frequency difference of the mixed signal (8.4) is dependent on the phase of RF source, seen from the sinusoidal terms, rewritten as $\sin(\Delta\omega_\pm t + \phi_m - \phi_s)\cos\phi_{RF} \pm \cos(\Delta\omega_\pm t + \phi_m - \phi_s)\sin\phi_{RF}$. The quadrature is obtained at $\phi_{RF} = 0$. If noise of RF source is taken into account, the variance of SL phase noise should be given by the sum of variances of ML and RF. If noises and drifts occurring in optics and electronics in the loop exist, some Langevin terms have to be added to Eq. (8.6).

The frequency and phase of slave laser can be modulated by tuning the RF source in heterodyne OPLL, $\omega_s(t) = \omega_m \pm \omega_{RF}(t)$ and $\phi_s(t) = \pm\phi_{RF}(t) - \varphi_0$ used as a precisely frequency swept laser or a frequency transferred laser [2]. In homodyne OPLL, frequency and phase modulations can be implemented by a phase modulated master laser with an external phase modulator inserted in its output path [1]. It is one of the important advantages and usages of OPLL. In heterodyne OPLL, mixing with RF brings about a function of beat signal amplification by using higher RF power. In electronics part of the loop the electrical PLL technology, especially the digital PLL, may be used to optimize loop's performance.

Generally applications require OPLL with the following performances. (1) High precision of locked frequency and phase for OPLL output with low noise and high stability; (2) Wide range of locked frequency and phase; (3) Fast locking speed responding to external disturbances or to the intentional adjustment. These performances are basically related to loop's band width and its sensitivity of frequency discrimination, contributed both from optics and electronics, and embodied in the loop gain factor K. Detailed analyses can be found in papers and monographs.

8.1.2 Injection Locking of Semiconductor Lasers

The optical injection locking (OIL) of semiconductor lasers is introduced in Sect. 5.1.4, which can be regarded as a direct OPLL without electrical components inserted in the loop. The frequency and phase of a narrow linewidth single frequency master laser can be duplicated in the slave laser when the injection locking is established. The injected optical wave of master laser (ML) plays a role of seed to the oscillation of slave laser (SL). Referring formula (5.31b) in 5.1.4, the difference between phases of SL and ML will reach a stable value in case of the loop locked, expressed with the subscripts used in this section as

$$\omega_s - \omega_m = \kappa \sin(\phi_s - \phi_m). \tag{8.7}$$

The injection rate $\kappa = E_m/(\tau_c E_s)$, where E_m is the injected field of ML, E_s is the intra-cavity field of SL, and τ_c is the round-trip propagation time in SL cavity. The OIL has a merit of fast response since the OE conversion and feedback control of laser driver are not needed in the loop. However, the original characteristics of ML and the injection rate have to meet certain conditions, as analyzed in Sect. 5.1.4.

The OIL can be used either to lock the main mode of SL, similar to homodyne OPLL; or to lock one of its side modes, similar to heterodyne OPLL [4, 5]. The locked mode may be either one of the longitudinal modes of F-P cavity LD working in continuous wave, even if not lasing in free running or one of the side bands in state of high frequency modulation. Figure 8.2a shows a schematic diagram of side band locking. The slave laser is biased by a DC driver and modulated by a RF source with a Bias-T. One of its side bands is located at the frequency of ML. When the OIL is established, the main mode and other side bands are locked into stable

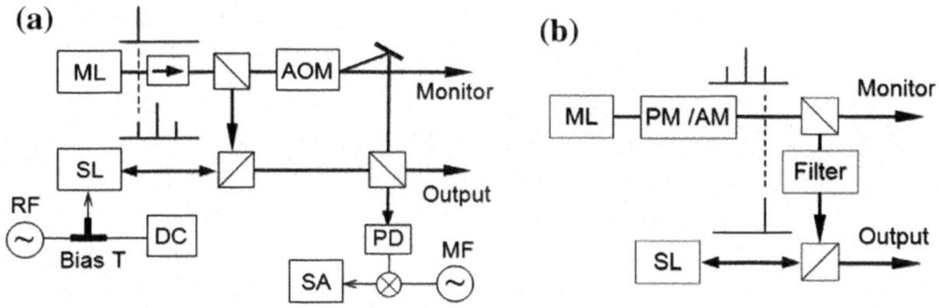

Fig. 8.2 Schemes of **a** SL side band injection locked. **b** SL main mode locked by ML side band

states. The other parts in the figure are used for characteristics measurement and monitoring. The ML beam with frequency shifted by the acousto-optic modulator (AOM) beats with the output of SL; the PD detected signal is mixed with a middle frequency source, and is analyzed by the spectrum analyzer (SA).

The main mode of SL can also be locked at one of the side bands of ML working at high frequency modulation, as shown in Fig. 8.2b. In the scheme the ML is usually modulated by an external phase modulator (PM) or amplitude modulator (AM) to ensure good quality of ML spectrum. The schemes shown in Fig. 8.2 gives wide frequency range of OPLL, and more flexible frequency selectivity, since the spacing between side bands of modulation is usually much smaller than the spacing between F-P modes of LD.

8.2 Applications of OPLL in Laser Frequency Transfer

OPLL has been used in many important areas, including coherent optical communication (COC), transportation of frequency and time standards in networks, microwave photonics and radio over fiber (ROF), physical researches, especially on laser cooling and cooled atoms, and the effect of electromagnetically induced transparency (EIT), and so on. These applications involve a wide scope of technologies and physical mechanisms; here we just give a brief introduction to those related to the single frequency lasers.

8.2.1 *Coherent Optical Communication*

Optical fiber communication (OFC) has been recognized as one of the foundation stones of modern information infrastructure of the world. In the OFC system the intensity modulation and direct detection (IM-DD) method and the dense wavelength division multiplex (DWDM) are used as the current technologies now. As the demands on information capacity and transportation speed increase continuously, the coherent optical communication (COC) is being developed rapidly, which uses not only the amplitude of optical pulses, but also its frequency, phase and polarization state as the carriers of information, and thus provides much larger capacity. COC is suitable especially to inter-satellite communications since the vacuum between them will not distort the optical frequency and phase. In fiber systems, its applications are developed fast. Several basic formats used in COC have been developed and matured, including amplitude shift keying (ASK), frequency shift keying (FSK), and phase shift keying (PSK). Coherent optical orthogonal frequency division multiplex (CO-OFDM) and polarization multiplex are promising technologies for multiplexing. Compared with DWDM, the multiplex density and spectral efficiency of COC are greatly increased. At the receiver side, correlation detection must be used for retrieving information contained in optical frequencies and phases. Obviously, COC must use a high coherent, high stable laser as its source; and its frequency and phase should be modulated and detected in high speed. Therefore OPLL is a must technology in COC [1–3, 6–9].

The transmitter of COC should meet two basic requirements. First, its frequency and/or its phase should be modulated, typically by using OPLL. Moreover, the data have to converted to a pair of in-phase and quadrature signals by an *I*/*Q* modulator [10]. Second, for the optical frequency division multiplex (OFDM) the transmitter should contain a series of frequencies, which is obtained usually by using single side band (SSB) modulators, whose function is introduced in Sect. 7.4.2. Reference [11] proposed and demonstrated an OFDM source composed of multiple narrow lines with smaller intervals by using a SSB modulator to shift the laser frequency multiply with a circulating loop.

At receiver terminals, the data carried by optical frequency and/or phase have to be demodulated, which are usually very weak, even drowned by noises. That is just one of the scopes of OPLL's ability [2, 3, 7]. Figure 8.3a shows a schematic structure of correlation detector based on OPLL.

The function of 180° coupler in the figure is to give "sum" and "difference" of the local oscillation (LO) and the data signal, respectively, expressed as

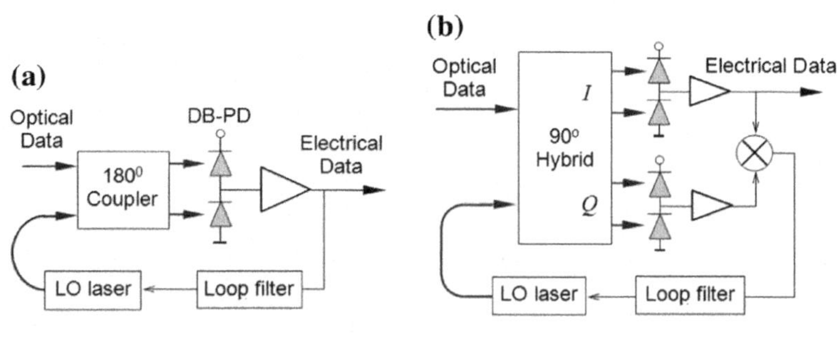

Fig. 8.3 **a** A balanced OPLL; **b** A Costas loop

$$
\begin{aligned}
E_1 &= (E_{\mathrm{LO}} + E_{\mathrm{sig}})/\sqrt{2} \\
E_2 &= (E_{\mathrm{LO}} - E_{\mathrm{sig}})/\sqrt{2}.
\end{aligned} \tag{8.8}
$$

The current signal of double balanced photodetector (DB-PD) is

$$
i_{PD} = \rho_{\mathrm{PD}}(|E_1|^2 - |E_2|^2) = 2\rho_{\mathrm{PD}}|E_{\mathrm{LO}}||E_{\mathrm{sig}}|\cos(\Delta\omega t + \Delta\phi), \tag{8.9}
$$

where $\Delta\omega = \omega_{\mathrm{LO}} - \omega_{\mathrm{sig}}$, $\Delta\phi = \phi_{\mathrm{LO}} - \phi_{\mathrm{sig}}$. After the loop filter it is reduced to the error signal to control the diver of local laser. When the OPLL is established, the frequency and phase of local laser will be locked on those of detected signal, and $i_{\mathrm{PD}} = 2\rho_{\mathrm{PD}}|E_{\mathrm{LO}}||E_{\mathrm{sig}}|$ is output as the electrical data of ASK format. It is seen that the OPLL technology is a perfect frequency selective device, so long as the local laser has linewidth narrow enough. It is very useful for OFDM system, since many different frequencies with narrow intervals are contained in the system.

The OPLL scheme of Fig. 8.3a is usually termed the balanced phase locked loop, since a balanced detector is used. Its shortcoming is that the FSK data and PSK data can not be retrieved by such a scheme; the data may be lost or distorted in the output electrical signal. The frequency modulation and/or phase modulation signal carried by the received beam should be retained in the electrical output, while the frequency of local oscillation is locked on one of the channels of received beam. Many schemes were developed for the purpose; among them Costas loop is often used [12], whose structure is shown in Fig. 8.3b schematically. Instead of the 180° coupler in the balanced loop, a 90° hybrid bridge is used, which gives 4 outputs, expressed respectively as

$$
\begin{aligned}
E_1 &= (E_{\mathrm{LO}} + E_{\mathrm{sig}})/\sqrt{2} \\
E_2 &= (E_{\mathrm{LO}} - E_{\mathrm{sig}})/\sqrt{2} \\
E_3 &= (E_{\mathrm{LO}} + jE_{\mathrm{sig}})/\sqrt{2} \\
E_4 &= (jE_{\mathrm{LO}} + E_{\mathrm{sig}})/\sqrt{2}.
\end{aligned} \tag{8.10}
$$

The in-phase (I) and quadrature (Q) beat signals are obtained from the two DB-PDs:

$$
\begin{aligned}
i_{\mathrm{PD1}} &= \rho_{\mathrm{PD}}(|E_1|^2 - |E_2|^2) = \rho_{\mathrm{PD}}|E_{\mathrm{LO}}||E_{\mathrm{sig}}|\cos(\Delta\omega t + \Delta\phi) \\
i_{\mathrm{PD2}} &= \rho_{\mathrm{PD}}(|E_3|^2 - |E_4|^2) = \rho_{\mathrm{PD}}|E_{\mathrm{LO}}||E_{\mathrm{sig}}|\sin(\Delta\omega t + \Delta\phi),
\end{aligned} \tag{8.11}
$$

where $\omega_{\mathrm{sig}} = \bar{\omega}_{\mathrm{sig}} + \delta\omega_{\mathrm{sig}}(t)$ and $\phi_{\mathrm{sig}} = \phi_{\mathrm{sig}}(t)$ contains the frequency modulated and/or phase modulated data. The mixed output of I and Q signals is taken as error signal of frequency deviation and used to control the local laser. When the OPLL is established, the LO frequency will be locked on one of the input frequencies, $\omega_{\mathrm{LO}} = \bar{\omega}_{\mathrm{sig}}$. Meanwhile, I (or Q) signal is used to be demodulated for retrieving transported data.

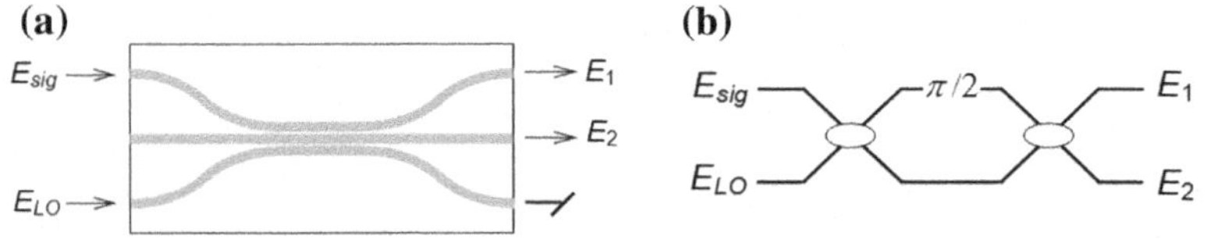

Fig. 8.4 Schemes of 90° bridge: **a** with a 3 × 3 fiber/waveguide coupler; **b** with a $\pi/2$ phase shifted Mach-Zehnder interferometer

The 180° and 90° bridges used in Fig. 8.3 can be fabricated either by block optical components, or by optical waveguides and fibers. Some of them are introduced here. Figure 8.4 shows two schemes of 180° coupler. Device (a) is based on a 3 × 3 waveguide coupler or fiber coupler. The three waveguides are arranged in a plane, and the middle waveguide is coupled with the other two by the evanescent field, whereas the coupling between the two side waveguides can be neglected. When the coupling coefficients and coupling length meet certain conditions, the transfer function between the inputs and outputs can be expressed as [13, 14]:

$$\begin{pmatrix} A \\ B \\ C \end{pmatrix} = \frac{1}{2}\begin{pmatrix} 1 & j\sqrt{2} & -1 \\ j\sqrt{2} & 0 & j\sqrt{2} \\ -1 & j\sqrt{2} & 1 \end{pmatrix}\begin{pmatrix} A_0 \\ B_0 \\ C_0 \end{pmatrix}. \tag{8.12}$$

In case the signal and LO input as shown in Fig. 8.4a, the two outputs are obtained as $E_1 = (E_{\rm sig} - E_{\rm LO})/2$ and $E_2 = j(E_{\rm sig} + E_{\rm LO})/\sqrt{2}$, giving the same relation as (8.8), except for the amplitude difference, which can be compensated by optical attenuator or by electrical processing.

Figure 8.4b is a $\pi/2$ phase shifted Mach-Zehnder interferometer (MZI) [15]. By referring formula (8.2), its transfer function is obtained as

$$\begin{aligned} E_1 &= \tfrac{1}{2}[(E_{\rm sig} + jE_{\rm LO})e^{j\pi/2} + j(jE_{\rm sig} + E_{\rm LO})] = \tfrac{j-1}{2}(E_{\rm sig} + E_{\rm LO}) \\ E_2 &= \tfrac{1}{2}[j(E_{\rm sig} + jE_{\rm LO})e^{j\pi/2} + (jE_{\rm sig} + E_{\rm LO})] = \tfrac{j-1}{2}(E_{\rm sig} - E_{\rm LO}). \end{aligned} \tag{8.13}$$

It is just the relation required by (8.8).

Various structures of I/Q bridge have been presented in journals [9, 16, 17], including schemes with optical waveguides and with block optics. Figure 8.5a shows a structure composed of two cross-connected MZIs with four 2 × 2 3 dB couplers [10]. Three of the four arms have identical optical paths except for phase shifts of integer multiplied 2π, whereas the other one has a different path from them by $\pi/2$ phase shift. Similarly, the four outputs are obtained by referring (8.2):

$$\begin{aligned} E_1 &= j(E_{\mathrm{sig}} + E_{\mathrm{LO}})/2 \\ E_2 &= (-E_{\mathrm{sig}} + E_{\mathrm{LO}})/2 \\ E_3 &= (jE_{\mathrm{sig}} - E_{\mathrm{LO}})/2 \\ E_4 &= (-E_{\mathrm{sig}} + jE_{\mathrm{LO}})/2. \end{aligned} \tag{8.14}$$

It coincides with (8.10) exactly. The 2 × 2 and 3 × 3 couplers used in structures of Figs. 8.4 and 8.5a can be fabricated either by planar waveguides, such as SiO_2 film on Si substrate, or by fused silica fiber. The theory and technical issues of their design and fabrications can be found in journals and monographs.

Figure 8.5b gives a schematic diagram of *I*/*Q* bridge composed of bulk optics [15, 16]. In the structure, the polarization of LO is converted to a circularly polarized beam by the $\lambda/4$ plate (QWP), which is the sum of two linearly polarized wave with $\pi/2$ phase shift between them. It is then divided into two beams by the beam splitter (BS), where the phase of reflected beam is shifted by $\pi/2$ respect to the transmission, according to Fresnel formulas. The signal beam E_{sig} is divided into two at BS also, and its reflection experiences $\pi/2$ phase shift. The same phase shift occurs at the two PBS. Therefore, E_a is the sum of once reflected signal wave and the transmitted LO; E_b is the sum of twice reflected signal wave and once reflected, circularly polarized LO; E_c is the sum of transmitted signal wave and once reflected, circularly polarized LO; E_d is the sum of once reflected signal wave and twice reflected, circularly polarized LO. In result, the four outputs $E_{a\text{-}d}$ will have the same phase relations as (8.14). A polarization controller is inserted in the input path to adjust the polarization of signal beam and get equal reflection and transmission at BS. Several other structure than the above were also proposed such as that by utilizing birefringence crystals [17].

The balanced loop and the Costas loop are the basic structure used in COC. To meet the requirements of different formats and for different performances, more advanced loops have been proposed and developed, such as SyncBit Loop, Dither Loop, and Decision-driven Loop [2, 18]. A lot of technical problems are involved in these schemes; details can be read in journals.

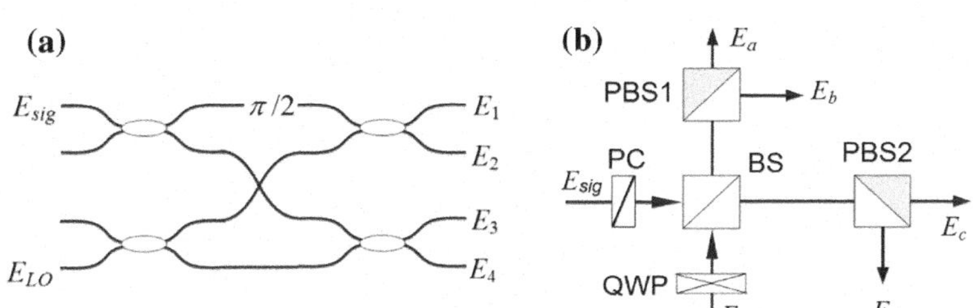

Fig. 8.5 Schemes of *I*/*Q* bridge: **a** with cross-connected MZIs; **b** composed of polarization beam splitters and combiners

8.2.2 Transportation of Time and Frequency Signals

The synchronization of different communication networks is a basic condition for information transportation. In the optical fiber communication networks, the transported data are encoded by pulses; the pulsed signals will be distorted after propagation in the fiber. At the receiver terminals, the amplitudes, the pulse waveforms, and its periods of coded data have to be recovered by regeneration, reshaping and retiming (3R). In relatively low speed systems 3R are realized by electrical processing of the signals. All-optical 3R are also developed in optical communication systems. The transportation of frequency and time standard is critical for retiming. Without the time standard the coded data will not be recognized. It is more and more important for the higher speed systems and the larger covered ranges [1, 19]. It is also critical for the global positioning system (GPS), where the precision and resolution of positioning are directly dependent on the precisions of frequency and time. The related frequency band is expanded from RF to the optical band. As the optical frequency comb and its applications are developing, optical waves with linewidth down to the order of Hz and the frequency stability up to 10^{15} or higher are being used; transportation of such high precision signals must be implemented [20, 21]. The laser frequency stabilization and optical phase locking are the basis of these developing technologies.

A lot of researches have been focused on the transportations and distributions of frequency and clock signal over long distance optical fiber networks [22–25]. When a single frequency optical wave is propagating in fiber, its phase will experience fluctuations due to the temporal and spatial variations of fiber index, and the thermal noise. The phase fluctuation of $\phi(t) = \int_0^L \delta\phi(z, t - \tau + z/v)$ will occur at the remote end, where $\delta\phi$ is the phase fluctuation per unit distance and $\tau = L/v$ is the single trip time delay, with light velocity of v in fiber; and a frequency fluctuation of $\delta\omega = \partial\phi(t)/\partial t$ occurs in result [26]. Such a distortion must be removed when the optical frequency itself is the single to be transported. Similar problem has been already met and solved in clock synchronization over space by using the Doppler-cancellation technique [27]. The basic concept is to let the delivered wave returned back to the local end, and to control the transmitter by PLL with the roundtrip distortion as feedback errors. It is actually a large phase locked loop, spanning the long fiber. The phase distortion in round trip is expressed as [28]

$$\phi_{RT}(t) = \int_0^L \{\delta\phi[z, t - 2\tau + z/v] + \delta\phi[z, t - z/v]\} \approx 2\int_0^L \delta\phi(z, t - \tau) \quad (8.16)$$

The last approximation is based on the assumption that the phase change in the roundtrip propagation equals twice of the single trip propagation, if the external disturbances are much slower than the roundtrip propagation of optical signal. By Fourier transform, the frequency spectrum of distortion is expressed respectively for the forward and roundtrip propagations as [26].

$$\tilde{\phi}_F(f) = \int_0^L e^{-j2\pi f(\tau - z/v)} \delta\tilde{\phi}(z,f), \tag{8.17a}$$

$$\tilde{\phi}_{RT}(f) = 2e^{-j2\pi f\tau} \int_0^L \delta\tilde{\phi}(z,f) \cos[2\pi f(\tau - z/v)]. \tag{8.17b}$$

Symbols '~' above stand for the respective Fourier transforms. Their power spectral densities (PSD) are deduced to be

$$S_F(f) = \left\langle \left|\tilde{\phi}_F(f)\right|^2 \right\rangle = \int_0^L \left\langle \left|\delta\tilde{\phi}(z,f)\right|^2 \right\rangle, \tag{8.18a}$$

$$S_{RT}(f) = 2S_F(f)[1 + \text{sinc}(4\pi f\tau)]. \tag{8.18b}$$

The deduction of (8.18b) is based on the assumption that the noise is independent of position. It is shown that the difference between PSD of the single trip and the roundtrip is related to the frequency of noises. At the low frequency limit, the ratio of two PSD reaches 4, whereas at high frequency band, it is towards 2; and the criterion depends on the fiber length.

Obviously, OPLL technology is the most powerful tool for compensating the phase distortions. Figure 8.6 shows a schematic diagram of phase noise compensation [19], where frequency shifts of Ω and $-\Omega$ are implemented respectively by the acousto-optic modulators AOM1 and AOM2. The returned wave contains phase noise of $2\omega_\Phi$. Two beams are combined at the beam splitter (BS) and detected by the photodiode (PD); the beat with frequency of $2\Omega + 2\omega_\Phi$ is thus obtained as the frequency discriminating signal. It is then used to control the VCO through an electrical PLL. AOM2 is driven by the half of VCO output frequency with a frequency divider ($f/2$). A large OPLL over local and remote ends is then built; and the phase noise at the remote end, generated in fiber propagation, will be suppressed when the loop is locked.

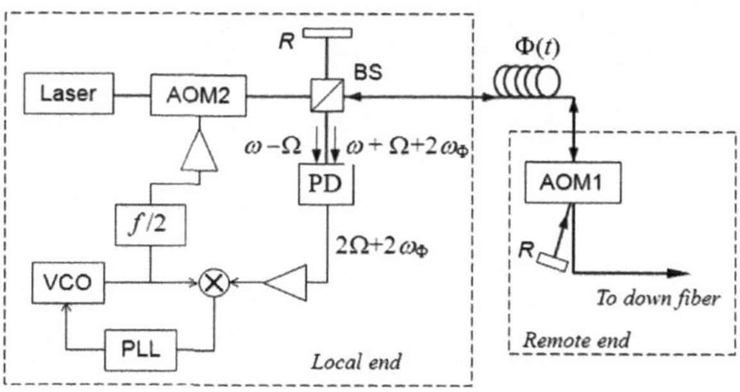

Fig. 8.6 Compensation of optical phase distortion generated in fiber propagation

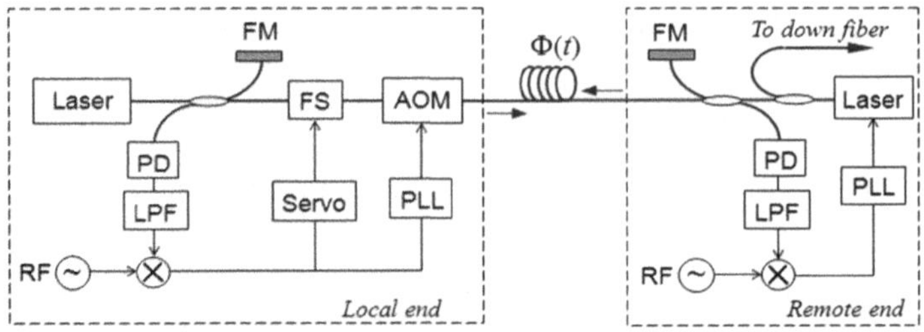

Fig. 8.7 System with local and remote ends locked mutually

Since the fiber index fluctuations are usually slower than the propagation speed of optical signal, the induced phase distortion can be compensated by fiber phase modulators, which are based on the photo-elastic effect and/or thermo-optic effect. References [22, 23] demonstrated a system with two sources placed at the local end and the remote end, locked mutually by OPLL, as shown in Fig. 8.7.

The local laser is a frequency stabilized narrow line laser, and its frequency is modulated by the AOM before sent to the remote. The AOM is driven by the signal generated by an OPLL. The wave returned from the remote passes through the AOM in direction counter to that of the local laser beam, through the fiber phase modulator (FS), and is combined with the local laser reflected by the Faraday rotation mirror (FM), giving beat signal as feedback errors to control the AOM driver. The role of FM is to avoid the external disturbance to the polarization state in the loop. The phase modulator FS is a fiber stretcher, driven by PZT and controlled by a serve. At the remote end, the laser is controlled by OPLL, in which the error signal is given by beating of forward propagating wave and the laser output. The positive and negative frequency shifts by two AOM in Fig. 8.6 are now replaced by forward and backward passing through the same AOM. The frequencies of remote laser beam and output from the local will be locked when the phase-locking is established. For transregional loops, especially with the commercial fiber lines utilized, quite a lot of technical issues have to be solved, such as delays and noises in different sections and different nodes. Even the earth spin induced Sagnac effect has also to be taken into account. Reference [28] presented related analysis and experimental results.

The joint transfer of time and frequency can be realized via ready-installed optical fiber communication networks for multiple users. The technical issues related to such a scheme were studied in Refs. [29, 30].

8.2.3 Microwave Photonics and Radio Over Fiber

The demand on communication has been growing increasingly all over the world. The combination of optical fiber communication and wireless communication becomes an inevitable trend. Its benefit is obvious: the wireless provides possibility of mobile communication, whereas the fiber network provides possibility of connection over long distances worldwide with huge capacity. Since the code speed of optical communications has reached RF and millimeter wave ranges, it is possible to process and transport RF signals by optical waves. On the contrary, the ordinary method of RF signal processing needs much bulky and expensive devices and components, being one of the bottlenecks in development of RF technology. Therefore, a new technology, termed the radio over fiber (ROF), emerged years ago [31, 32].

The ROF system is basically composed of several parts. (a) Conversion of RF signals to optical signals, i.e. generation of optical wave carrying RF signals; (b) Transportation of the optical carrier in optical fibers; (c) Processing of RF signals by optical devices and components; and (d) Conversion the optical signal back to electrical domain in RF band.

To ensure conversions between RF and optical domain with high fidelity, the process must utilize high coherent, low noise optical sources, and high speed, flexible frequency transfer and phase control. Problems of signal distortion and additive noises in the conversions and fiber propagating must be solved [31–34].

The optical carrier of RF signals can be realized by using the beating of a pair of narrow line laser beams with their frequency difference coincident with the RF frequency. That is,

$$\begin{aligned} E &= E_1 \cos(\omega_1 t + \phi_1) + E_2 \cos(\omega_2 t + \phi_2) \\ &= \bar{E} \cos(\bar{\omega} t + \bar{\phi}) \cos(\Delta\omega t + \Delta\phi) + \Delta E \sin(\bar{\omega} t + \bar{\phi}) \sin(\Delta\omega t + \Delta\phi), \end{aligned} \tag{8.15}$$

where $\bar{\omega} = (\omega_1 + \omega_2)/2$, $\Delta\omega = (\omega_1 - \omega_2)/2$, $\bar{E} = (E_1 + E_2)/2$, and $\Delta E = (E_1 - E_2)/2$. This is an optical carrier with frequency of $\bar{\omega}$ and the carried signal is in band of $\omega_1 - \omega_2$. Obviously, the two laser beams must have narrow linewidth and high stability, and their frequency and phases must be locked with each other. Any fluctuations of frequencies and phases will be transformed to be noises of RF wave. Therefore it is necessary to use OPLL technology.

Figure 8.8a shows an ROF transmitter, composed of a master laser and a slave laser, locked with each other by heterodyne OPLL [35, 36]; and the frequency

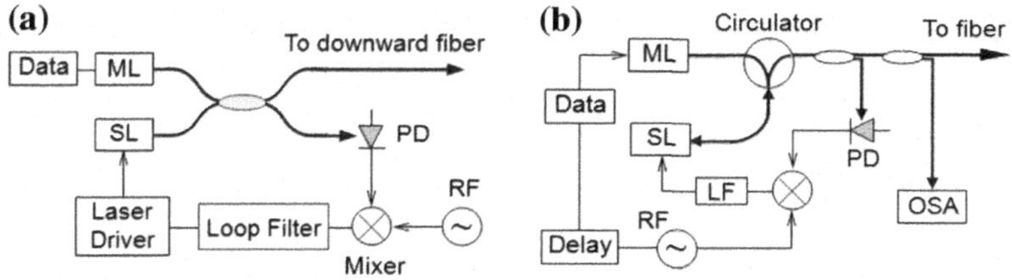

Fig. 8.8 **a** ROF transmitter with OPLL lasers; **b** ROF transmitter with OPLL and injection locking

difference between them is determined by RF source. The master laser and frequency locked slave laser are combined at the fiber coupler to downward fiber, carrying the data contained in the RF signal.

The scheme of Fig. 8.8a has some shortcomings. A certain delay must exist in the loop; it is actually a phase bias between the master and slave lasers and corresponds to a phase noise or phase drift of the RF signal to be carried. Besides, the RF frequency is limited by the locking range of OPLL. Another scheme, called the optical injection phase lock loop, was proposed to solve the problems [36–38], which combines the injection locking and OPLL together, as shown in Fig. 8.8b. Different from the conventional OPLL, one of the slave LD modes is locked by injection of the master laser, while the other mode is locked by OPLL through the feedback control of its driver. The frequency limitation by the RF source is thus broken through by the side mode injection locking; and the scheme can be used for millimeter wave signal transportation. In addition, propagations of ML and SL beams are synchronized at the output facet of SL without a phase bias, so that the phase noise is lowered greatly. A variable delay is inserted in the loop to compensate the delay of OPLL. The data to be transported is now carried by the master laser.

The high speed single sideband (SSB) *I*/*Q* modulator is one of the key devices in ROF systems. A scheme with two-stage *I*/*Q* modulator and two-stage Mach-Zehnder modulator was reported [39] and optical carriers with spacing of 4-eightfold RF frequency was obtained.

Multiple channel transmitter and receiver are key components of communication systems. References [40, 41] proposed a scheme of multi-channel transmitter by using injection locking method, as shown in Fig. 8.9. Multiple side bands of DFB laser are generated by phase modulation, which are used as seeds for injection locking of a mode locked LD (MLLD). The ±1-order side bands of DFB laser is used to lock the modes of MLLD; thus the RF frequency should be adjusted to be half of MLLD mode spacing, $f_{RF} = \Delta f/2$. The output beam is amplified by erbium-doped fiber amplifier (EDFA) and divided in pairs by an arrayed waveguide grating (AWG); and the divided outputs are then used to injection-lock the F-P cavity LDs (FPLD). As result, a series of transmitters with different optical frequency but identical RF frequency are obtained. The data to be carried are loaded to every FPLD through their drivers.

Applications of frequency swept lasers, such as coherent optical analyzer, coherent optical communications, and FMCW lidar, require crucially the source

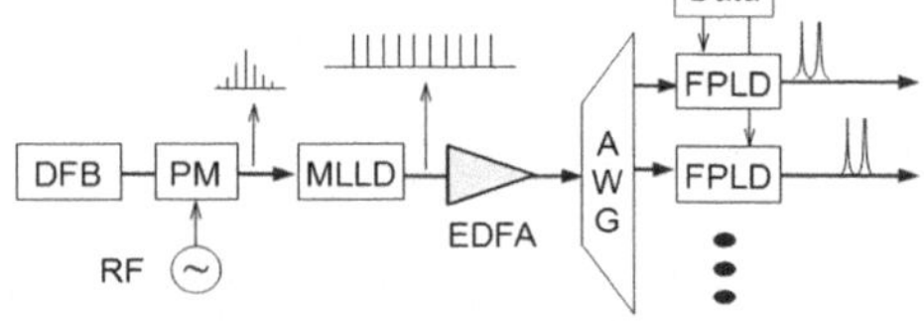

Fig. 8.9 Frequency division multiplex ROF transmitter

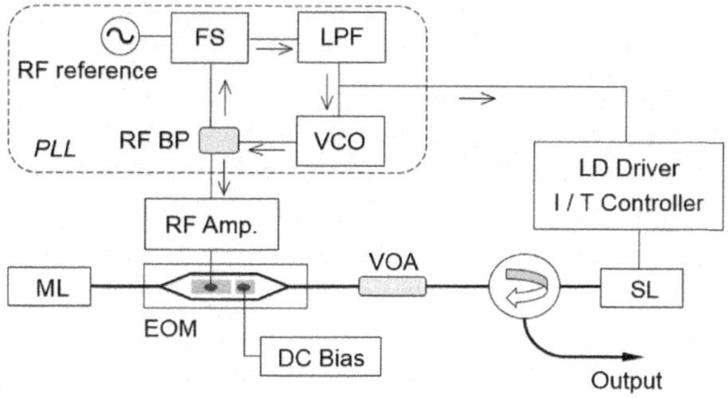

Fig. 8.10 Linearized frequency swept laser source based on sideband injection-locking. *ML*: master laser; *SL*: slave laser; FS: frequency synthesizer; *RF PS*: *RF* power splitter; *LPF*: loop filter

with high spectral purity and broad sweeping range. Reference [42] proposed and demonstrated a scheme based on high-order modulation injection-locking, as shown in Fig. 8.10.

In the experimental setup a low noise external cavity laser with linewidth less than 3 kHz is used as the master laser (ML), and a DFB LD is used as the slave laser (SL). The ML is modulated by an EOM, which is driven by an RF source controlled by a voltage controlled oscillator (VCO) with the base band suppressed and the high-order sidebands enhanced. One of the high-order sidebands (fifth in the demonstration) is tuned to the SL mode, whose driver provides precise current and temperature controlling. The RF frequency is stabilized to a reference by the phase-locked loop, as depicted by a dashed line frame in the figure. When the RF frequency is swept by $\Delta\omega$, the swept range of high-order will reach $n\Delta\omega$. In the experiment a tuning range of 15 GHz and sweeping rate of 2.5 THz/s were obtained by the 5th order side band injection locking. Linewidth of 2.5 kHz was measured during frequency sweeping, with the ML coherence maintained perfectly.

At the receiver end of ROF system, the heterodyne detection and OPLL are usually used, similar to the receiver of coherent optical communications. The high speed photodetector is one of the key components in conversion of optical wave to microwave. A special PD, called the uni-traveling-carrier waveguide photodiode (UTC-PD) has been developed for detection of optical signal and as RF emitter in a whole [43]. The OPLL technology plays important roles in fields of ROF technology, microwave photonics and lightwave radar (lidar) [44].

8.2.4 Researches on Atomic Physics

Physics of Cooled Atoms. Laser is an important and powerful tool not only for revealing material structures and physical processes inside, but also for controlling movements of atoms and molecules. Researches of laser cooling and trapping promote development of many new important technologies, especially the high precision frequency standard and time standard [45–48]. The physical picture of laser cooling may be simply understood as that a moving atom can be slowed down by collision with a photon moving in counter direction. It must be an inelastic collision, i.e. the photon is absorbed by the atom, so that the whole momentum is thus decreased due to the conservation law of momentum. Therefore the photon's energy, i.e. its frequency, must match with the atom's energy levels precisely. The excited atom will transit down to its ground states, which is usually composed of two sub-levels for different spin states. Therefore two laser beams with frequencies matching them are needed, called the pump beam and the re-pump beam [48], respectively. Many other important physical effects are involved in laser cooling process and in its applications, such as the Doppler frequency shift. The lasers used in this field must have high coherence and flexible tunability.

Typical atoms used in laser cooling and in frequency standard include rubidium (Rb), cesium (Cs), and strontium (Sr) etc. The related wavelength bands are 780 nm, 852 nm, and 689 nm, respectively, where semiconductor lasers are available. The D_2 absorption line of Rb is at 780.02 nm, corresponding to the transition of $5S_{1/2}$–$5P_{3/2}$. The frequency difference between pump and re-pump beams is required to be 6.567 GHz. OPLL and injection locking are often used to provide laser pairs for the purpose. For example, Ref. [49] utilized the technology of side band injection locking to obtain a pair of laser beams with frequencies stabilized at the two sub-levels of Rb.

For the cooled Cs atoms, the absorption line corresponding to the transition from $6S_{1/2}$ ($F = 3$) to $6S_{3/2}$ ($F' = 4$) is mostly used. The gap of sub-levels of the ground state is 8.9536 GHz. Reference [50] presented a scheme to obtain a pair of lasers, as shown in Fig. 8.11. The master laser is an AlGaAs Littrow ECDL working at 852 nm with frequency stabilized at one of the sub-levels by the saturated absorption spectroscopy (Cs-SAS), being used as a pump beam. The slave laser is modulated by RF frequency of 8.9536 GHz; one of its first order side bands is locked by the master laser, being used as a re-pump beam. In the set-up, the $\lambda/2$ wave-plate is used to rotate the polarization and to adjust the power for injection, and the polarization beam splitter (PBS) is used to ensure the polarization direction matching the optical isolator (OI) and collimating the mode of slave laser. It was

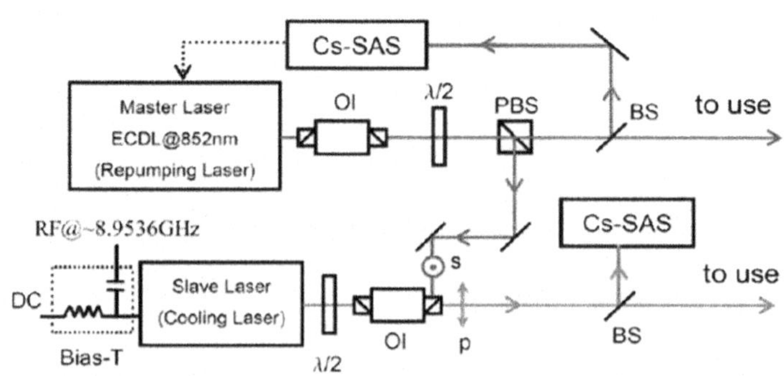

Fig. 8.11 Injection locked lasers for cooled atom researches. Reprinted from [50] with permission

reported that the injected power to SL is only 0.1 mW, whereas its output reaches 100 mW. The frequency difference between ML and SL can be tuned by tuning RF source. The set-up was reported being used to study the magneto-optical trap (MOT) in laser cooling.

Quite many researches are concentrated on the physics of cooled atoms. Reference [51] presented a scheme with two slave lasers used for pumping and re-pumping; one of the SLs was injection locked by an ECDL with linewidth of 100 kHz directly, and the other is locked by the first order side band of EOM modulated master laser beam with modulation frequency of 9.19 GHz. In the scheme one of the side bands is used as an injecting seed with the others filtered out. The work adopted a high finesse F-P cavity as the filter; when the base band of ML is tuned to one of the F-P modes, it will be transmitted totally through the F-P, whereas the sidebands beams are obtained from the reflection. High speed switching between different frequencies is also needed in researches of cooled atom physics. Reference [51] demonstrated an experiment of frequency switching in 20 μs by using a tunable F-P interferometer, which was controlled by a serve.

Electromagnetically Induced Transparency (EIT). EIT is a coherent optical nonlinearity with manifestation of a narrow width transparency window within an absorption line of medium under certain pump conditions. It is one of the hot topics in atomic physics and quantum mechanics, and is with practical interests of application. The coherent population trapping (CPT) used in laser cooling of atoms is a special case of EIT. In the researches of EIT and CPT the lasers with narrow linewidth and stable frequency matching with a group of energy levels are indispensable. Reference [52] reported a set-up for researches of velocity selective CPT, where the SL is injection locked by one of the side bands of RF modulated ML, giving a pair of lasers with frequency spacing of 9.2 GHz at 852 nm to match with the levels of Cs.

References [53–56] presented researches on EIT of rubidium with effect of hyper fine energy level structures revealed, and on the effect of laser's linewidth narrowing, which comes from the slow light effect due to the high group dispersion at the narrow transparent widow of EIT. As an example, Fig. 8.12 shows the experimental set-up used in Ref. [53] to study the linewidth reduction by intra-cavity EIT, where two ECDLs are used as the coupling laser and the probe laser, both of them are stabilized by the saturated absorption spectroscopic system (SAS) with a Rb cell as the frequency discriminator, and can be tuned separately; the laser's powers can be adjusted by combinations of half wavelength plates (HWP) and

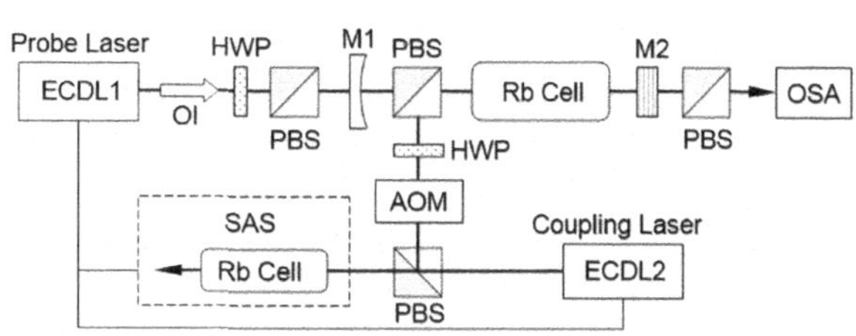

Fig. 8.12 Experimental set-up for research of intra-cavity linewidth reduction by EIT

polarization beam splitters. With *V*-type level structure of ^{85}Rb, EIT is realized in Rb cell set in the cavity composed by mirrors of M1 and M2. The cavity linewidth was measured 1.2 MHz, which is sixth of the empty cavity linewidth; in addition, the narrow linewidth is kept in the tuning range of 100 MHz. A narrower linewidth of 250 kHz was obtained in Ref. [55], and a low frequency fluctuation of 60 kHz in the observation time of 2000 s was demonstrated in Ref. [56].

Optical Tweezers. The mechanical force of optical waves can also be used to move atoms and micro particles. Based on the effect, a special technology emerged then, called the optical tweezers. More precise performances of lasers are required for the application of optical tweezers; besides the frequency and phase, the spatial distribution of optical field is also concerned [57]. There is a broad space for applications of single frequency semiconductor lasers.

8.3 Optical Frequency Comb and Its Characteristics

Optical frequency comb is a special laser source; in the time domain it is a periodic ultra-short pulsed laser, usually with pulse-width in the order of femto-second (fs = 10^{-15} s); in the spectral domain it is a frequency comb with equal spacing, narrow frequency width, and an extremely wide envelope. It is a high precision ruler in the optical spectrum. Its applications cover many areas, include the broad band optical frequency standard, the optical clock, optical frequency synthesis, precision measurements of frequency and distance, time and frequency transportation, coherent control, atto-second (10^{-18} s) sciences, space technology, astronomy, and fundamental physical researches. Because of the importance of optical frequency comb and laser-based spectroscopy, the contributors and inventors, John L. Hall and Theodor W. Hänsch were awarded the Nobel Prize in Physics 2005 [58]. By the way, Hall is one of the inventors of PDH method introduced in Sect. 6.4, and Hänsch is the pioneer of polarization spectroscopy introduced in Sect. 6.3.

The frequency comb involves a lot of physical fundamentals and advanced technologies, and is expounded in papers and monographs [58–66]. Here we just give a brief introduction to its basic characteristics and applications related to the single frequency lasers.

8.3.1 Principle of Mode-Locked Laser

The optical frequency comb is generated based on the mode-locked laser [67, 68]. The field of a multiple longitudinal mode laser is written as

$$E(t) = \sum_m A_m \exp[-j(2\pi v_m t + \phi_m)] + \text{c.c.}, \tag{8.19}$$

where A_m is the field amplitude of mth mode, and ϕ_m is its phase. The mode spacing is the same as that of an ordinary laser: $\Delta v = c/(2\eta_g L) = v_g/(2L)$, with here cavity length L, group index $\eta_g = \eta(1+\rho)$, where $\rho = (v/\eta)\partial\eta/\partial v$ is the dispersion coefficient, and group velocity of light v_g. The mode spacing is the reciprocal of roundtrip delay $\tau = 2L/v_g$. If the index η is a constant in spectrum, $\rho = 0$, the mode spacing Δv will be a constant, and the mode frequency can be written as $v_m = v_c + m\Delta v$, where v_c is an offset optical frequency. If the phases of N modes is locked-in by some method to be the same (with difference of integer multiplied 2π), and thus can be taken as zero $\phi_m = 0$, a mode locked laser is then built, whose field is written as

$$E(t) = e^{-j2\pi v_c t} \sum_{m=0}^{N-1} A_m \exp(-j2\pi m\Delta v t) + \text{c.c.} \tag{8.20}$$

In case the mode amplitudes are the same, the envelope amplitude is expressed analytically:

$$A(t) = A\sum_{m=0}^{N-1} e^{-j2\pi m\Delta v t} = A\frac{\sin(\pi N\Delta v t)}{\sin(\pi\Delta v t)} e^{-j[2\pi v_c + \pi(N-1)\Delta v]t}. \tag{8.21}$$

Its intensity is then obtained as

$$I(t) = A^2 \frac{\sin^2(\pi N\Delta v t)}{\sin^2(\pi\Delta v t)}. \tag{8.22}$$

It is a pulse train with repetition frequency of $f_{\text{rep}} = 1/\tau = \Delta v$, pulse width of $\Delta\tau \propto \tau/N$, and pulse peak intensity of $I(t) \propto N^2A^2$. If the mode amplitudes are not equal with each other, such as in a Gaussian envelope approximately, the analytical expression may not be obtained, but the laser beam is still a pulse train. It is shown that the larger the number of locked modes, the narrower the pulse width is, and the higher the peak intensity is.

In practice, the dispersion is inevitable. According to the universal Kramers–Kronig relation, the gain spectrum of laser medium causes a correspondent index spectrum. The dispersion will affect the performances of mode-locked laser. The number of locked modes will be limited, and the pulse width will be broadened. For building a mode-locked laser the main task is to find an effective mode locking mechanism and to expand the spectral range of mode locking. Generally two methods of mode-locking have been developed. One is the active mode-locking, in which gain and/or cavity loss of the laser are modulated actively with modulation frequency equal to the mode spacing. For semiconductor lasers, the active mode-locking can be realized by modulating the pumping current; or by inserting a modulator in the extended cavity. The other is the passive mode-locking, usually with a saturable absorber inserted in laser cavity. Once some power fluctuation appears, the saturable absorber will strengthen the fluctuation and suppress its temporal width; by multiple roundtrips a pulse train will be formed with the repetition frequency matching the roundtrip delay self-consistently.

Due to the great afford in the field, excellent mode locked lasers are now available. One of the mature devices is the titanium-doped sapphire mode-locked laser. Typical Ti–sapphire femto-second lasers are working at 800 nm band, with pulse width of 10–100 fs, repletion frequency of 0.1–1.0 GHz, and covered spectral range in the order of 10 THz, corresponding to tens nanometers in the near infrared band. Erbium and ytterbium doped mode-locked fiber lasers are also developed, whose central wavelengths are at 1560 nm band and 1040 nm band respectively.

8.3.2 Application of Four Wave Mixing in Frequency Trimming

The mode-locked laser is the basis of ultrahigh intensity, ultrashort pulse laser source for many important advanced researches. People also intend to utilize it as a frequency ruler for high precision metrology applications, in which the mode spacing should be exactly the same, every frequencies should be very stable, and the total number of frequencies is extremely high, even up to millions. For such purposes, the nonlinear optical effects (NLO) must be utilized, especially the four wave mixing (FWM) [67, 68].

In linear medium, the dielectric constant is independent of the intensity of incident optical beam. When the intensity increases to a certain level, the dielectric constant becomes dependent on the intensity; the electric polarization will be proportional to a power series of the electric field:

$$\boldsymbol{P} = \chi^{(1)}\boldsymbol{E} + \chi^{(2)}:\boldsymbol{EE} + \chi^{(3)}:\boldsymbol{EEE} + \cdots,$$

no longer a linear function. When two or more optical beams with different frequencies propagate in the medium simultaneously, new frequency waves will generate due to the product of electric fields. FWM is based on the third nonlinear effect with coefficient $\chi^{(3)}$. If the incident optical wave contains three frequencies: f_1, f_2 and f_3, a new frequency will generate, such as $f_4 = f_1 + f_2 - f_3$ and $f_4 = f_1 - f_2 + f_3$. The FWM occurs also in case of two frequencies, generating the third frequency of $f_3 = 2f_1 - f_2$ and $f_3 = 2f_2 - f_1$, called the degenerate FWM. It is noticed that the generated frequency is determined by the difference of original frequencies, independent of the medium dispersion. Furthermore, the generated frequency will mix with the original frequencies, generating more new frequencies. Therefore, if FWM occurs continuously, a series of frequencies with exactly equal spacing will generate.

However, as the new frequencies generate, their intensities will be lowered; and the efficiency of nonlinear effect is thus declined correspondingly. Therefore a sustained FWM process must be assisted by certain power amplification. Fortunately, high nonlinear optical fibers have been developed, such as photo crystal fibers (PCF) and micro-structured fibers, where strong nonlinear effects will occur, including FWM, optical parameter amplification (OPA), Raman amplification, and other nonlinear effects; and supercontinuum has been realized based on the high nonlinear optical fibers. The details can be read in journals and monographs [67–70].

Incorporated with FWM, optical frequency combs are now built based on mode-locked lasers. The mode spacing will be trimmed to equal with each other by the FWM in a propagation distance long enough, and number of frequency teeth will be increased more and more. The necessary condition is the FWM process sustained long enough. The PCF with high NLO coefficients can perfectly meet the requirements [71, 72]. Another scheme is with a resonant cavity, in which FWM effect will be enhanced greatly [73, 74]. A frequency comb is then generated.

8.3.3 Control of Carrier-Envelope Phase

An ideal frequency comb is an exactly periodical pulse train based on a mode locked laser with repetition frequency locked on the frequency standard in RF band. In spectrum domain it can be described by a Fourier series of frequency components with exactly equal spacing of [71, 72].

$$f_n = f_0 + nf_{\mathrm{rep}}, \tag{8.23}$$

where f_0 is the carrier-envelope offset frequency; $f_{\mathrm{rep}} = 1/\tau$ is the frequency spacing, i.e. the modulation frequency of active mode-locked laser. The range of integer n is determined by the gain spectrum; it may be written as $n_0 \leq n \leq n_0 + N - 1$ for total N modes.

Since the optical frequencies are not necessary an integer multiple of f_{rep}, existence of a carrier-envelope offset frequency f_0 is inevitable. The offset frequency will cause a phase difference between the pulse peak and the peak of optical wave, as shown in Fig. 8.13. The optical field is now expressed as

$$E(t) = \mathrm{e}^{-j2\pi f_0 t} \sum_{n=n_0}^{n_0+N-1} A_{\mathrm{n}} \exp(-j2\pi n f_{\mathrm{rep}} t) + \mathrm{c.c.} \tag{8.24}$$

In time domain, the pulse peaks are located at $t = M/f_{\mathrm{rep}} = M\tau$. With $m = n - n_0$ used as the ordinal number of the summation, the peak field at the peak moments is written as:

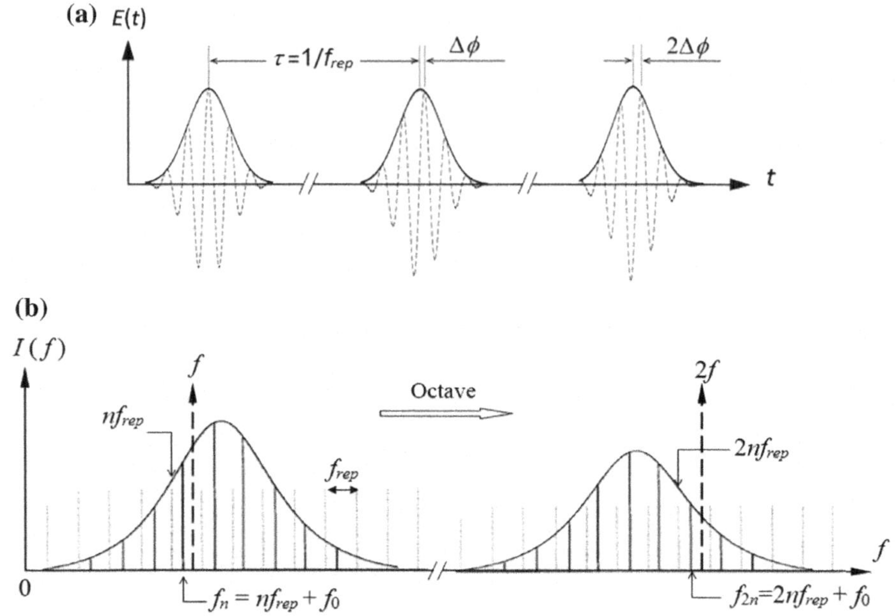

Fig. 8.13 **a** Waveform of femto-second pulse train; **b** Spectral relations of comb and its octave

$$E(M\tau) = \mathrm{e}^{-j2\pi(f_0 + n_0 f_{rep})M\tau} \sum_{\mathrm{m}=0}^{N-1} A_{\mathrm{m}} \mathrm{e}^{-j2\pi m f_{\mathrm{rep}} M\tau} + \mathrm{c.c.} = \mathrm{e}^{-j2\pi f_0 M\tau} \sum_{\mathrm{m}=0}^{N-1} A_{\mathrm{m}} + \mathrm{c.c.} \tag{8.25}$$

The phase difference between the adjacent pulse peaks, Mth and (M + 1)th, is then obtained as $\Delta\phi = 2\pi f_0 \tau$, which is called the carrier-envelop phase, as shown in Fig. 8.13a [59, 60].

One of the important applications of frequency comb is to establish a high precision frequency standard and time standard, i.e. an optical clock. The modern time standard affirmed by international organizations is based on the transition between two ultra-fine levels of ^{133}Cs; that is, one second equals 9,192,631,770 periods of the radiation generated by the transition. Because the optical frequency is 3–4 orders higher than microwave frequency, the precision of timing by using optical frequency will be much higher than that by microwave, if the precision of frequency counting is the same. For the purpose, the measurement of optical frequency must be linked with the microwave frequency of radiation of ^{133}Cs or other atoms. If the mode spacing of optical frequency comb is equal to the frequency standard in RF band, the comb will play a role of time standard in optical band.

The precision of frequency counting is affected by the carrier-envelop phase $\Delta\phi$. Such a phase exists almost in all mode-locked lasers, but it does not lead to serious effects for wider pulses. However, in applications of time standard, the pulse width is decreased to the order of 10 fs, in which only 2–3 period of optical oscillation is involved. The variation of carrier-envelop phase becomes an important noise of pulse peak position measurement, and brings about a serious influence on the quantitative relationship between the optical frequency and the microwave frequency. In applications, it is needed to transport the comb through optical fibers; the fiber dispersion will further enlarge the carrier-envelope phase. The length and the index of fiber and other medium are susceptible to temperature fluctuation and mechanical vibrations. Due to such noises the carrier-envelope phase will fluctuate randomly.

Therefore the carrier-envelope phase must be measured and removed, especially for the comb applications with metrological significations, such as atomic clocks. A special technology, called self-referencing is developed for the purpose [60–62]. Its basic concept is to compare the comb spectrum $f_n = nf_{\text{rep}} + f_0$ at the longer wavelength band with its optical octave $f_{2n} = 2nf_{\text{rep}} + f_0$ at the short wavelength band; the offset frequency can then be obtained: $f_0 = 2f_n - f_{2n}$. The frequency span for producing f_{2n} is so wide that the supercontinuum has to be utilized, which combines FWM, the optical parameter amplification (OPA) and other nonlinear effects together in high nonlinear photonic crystal fibers [69]. In the process the comb frequency can be duplicated with equal spacing, while the offset frequency keeps unchanged. Besides, two reference frequencies f and $2f$ are needed to measure frequencies f_n and f_{2n} by beating, as illustrated in Fig. 8.13b.

Figure 8.14 shows a schematic diagram of typical setup for carrier-envelop offset frequency measurement [60–63]. The optical frequencies of comb must be traced to the frequency standard in RF range for its applications as the frequency ruler. The repetition frequency of femto-second laser is stabilized on the reference frequency given by the synthesizer traced to a Cs clock by using PZT to control its cavity length. The frequency comb is obtained in transmitted beam of fs laser from the nonlinear fiber (NLO). A frequency stabilized laser is used to generate reference frequencies of f and $2f$, by using the second harmonic generation (SHG). For example, an Nd–YAG laser with frequency stabilized at one of the absorption lines of iodine (I_2) was used in Ref. [60]. Two groups of frequency pairs are resolved by

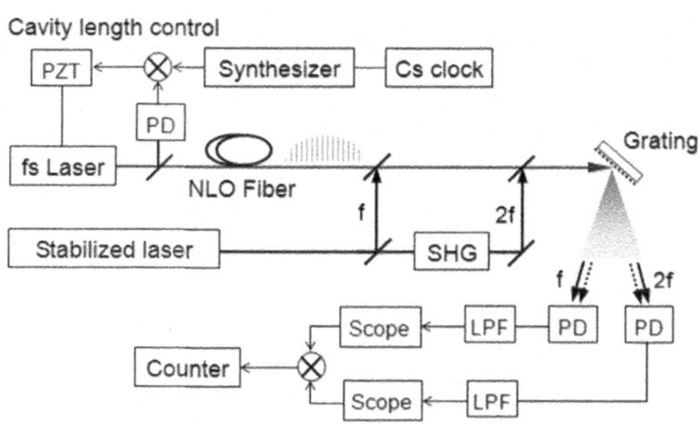

Fig. 8.14 Measurement setup of carrier-envelope offset frequency

the grating, and correlated at the two PDs respectively. The beat frequencies are measured by a frequency counter:

$$\begin{aligned} \Delta f_1 &= \left| f - (nf_{\text{rep}} + f_0) \right| \\ \Delta f_2 &= \left| 2f - (2nf_{\text{rep}} + f_0) \right|. \end{aligned} \tag{8.26}$$

The carrier-envelope offset frequency f_0 can then be obtained from $2\Delta f_1 - \Delta f_2$, after the possible sign uncertainty in the absolute arguments above is cleared [60]. The measured offset frequency is then used to determine the frequencies of comb teeth. The process of such measurements and controlling involves a series high precision technologies and deep physical mechanisms. Details can be read in the references.

8.4 Applications of Optical Frequency Comb in Laser Frequency Transfer

As stated above, the applications of frequency comb cover many areas. Among them, laser frequency stabilization and frequency transfer are mostly attractive and useful.

8.4.1 Optical Frequency Locking and Frequency Synthesizer

The optical frequency comb provides an absolute frequency standard in an extremely wide band to be used for the frequency stabilization of lasers in all over the band. In applications, a source with an arbitrary accurate frequency is often required, called the optical frequency generator or the optical frequency synthesizer (OFS). The comb contains 10 thousands to millions frequencies. However, limited by the total output power, the power of each comb tooth is very small, unable to be used simply by just spectrally decomposition. The mostly used method for OFS is to use one of the comb frequencies as a reference frequency to stabilize a laser, so that the laser will give output not only with an accurate frequency, but also with higher power.

Figure 8.15 shows a schematic diagram of frequency stabilized laser based on comb [75]. The single frequency laser can be correlated with all of the comb teeth, giving a series beat frequencies; the lowest beat frequency corresponds to the nearest comb tooth. A narrow linewidth low pass filter (LPF) is used to select one of them as the frequency discriminating signal. It is usually converted from analog to digital and feedback to the laser driver via PPL to build an OPLL. When the OPLL is established closely, the laser frequency will be stabilized to the selected comb

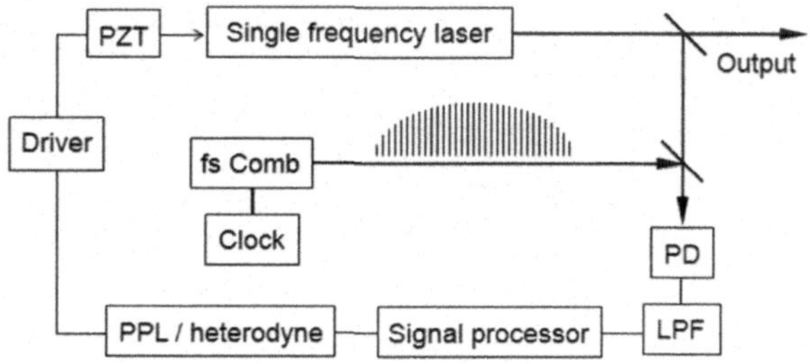

Fig. 8.15 A laser with frequency locked at accurate standard given by comb

tooth. By using a heterodyne PLL, the laser frequency can be locked to some position between two adjacent teeth. Good results were demonstrated by the scheme in the experiments reported by Ref. [75], where a comb with spectral width of 520–1100 nm was used, and the signal-to-noise ratio of output reached 40 dB, measured with 100 kHz detection bandwidth.

The combs are also used for the injection locking of semiconductor lasers. By the method multiple diode lasers can be locked simultaneously with absolutely standardized frequencies. Figure 8.16 shows a schematic diagram of such a system [76, 77], where the femto-second comb is used as the master laser, the scanning F-P interferometer is used to select a group of comb teeth, which are resolved spatially by the grating. On the part of slave lasers, an optical isolator (OI) based on Faraday rotation is used, and the locked laser beam outputs from the polarization beam splitter. A taper waveguide semiconductor amplifier (SOA) may be used for those comb teeth with low powers. The system contains monitoring and controlling elements for optimization of working state, which is not depicted in the figure for simplicity.

Since the comb has functions of tracking frequencies to the frequency standard, it is a powerful tool for measuring frequencies with absolute metrological significance. In the experiment reported by Ref. [78], one of the comb teeth was selected by an arrayed waveguide grating (AWG), which is used widely in fiber communication systems, and used to stabilize the frequency of DFB laser by injection locking. Its output was then mixed with another laser, which is stabilized by P16

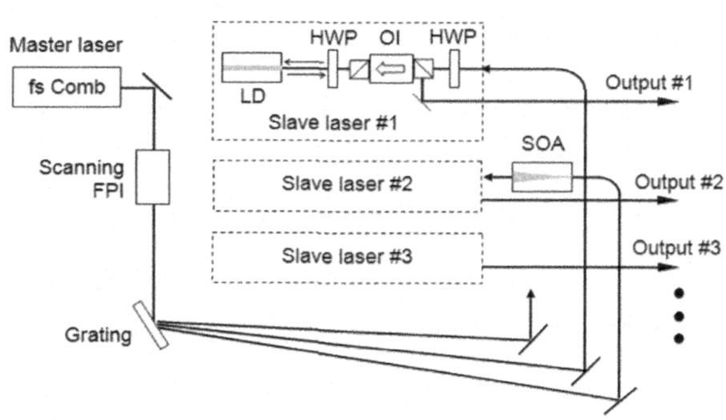

Fig. 8.16 Injection locked laser system based on frequency comb

absorption line of acetylene ($^{13}C_2H_2$). From the beat frequency, the frequency of P16 transition is then precisely measured to be 194,369,569,385.7 kHz.

8.4.2 High Precision Optical Frequency Sweeping

Frequency swept lasers are often required to give quantitatively absolute frequency reading in some applications, such as a frequency (wavelength) meter. The frequency comb makes it possible. However, if the frequency is located at the middle of two teeth, $(f_n + f_{n+1})/2$, an uncertainty will occur both in case of frequency locking and in frequency sweeping, especially when the sweeping range is over two or more comb teeth. To solve the uncertainty, several methods have been proposed. The basic idea is to adjust the repetition rate of comb controllably.

Figure 8.17 shows a scheme proposed by Ref. [79]. In the system a part of output beam of the frequency swept laser is modulated by an acousto-optic modulator; its 0th order component, i.e. the laser frequency f_{cw}, and the first order component $(f_{cw} + f_{RF})$ are beaten with the comb output, and detected by PD1 and PD2 respectively. The comb output is also correlated with an idiom stabilized laser with beat signal detected by PD3 for tracking the frequency standard. The three beat signals are processed in the digital servo, which provides three signals to control the frequency swept laser, the AOM, and the repletion rate of comb, respectively. When the tuned frequency approaches and crosses the middle point between the teeth, the modulation frequency of AOM will be adjusted or switched to maintain the continuous tuning. All the frequencies are locked to the atomic clock. A sweeping speed of 30 GHz/s and a frequency stability of $\sim 2 \times 10^{-14}$ were reported in this system.

Because the power of a single tooth is very low, the frequency stabilized laser utilizing only one comb tooth has a shortcoming of low SNR. A scheme of using multiple comb teeth was proposed [80], as shown in Fig. 8.18. A comb based on fs mode-locked fiber laser at 1560 nm band was used in the experiment. The system has the following features.

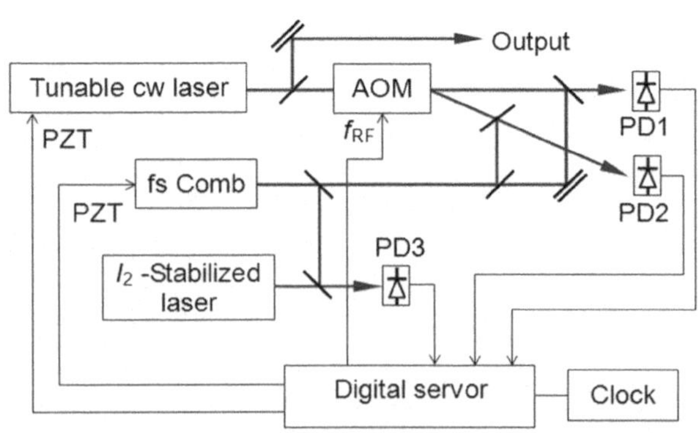

Fig. 8.17 A frequency swept laser locked to a comb

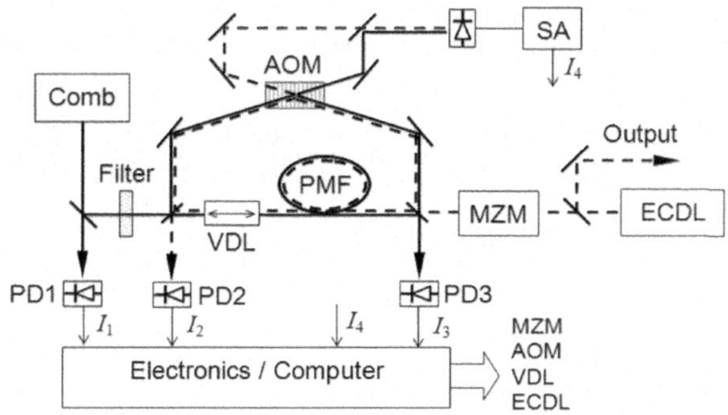

Fig. 8.18 A frequency stabilized laser by using comb. *Solid line*: comb path; *doted line*: ECDL path; *VDL* *Variable delay line*; *MZM* M-Z modulator; *SA* Spectrum analyzer

An imbalanced Mach-Zehnder interferometer is used as the frequency discriminator, with a 10 m long PMF inserted in one arm to shift the interference signal to a bias frequency of ν_{aom}. The output of MZI for comb light, depicted by solid lines in the figure, is detected by PD3. The output of MZI for ECDL laser, depicted by dashed lines, is detected by PD2. To distinguish the signals of comb and those of ECDL, a double sideband Mach-Zehnder modulator (MZM) is used to shift the frequency of ECDL by ν_{mod}. The beat signals of AOM transmissions of comb and ECDL are analyzed by the spectrum analyzer. The interferometer works in the way of common mode for the two light waves to mitigate external disturbances.

The conducive bandwidth of MZI is related to the second order dispersions of fiber, $\beta_2 = \left|\partial^2\beta/\partial\omega^2\right|$, which induces a dephasing factor $\Delta\phi = \beta_2 L\Delta\omega^2/2$ for path difference L. If a dephasing of 1 rad is taken for estimation, the conducive bandwidth is obtained to be $\Delta\nu_{\text{comb}} = 2/(\pi\sqrt{2\beta_2 L})$. The PMF used here is helpful to avoid the polarization fading. The number of comb lines in the bandwidth reached 10 thousands in the experiment, whose power is much higher than only one line when used for discriminating the frequency of ECDL. To match with the bandwidth a filter is used to get rid of comb lines out of the band, which will otherwise reduce the visibility of interferometer.

The optical path difference is fine-adjusted by the variable delay line (VDL) to obtain synchronization of pulses. The detected signals of three PDs, I_1, I_2 and I_3, contain interference signals with different frequencies. By mixings of them in data processing, the error signal, $\propto \sum_n \cos[2\pi(\nu_{\text{ecdl}} - \nu_n)t + \phi]$, can be obtained and is used to lock the ECDL frequency. Detailed analyses of this scheme are given in Ref. [80], and experimental results are demonstrated with frequency stability of $\Delta\nu/\nu = 3 \times 10^{-9}$.

For applications of comb in the fiber communications, Ref. [81] proposed a scheme of frequency sweeping by means of tuning the repetition frequency of fs mode-locked laser; and arbitrarily customized frequencies were obtained in the C-band covering 192–196 THz. In the experiment a comb based on Ti–sapphire fs laser was used, whose frequencies were locked on the standard given by hydrogen-maser, and frequencies of external cavity diode lasers were locked on the comb lines. The frequency sweeping range of several GHz was reported. Reference

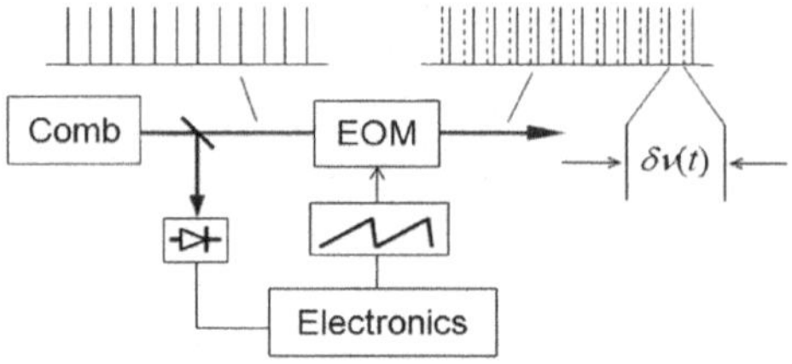

Fig. 8.19 Scheme of externally tuning comb

[75] reported experiment on fast frequency switching, showing a response rate of 8 MHz per millisecond.

In the above schemes of tuning comb the basic idea is to adjust the repetition rate, so that different comb teeth will be tuned in different rate because the spacing between them is changed. Reference [82] proposed a different scheme, which is based on the external modulation, which will tune all the comb teeth in the same rate with the spacing kept unchanged. Such a tuning way is required in some applications. Figure 8.19 shows a schematic diagram of the external modulation of comb. The external modulation here is different from that described in Sect. 7.3 for tuning a single frequency laser working in continuous wave mode (cw). The external modulation of comb is for a pulse train; the driving voltage of modulator must be synchronized with the pulse repetition rate.

Referring to formula (7.17), the phase variation is a function of wavelength, written as

$$\Delta\phi = \frac{\pi\gamma(\lambda)n_0^3(\lambda)}{\lambda}\frac{V(t)L}{d} \tag{8.27}$$

Obviously, it is not only inversely proportional to the wavelength, but also determined by the dependence of the electro-optic coefficient of crystals on wavelength; and the dispersion of crystal has to be taken into account also. It is much different from the case of cw single frequency laser, since the spectrum of comb covers an extremely large range. To overcome the problems, real-time monitoring and feedback control are necessary. Detailed experimental setup and results are given in Ref. [82].

The optical frequency comb plays important roles in many areas, especially in researches of the cooled atom physics and the quantum optics [83–86]. Detailed reviews are given in Refs. [87, 88].

References

1. Ristic S, Bhardwaj A, Rodwell MJ et al (2010) An optical phase-locked loop photonic integrated circuit. J Lightwave Technol 28(4):526–538
2. Herzog FT (2006) An optical phase locked loop for coherent space communications. Dissertation of Swiss Federal Institute of Technology, Zurich

3. Ferrero V, Camatel S (2008) Optical phase locking techniques: an overview and a novel method based on single side sub-carrier modulation. Opt Express 16(2):818–828
4. Kashima N, Yamaguchi S, Ishii S (2004) Optical transmitter using side-mode injection locking for high-speed photonic LANs. J Lightwave Technol 22(2):550–557
5. Snadden MJ, Clarke RBM, Riis E (1997) Injection-locking technique for heterodyne optical phase locking of a diode laser. Opt Lett 22(12):892–894
6. Ip E, Lau APT, Barros DJF et al (2008) Coherent detection in optical fiber systems. Opt Express 16(2):753–791
7. Herzog F, Kudielka K, Erni D et al (2005) Optical phase locked loop for transparent inter-satellite communications. Opt Express 13(10):3816–3821
8. Yoshida M, Goto H, Kasai K et al (2008) 64 and 128 coherent QAM optical transmission over 150 km using frequency-stabilized laser and heterodyne PLL detection. Opt Express 16(2):829–840
9. Shieh W, Bao H, Tang Y (2008) Coherent optical OFDM: theory and design. Opt Express 16(2):841–859
10. Clark TR Jr, O'Connor SR, Dennis ML (2010) A phase-modulation I/Q-demodulation microwave-to-digital photonic link. IEEE Trans Microwave Theory Tech 58(11):3039–3058
11. Li J, Zhang X, Tian F et al (2011) Theoretical and experimental study on generation of stable and high-quality multi-carrier source based on re-circulating frequency shifter used for Tb/s optical transmission. Opt Express 19(2):848–860
12. Hodgkinson TG (1986) Costas loop analysis for coherent optical receiver. Electron Lett 22(7):394–396
13. Fang Z, Chin K, Qu R et al (2012) Fundamentals of optical fiber sensors. Wiley
14. Stephens TD, Nicholson G (1987) Optical homodyne receiver with a six-port fiber coupler. Electron Lett 23(21):1106–1108
15. Wang Y, Leeb WR (1987) A 90° optical fiber hybrid for optimal signal power utilization. Appl Opt 26(19):4181–4184
16. Garreis R, Zeiss C (1991) 90° optical hybrid for coherent receivers. Proc SPIE 1522:210–219
17. Langenhorst R, Wenke G (1989) Compact bulk optical 90° hybrid for balanced phase diversity receivers. Electron Lett 25(22):1518–1519
18. Herzog F, Kudielka K, Erni D et al (2006) Optical phase locking by local oscillator phase dithering. IEEE J Quantum Electron 42(10):973–985
19. von Lerber T, Honkanen S, Tervonen A et al (2009) Optical clock recovery methods: review. Opt Fiber Technol 15:363–372
20. Ma L, Jungner P, Ye J et al (1994) Delivering the same optical frequency at two places: accurate cancellation of phase noise introduced by an optical fiber or other time-varying path. Opt Lett 19(21):1777–1779
21. Foreman SM, Holman KW, Hudson DD et al (2007) Remote transfer of ultrastable frequency references via fiber networks. Rev Sci Instrum 78:021101(1–25)
22. Musha M, Hong F, Nakagawa K et al (2008) Coherent optical frequency transfer over 50-km physical distance using a 120-km-long installed telecom fiber network. Opt Express 16(21):16459–16466
23. Lopez O, Haboucha A, Chanteau B et al (2012) Ultra-stable long distance optical frequency distribution using the internet fiber network. Opt Express 20(21):23518–23526
24. Lopez O, Kanj A, Pottie PE et al (2013) Simultaneous remote transfer of accurate timing and optical frequency over a public fiber network. Appl Phys B 110:3–6
25. Calonico D, Bertacco EK, Calosso CE et al (2014) High-accuracy coherent optical frequency transfer over a doubled 642-km fiber link. Appl Phys B 117:979–986
26. Williams PA, Swann WC, Newbury NR (2008) High-stability transfer of an optical frequency over long fiber-optic links. J Opt Soc Am B 25(8):1284–1293
27. Eckstein JN, Ferguson AI, Hänsch TW (1978) High-resolution two-photon spectroscopy with picosecond light pulses. Phys Rev Lett 40(13):847–850
28. Calosso CE, Bertacco EK, Calonico D et al (2015) Doppler-stabilized fiber link with 6 dB noise improvement below the classical limit. Opt Lett 40(2):131–134

29. Chen W, Liu Q, Cheng N et al (2015) Joint time and frequency dissemination network over delay-stabilized fiber optic links. IEEE Photon J 7(3):7901609(1–10)
30. Cheng N, Chen W, Liu Q et al (2016) Joint transfer of time and frequency signals and multi-point synchronization via fiber network. Chin Phys B 25(1):014206(1–8)
31. Seeds AJ (2002) Microwave photonics. IEEE Trans Microwave Theory Tech 50(3):877–887
32. Yao J (2009) Microwave photonics. J Lightwave Technol 27(3):314–335
33. Stephens WE, Joseph TR (1987) System characteristics of direct modulated and external modulated RF fiber-optic links. J Lightwave Technol 5(3):380–387
34. Yu J, Chang GK, Jia Z et al (2006) A ROF downstream link with optical mm-wave generation using optical phase modulator for providing broadband optical-wireless access service. Optical fiber communication conference, paper-OFM3
35. Gliese U, Nielsen TN, Bruun M et al (1992) A wideband heterodyne optical phase-locked loop for generation of 3–18 GHz microwave carriers. IEEE Photon Technol Lett 4(8):936–938
36. Bordonalli AC, Walton C, Seeds AJ (1999) High-performance phase locking of wide linewidth semiconductor lasers by combined use of optical injection locking and optical phase-lock loop. J Lightwave Technol 17(2):328–342
37. Ramos RT, Gallion P, Erasme D et al (1994) Optical injection locking and phase-lock loop combined systems. Opt Lett 19(1):4–6
38. Johansson LA, Liu CP, Seeds AJ (2002) A 65-km span unamplified transmission of 36 GHz radio-over-fiber signals using an optical injection phase-lock loop. IEEE Photon Technol Lett 14(11):1596–1598
39. Zhao Y, Zheng X, Wen H et al (2009) Simplified optical millimeter-wave generation configuration by frequency quadrupling using two cascaded Mach-Zehnder modulators. Opt Lett 34(21):3250–3252
40. Lin CT, Shih PT, Jiang WJ et al (2009) A continuously tunable and filterless optical millimeter-wave generation via frequency octupling. Opt Express 17(22):19749–19756
41. Hu WW, Inagaki K, Tanaka T et al (2004) Millimetre-wave band (50 GHz) multi-carrier generation using injection-locking technique for radio-over-fiber WDM communication system. Electron Lett 40(23):1505–1506
42. Wei F, Lu B, Wang J et al (2015) Precision and broadband frequency swept laser source based on high-order modulation-sideband injection-locking. Opt Express 23(4):4970–4980
43. Ito H, Furuta T, Nakajima F et al (2005) Photonic generation of continuous THz wave using uni-traveling-carrier photodiode. J Lightwave Technol 23(12):4016–4021
44. Gao S, Hui R (2012) Frequency-modulated continuous-wave lidar using I/Q modulator for simplified heterodyne detection. Opt Lett 37(11):2022–2024
45. Chu S (1998) The manipulation of neutral particles. Rev Mod Phys 70(3):685–703
46. Cohen-Tannoudji, Dupont-Roc, Grynberg (1992) Atom-photon interaction: basic processes and applications. Wiley
47. Phillips WD (1998) Laser cooling and trapping of neutral atoms. Rev Mod Phys 70(3):721–741
48. Wang Y (2007) Laser cooling and trapping of atoms. Peking University Press (in Chinese)
49. Kowalski R, Root S, Gensemer SD et al (2001) A frequency-modulated injection-locked diode laser for two-frequency generation. Rev Sci Instrum 72:2532–2534
50. Diao W, He J, Liu Z et al (2012) Alternative laser system for cesium magneto-optical trap via optical injection locking to sideband of a 9-GHz current-modulated diode laser. Opt Express 20(7):7480–7487
51. Szymaniec K, Ghezali S, Let Cognet et al (1997) Injection locking of diode lasers to frequency modulated source. Opt Commun 144:50–54
52. Park SE, Kwon TY, Lee HS (2003) Production of Raman laser beams using injection-locking technique. IEEE Trans Instrum Measur 52(2):277–279
53. Ying K, Niu Y, Chen D et al (2014) Realization of cavity linewidth narrowing via interacting dark resonances in a tripod-type electromagnetically induced transparency system. J Opt Soc Am B 31(1):144–148

54. Ying K, Niu Y, Chen D et al (2014) Cavity linewidth narrowing by optical pumping-assisted electromagnetically induced transparency in V-type rubidium at room temperature. J Mod Opt 61(4):322–327
55. Ying K, Niu Y, Chen D et al (2014) Observation of multi-electromagnetically induced transparency in V-type rubidium atoms. J Mod Opt 61(8):631–635
56. Ying K, Niu Y, Chen D et al (2014) Laser frequency offset locking via tripod-type electromagnetically induced transparency. Appl Opt 53(12):2632–2637
57. Thom T, Wilpers G, Riis E et al (2013) Accurate and agile digital control of optical phase, amplitude and frequency for coherent atomic manipulation of atomic systems. Opt Express 21(16):18712–18723
58. Hänsch TW (2006) Nobel lecture: passion for precision. Rev Mod Phys 78(4):1297–1309
59. Xu L, Spielmann C, Poppe A et al (1996) Route to phase control of ultrashort light pulses. Opt Lett 21(24):2008–2010
60. Diddams SA, Jones DJ, Ye J et al (2000) Direct link between microwave and optical frequencies with a 300 THz femtosecond laser comb. Phys Rev Lett 84(22):5102–5105
61. Holzwarth R, Udem T, Hänsch TW et al (2000) Optical frequency synthesizer for precision spectroscopy. Phys Rev Lett 85(11):2264–2267
62. Jones DJ, Diddams SA, Ranka JK et al (2000) Carrier-envelope phase control of femtosecond mode-locked lasers and direct optical frequency synthesis. Science 288:635–639
63. Diddams SA, Jones DJ, Ma LS et al (2000) Optical frequency measurement across a 104 THz gap with a femtosecond laser frequency comb. Opt Lett 25(3):186–188
64. Ma LS, Bi Z, Bartels A et al (2004) Optical frequency synthesis and comparison with uncertainty at the 10^{-19} level. Science 303:1843–1845
65. Jones RJ, Moll KD, Thorpe MJ et al (2005) Phase-coherent frequency combs in the vacuum ultraviolet via high-harmonic generation inside a femtosecond enhancement cavity. Phys Rev Lett 94(19):193201(1–4)
66. Ye J, Cundiff ST (2005) Femtosecond optical frequency comb: principle. Springer, Operation and Applications
67. Yariv A (1997) Optical electronics in modern communications, 5th edn. Oxford University Press
68. Saleh BEA, Teich MC (2007) Fundamentals of photonics, Wiley
69. Dudley J, Genty G, Coen S (2006) Supercontinuum generation in photonic crystal fiber. Rev Mod Phys 78:1135–1184
70. Agrawal GP (2004) Nonlinear fiber optics. Elsevier Science
71. Sefler GA, Kitayama K (1998) Frequency comb generation by four-wave mixing and the role of fiber dispersion. J Lightwave Technol 16(9):1596–1605
72. Boggio JMC, Moro S, Windmiller JR et al (2009) Optical frequency comb generated by four-wave mixing in highly nonlinear fibers. Conference of Laser and Electro-Optics (CLEO), CMN7
73. Kourogi M, Enami T, Ohtsu M (1994) A monolithic optical frequency comb generator. IEEE Photon Technol Lett 6(2):214–217
74. Del'Haye P, Schliesser A, Arcizet O et al (2007) Optical frequency comb generation from a monolithic microresonator. Nature 450:1214–1217
75. Jost JD, Hall JL, Ye J (2002) Continuously tunable, precise, single frequency optical signal generator. Opt Express 10(12):515–520
76. Park SE, Kim EB, Park YH et al (2006) Sweep optical frequency synthesizer with a distributed-Bragg-reflector laser injection locked by a single component of an optical frequency comb. Opt Lett 31(24):3594–3596
77. Kim YJ, Jin J, Kim Y et al (2008) A wide-range optical frequency generator based on the frequency comb of a femtosecond laser. Opt Express 16(1):258–264
78. Ryu HY, Lee SH, Lee WK et al (2008) Absolute frequency measurement of an acetylene stabilized laser using a selected single mode from a femtosecond fiber laser comb. Opt Express 16(5):2867–2873

79. Schibli TR, Minoshima K, Hong FL et al (2005) Phase-locked widely tunable optical single-frequency generator based on a femtosecond comb. Opt Lett 30(17):2323–2325
80. Benkler E, Rohde F, Telle HR (2013) Robust interferometric frequency lock between cw lasers and optical frequency combs. Opt Lett 38(4):555–557
81. Ahtee V, Merimaa M, Nyholm K (2009) Single-frequency synthesis at telecommunication wavelengths. Opt Express 17(6):4890–4896
82. Benkler B, Rohde F, Telle HR (2013) Endless frequency shifting of optical frequency comb lines. Opt Express 21(5):5793–5802
83. Margolis HS, Huang G, Barwood GP et al (2003) Absolute frequency measurement of the 674-nm $^{88}Sr^{+}$ clock transition using a femtosecond optical frequency comb. Phys Rev A 67:032501(1–5)
84. Fortier TM, Coq YL, Stalnaker JE et al (2006) Kilohertz-resolution spectroscopy of cold atoms with an optical frequency comb. Phys Rev Lett 97:163905(1–4)
85. Thorpe MJ, Hudson DD, Moll KD et al (2007) Cavity-ringdown molecular spectroscopy based on an optical frequency comb at 1.45–1.65 μm. Opt Lett 32(3):307–309
86. Pysher M, Miwa Y, Shahrokhshahi R et al (2011) Parallel generation of quadripartite cluster entanglement in the optical frequency comb. Phys Rev Lett 107:030505(1–4)
87. Ye J, Schnatz H, Hollberg LW (2003) Optical frequency combs: from frequency metrology to optical phase control. IEEE J Sel Top Quantum Electron 9(4):1041–1057
88. Diddams SA (2010) The evolving optical frequency comb. J Opt Soc Am B 27(11):B51–B62

Chapter 9
Applications of Single-Frequency Semiconductor Lasers

9.1 Applications in Laser Cooling and Related Technologies

Laser cooling of atoms is one of the most attractive progresses in laser technology. The cold atomic clock and optical clock are developed based on the achievement. The single-frequency semiconductor laser is a key device in R&D of this field.

9.1.1 Time Standard and Atomic Clock

Time is one of the most important physical quantities, not only for people's daily life, but also for development of sciences and technologies. People began to measure the time thousands years ago by sundial, water-drop counting, and other methods. The pendulum clock, spring clock, and watch were invented long time ago, still used nowadays. Timing with error less than one second per day may meet the demand of daily life for ordinary persons, but far away from the requirement of modern technologies. Although the precision of quartz clock has reached 10^{-4} s per day in the past century based on the achievement of electronics, the frontier scientific researches and the more advanced technologies require higher precision of several orders. Noticing the stability of atom's absorption spectra, people made efforts to develop a new frequency standard and time standard based on the atom's absorption lines. A cesium (^{133}Cs) atomic clock was developed, indicating the emergence of a new era of time counting, the atom's time [1]. Based on the successful development of atomic frequency standard, one second was defined as the duration of 9,192,631,770 cycles of the electromagnetic radiation corresponding to the transition between the two fine levels of cesium in the Thirteenth International Conference of Metrology 1967. The new standard replaces the

Z. Fang et al., *Single Frequency Semiconductor Lasers*, Optical and Fiber Communications Reports 9, DOI 10.1007/978-981-10-5257-6_9

original standard based on the moving period of the sun; the astronomical time unit was then measured with the atom frequency standard [2, 3].

The basic structure of atomic clock is composed of a frequency transfer chain, where the frequency of crystal oscillator is locked to the resonant frequency of a resonator containing the atom's vapor. The mostly used atoms include cesium (Cs), rubidium (Rb), and hydrogen (H-maser). Figure 9.1a shows a schematic diagram of atomic clock, where the microwave resonant cavity filled with the atom's vapor plays a role of frequency discriminator. The beat signal given by mixing of the atom's transition frequency and the crystal oscillation is used as feedback error signal to control the frequency of crystal oscillator. A frequency synthesizer is used to cascade the oscillation frequency matching with the transition frequency, and PLL is used to process the error signal.

The precision of atomic clock is increased with the sustaining time of reaction between microwave and atoms, which move in a straight line. Since the velocity of atom's movement is very fast in room temperature, a long cavity is needed. However, some unfavorable factors occur in a long cavity, related to the uniformity of microwave field, the mechanical stability, and others. To avoid those difficulties N.F. Ramsey invented a special cavity after his name, as shown in Fig. 9.1b schematically. It is a cavity composed of two symmetrically and separately arranged sections with microwave phases locked with each other. With Ramsey cavity the performances of atomic clock were improved greatly and the atomic clock was developed successfully. Due to this contribution, Ramsey won the Nobel Prize in physics 1989 [4]. In the structure of Fig. 9.1b, the pump laser is used to select one single state from the fine level structures of the atoms; the probe laser is used to excite photoluminescence of the state reacted with the microwave, and detected by a photodiode. When the microwave frequency is tuned, the PD signal will show an interference pattern of atom-microwave reactions, called Ramsey fringes, which have linewidth much less than that of ordinary resonant cavity. The interferometric data can also be measured by detecting the number of reacted atoms by using the effect of magnetic deflection. Various atomic clocks, e.g., Rb-clock and H-maser, have been used widely in many important areas of advanced technologies, such as in telecommunication networks for synchronization of nodes, and in global navigation systems.

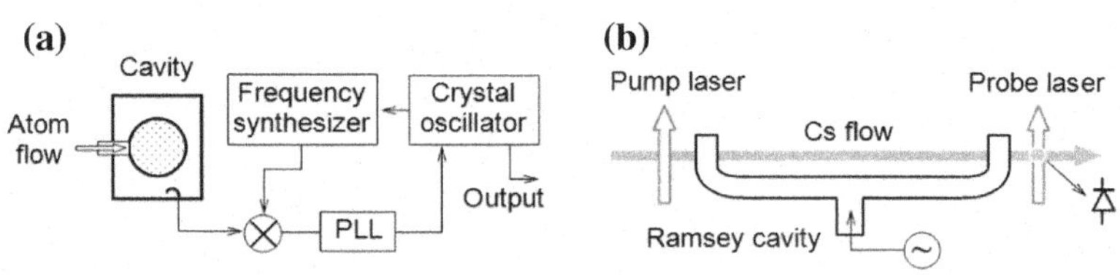

Fig. 9.1 **a** Basic structure of atom clock. **b** Ramsey cavity of atom clock

9.1.2 Laser Cooling of Atoms and Cold Atomic Clock

The precision of atomic clock is related mainly to two factors: the frequency width of atom's absorption line, and the duration length of reaction between atoms and microwave. The former is determined by the atom's intrinsic property, but it will be broadened by thermal movement, i.e., Doppler Effect, and the effect of external electromagnetic field, i.e., Zeeman Effect and Stark Effect. Therefore, cool down the atoms is the basic and necessary method for enhancing the precision of atomic clock. The accompanied advantage of using cooled atoms is that the reaction duration of microwave with the atoms will be extended, making the noises of atomic clock reduced.

After the invention of lasers, it was expected to cool the atoms by utilizing the exchange of momentums between atoms and photons [5]. The physics of cooled atoms and the cold atomic clock become hot R&D topics of advanced technology now. The atomic clock with cooled atom fountain came out successfully in 1990s [6]. For development of methods to cool and trap atoms with laser light, Steven Chu, Claude Cohen-Tannoudji, and William D. Phillips won the Nobel Prize in physics 1997 [7, 8, 9]. The precision of Cs fountain clock was reported 4–5 $\times$ 10^{-16} in 2010 [10]. As the development of space technology, especially the successes of global positioning system, a new target is to use the cold atomic clock in space [11]; it is expected that the frequency stability of cold atomic clock will be enhanced by one order in the environment without gravity. The basic concepts of laser cooling and related topics are explained qualitatively below.

(1) **Laser Cooling and Trapping of Atoms**

Cooling technology has played important roles in people's daily life, industries, and scientific activities. The classical method is based on the thermal dynamics; it has pushed the low temperature record down to far below the condensation point of helium. However a limit exists for further approaching the absolute zero point at −273.15 °C before the laser cooling.

It is well known that photons have not only energy of $hv = mc^2$, but also momentum of $mc = hv/c = \hbar k$. Photons will thus give a pressure to an object which they hit, called the light pressure. The tail of comets is just explained by the effect. The process of laser cooling is described by Fig. 9.2a, where the atom moving in the direction toward left absorbs a photon propagating in the direction toward right. According to the momentum conservation law, the atom velocity will be decreased, while its energy is increased, expressed as:

$$E_1 + \frac{1}{2}\mu v_1^2 + \hbar\omega = E_2 + \frac{1}{2}\mu v_2^2, \tag{9.1}$$

$$\mu \boldsymbol{v}_1 + h\boldsymbol{k} = \mu \boldsymbol{v}_2, \tag{9.2}$$

where $E_{1,2}$ are the energies of atom before and after the transition, $\boldsymbol{v}_{1,2}$ are the correspondent velocities, μ is the atom's mass. If $|\boldsymbol{v}_2| < |\boldsymbol{v}_1|$, the atom is slowed

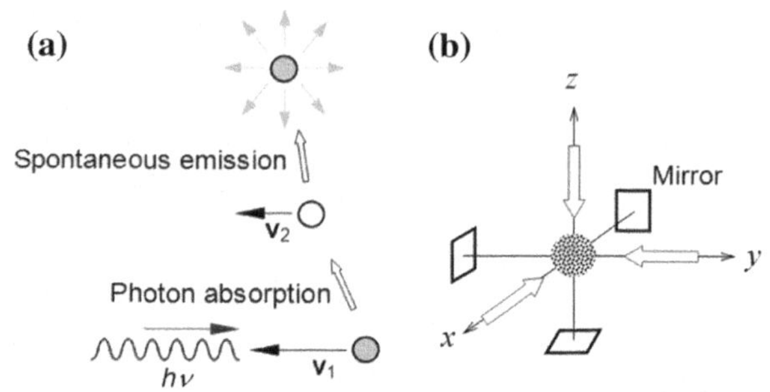

Fig. 9.2 **a** Schema of laser cooling of atom. **b** Formation of cooled atom molasses

down by a certain amount. It is therefore required that the light frequency must match with the absorption line. That is why the laser beam with monochromaticity good enough must be used. The atom at the excited state will eventually return back to the ground state by spontaneous emission in directions distributed randomly. If the process is accumulated, the atoms will be cooled as a whole. For a continuous laser cooling, the following factors must be taken into consideration.

Doppler Effect. As discussed in Chap. 6, the photon frequency observed by a moving atom will deviate from that of a static atom by Doppler shift of $\Delta\omega_D = \boldsymbol{k} \cdot \boldsymbol{v}$. Therefore, the cooling photon should have a negative detuning related to the peak of absorption line. Such a detuning is decreased as the atom is cooled down. Therefore, a precisely stabilized laser with controllable tunability is a basic condition for laser cooling. However, the laser frequency swept in serrated waves cannot meet the requirement of continuous cooling. To solve the problem, Stark Effect or Zeeman Effect is utilized [12]. If the cooling occurs in a spatially varied electric field or magnetic field in the laser beam direction, the absorption frequency of atoms will be changed correspondingly, and match with the frequency stabilized laser self-consistently along with the cooling process. A precisely designed cavity is then needed.

Re-pumping. As stated above, the atom at the excited state will return back to the background state by spontaneous emission. However, the background state is generally split into two levels with a small gap due to the electron spin. Some fraction of atoms exited from one of the two levels will return to the other level; and the number of atoms at that level will increase more and more. Another laser beam with frequency corresponding to the latter level is thus necessary for a continuous cooling. The latter process is called re-pumping.

Trapping of Cooled Atoms. Obviously it is necessary to cool down the atoms in the three-dimensional space, so that 6 laser beams propagating in *x*, *y*, and *z* directions oppositely are needed. Figure 9.2b shows a schematic diagram, where the opposite beams are obtained by three mirrors. The cooled atoms are concentrated around the middle point, forming the so-called molasses, which contains number of cooled atoms more enough for research and application. However, the atoms in molasses will diffuse off, and go down by the gravity. To keep the cooled atoms in a small volume, some trapping technologies are developed. Among them, the magneto-optical trap (MOT) is one of the mostly used methods. MOT uses a magnetic field with a designed gradient to confine the atoms based on Zeeman Effect. Other methods include bipolar force trapping, etc. [12].

Limitation of Lowest Temperature. The cooling process described above is termed Doppler cooling since decreasing of Doppler frequency shift is accompanied in the process. When the velocity is decreased to a certain degree, the absorbed photon energy will transfer to the thermal energy. The atoms will get a momentum in randomly distributed directions when they transit from the excited state down to the ground state by spontaneous emissions. In other words, a heating effect becomes dominant at a certain low temperature; and a limitation of lowest cooled temperature exists. Theoretical analyses indicate that the limitation is related to the natural linewidth δv of the atom, expressed as $T_{\min} = h\delta v/2k_{\mathrm{B}}$, where k_{B} is the Boltzmann constant. It is reported that the limited temperature is 124 and 145 μK for cesium and rubidium, respectively [12]. To overcome the limitation, sub-Doppler cooling methods are developed, such as with polarization gradient, with magnetic induction, and with velocity selective coherent population trapping (CPT).

(2) **Fountain Clock with Cooled Atoms**

The cooled atom will extend the reaction time with the microwave in the cavity when utilized to build an atomic clock. However, the slow the atoms move, the more serious the deflection of atom beam occurs in the gravity field, except for the case of vertically moving atoms. It means that the cavity should be arranged in vertical direction, so that the atomic beam will be moving like the water fountain. In the atomic fountain as shown in Fig. 9.3a, the atoms will react with the microwave field twice in a roundtrip; the cavity can be regarded as a folded Ramsey cavity [4, 6]. The equipment is surrounded by a magnetic shield to prevent the influence of

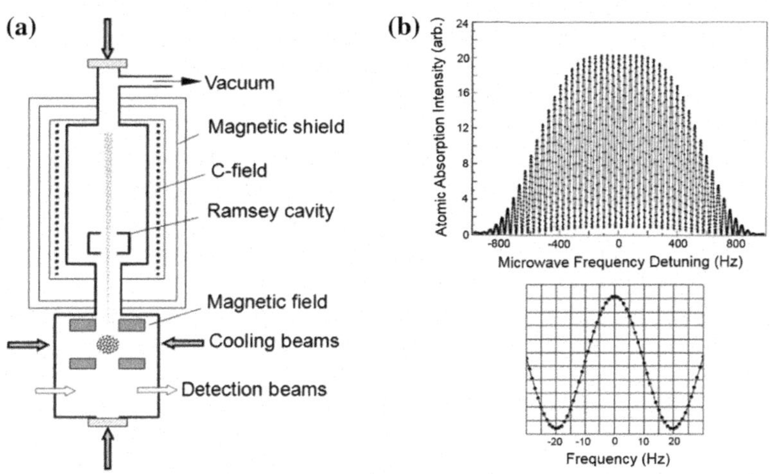

Fig. 9.3 **a** Schema of cold atomic fountain clock. **b** Ramsey fringes (*upper*) and one middle fringe (*down*). Reprinted from Ref. [15] with permission

the earth magnetic field; C-filed in the figure is a controllable weak magnetic field, playing a role of removing degenerate of Zeeman levels. A molasses is formed by six laser beams (including pump and re-pump lasers), and trapped in the MOT built by the magnetic field. A certain fraction of atoms in the MOT will be tossed upward by switching one of the lasers off, and then return back. The changes of atomic states are measured by the detection beam below the MOT.

The integrating sphere cold atom clock is a compact atomic clock that uses diffuse laser cooling [15]. Fig. 9.3b shows the measured Ramsey fringes of the clock, where the whole spectrum is given in the upper, and a single fringe at the middle part is shown in the below. The frequency width down to 20Hz was measured, much less than the microwave cavity width of about 1000Hz.

The above introduction to laser cooling is just a simple and qualitative description. The movement of atoms should be regarded as an ensemble, instead of individuals. It is necessary to analyze the process from viewpoint of the interaction between atoms and laser field, based on the theories of quantum mechanics and laser physics [12–14]. The mechanical force of laser field should be described as a stress tensor:

$$\boldsymbol{T} = \varepsilon_0(\boldsymbol{EE} - E^2\boldsymbol{I}/2) + \mu_0(\boldsymbol{HH} - H^2\boldsymbol{I}/2), \tag{9.3}$$

where $\boldsymbol{I}$ is a 3×3 unity tensor. The force of laser field applied on an object is $\boldsymbol{f} = \boldsymbol{n} \cdot \boldsymbol{T}$, where $\boldsymbol{n}$ is the normal vector. The atomic system should be defined by a certain Hamiltonian. The behavior of atoms in such a field obeys quantum mechanical equations. Further description and theoretical analysis are beyond the scope of this book. The mechanisms and theories of physical processes and related technologies can be found in journals and monographs in details.

Since the requirement to laser's power is not high for research and development in this field, the semiconductor laser is regarded as the best candidate. Obviously, a high coherence semiconductor laser with frequency stabilized on absorption lines and with fast and precise tunability is one of the key devices.

A further splendid target of the cold atom clock is to march in space with little tiny gravity, where the fountain effect becomes unwanted, and the microwave cavity can be designed more freely. It is believed that the precision of global positioning system will then be greatly enhanced. People are also pursuing the optical clock, in which atoms with highly stable energy levels, such as strontium (Sr), are trapped in the optical lattice, and incorporated with the femtosecond optical frequency comb as the accurate frequency measurement [15–19]. It is reported that the optical clock will provide the most accurate frequency standards with precision of one second per 300 million years [20].

(3) **Other Topics of Atomic Physics and Related Technologies**

Electromagnetically Induced Transparency (EIT). The typical picture of EIT is a narrow width transparency window within an absorption line of material under certain pump conditions [21–24]. Its mechanism involves a group of energy levels,

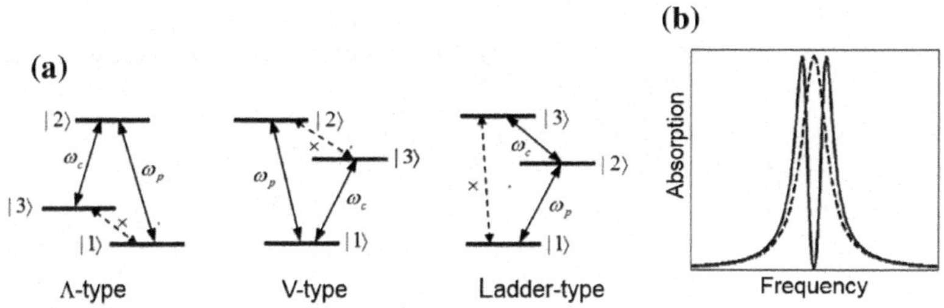

Fig. 9.4 **a** Three types of level configurations; **b** EIT spectrum

basically in three configurations: Λ-type, V-type, and ladder type, as shown in Fig. 9.4a, where ω_p is the frequency of probe light, ω_c is the frequency of coupling light; the transitions with symbol "x" are forbidden. The absorption is prevented by the destructive interference of the transitions between related energy levels when the material is pumped by the coupling light; the effect of EIT can then be measured by the probe. With the coupling light on, a transparency window will generate at the central part of absorption line, though the population at the ground state is maintained much larger than that at the excited state. Figure 9.4b shows two spectra for the case with coupling beam by the solid line and for the case without coupling beam by the dashed line. More levels may involve in the process to form a tripod-type configuration or others. Sub-levels of ultra-fine structure will also do so, showing multiple peaks in the EIT spectrum.

The mechanism of EIT is a destructive interference of wave functions related to the two transitions, described by Rabi oscillations in the quantum mechanics. Theoretically, the interaction of the atom and the coupling optical field can be described by the picture of dressed state. For example, in Λ configuration level E_2 will be split into two, leaving the original level as a transparent hole in spectrum. In the ideal case, the absorption line shape measured by a weak probe with frequency of ω_p can be expressed as [23]

$$L(\delta\omega_\mathrm{p}) = \frac{4}{\pi} \frac{\delta\omega_\mathrm{p}}{\Omega_c^2 - 4\delta\omega_\mathrm{p}^2 + j2\gamma_{12}\delta\omega_\mathrm{p}}, \tag{9.4}$$

where $\delta\omega_\mathrm{p} = \omega_\mathrm{p} - \omega_{12} = \omega_\mathrm{p} - (E_2 - E_1)/\hbar$, γ_{12} is the linewidth of $|2\rangle \to |1\rangle$ spontaneous transition, Ω_c is Rabi frequency. It is shown that the transparency occurs at $\omega_\mathrm{p} = \omega_{12}$, and the absorption peak moves to its two sides at $\omega_\mathrm{p} = \omega_{12} \pm \Omega_c$.

Such a state of an atom is also called the dark state, with no photon absorbed, neither emitted. For a collection of atoms, the interaction with the coupling optical field will make the system into a trapped state, called coherent population trapping (CPT). Taking the kinetic energy into Hamiltonian, the velocity selective CPT is used to overcome the Doppler limitation in the laser cooling.

Obviously, the transparent dip within the absorption line is very narrow, resulting in a strong dispersion of refractive index based on Kramers–Kronig relations. The light velocity at the region is lowered greatly. By using such a slow-light effect in lasers, the intra-cavity EIT, the laser line will be narrowed effectively [25–27].

Optical Tweezers. The mechanical force of laser field cannot only slow down the atoms, but also trap a nanometer- and micrometer-size dielectric particle in a highly focused optical beam. The particle can then be moved by the optical beam. Based on such an effect a special technology termed the optical tweezers is developed [28–30]. According to Maxwell electromagnetic field theory, the light pressure is described by the momentum of field. At the focused point of laser beam, the strong electric field gradient at the beam waist will generate a centripetal force, and a particle can be trapped there under certain conditions. The laser beam for optical tweezers is required not only with higher coherence, but also with controllable beam shape, special polarization states, and precisely movable focus. With the controllable orbital angular momentum of light, the micro particles can also be rotated [31]. The optical tweezers are very suitable to manipulate biological objects, becoming an important technology in related areas. The theory and technology are beyond the scope of this book. Interested readers can find them in literature.

9.1.3 Data for Atomic Physics and Related Spectroscopy

This section provides data of elements used mostly in researches of cooled atoms, and in laser spectroscopy. Some of the basic data are referred from Ref. [32].

(1) **Rubidium (Rb)**

Atomic number $Z = 37$. Mass number 85 (abundance 72.15%); 87 (abundance 27.85%). Melting point: 38.89 °C. Boiling point: 688 °C. Specific gravity: 1.532 g/cm^3 (solid state) and 1.475 g/cm^3 (liquid state). Vapor pressure at 300 K: $\sim 10^{-6}$ mmHg. Electron configuration: 1s^2; 2s^2, 2p^6; 3s^2, 3p^6, 3d^{10}; 4s^2, 4p^6, 5s^1. Valence = 1. The internal electron configuration is the same as the inertial gas krypton (Kr).

The transition related to laser cooling is from the ground state $5S_{1/2}$ to the first excited state $5P_{3/2}$ of the valence electron, called D_2 line. The transition from $5S_{1/2}$ to $5P_{1/2}$ is called D_1 line. Figure 9.5 shows the fine structure of D_2 line of its two isotopes ^{85}Rb and ^{87}Rb [12, 33]. The quantum number F is the total angular momentum, including the spin momentum I of atomic nucleus, which is 5/2 and 3/2 for ^{85}Rb and ^{87}Rb, respectively. The working frequency of ^{87}Rb clock is 6,834,682,610.904324 Hz; the working wavelength of pump laser is 780.02 nm [19].

Relations between frequency and wavelength: Wavelength 780 nm corresponds to 384 THz. $\Delta\lambda/\Delta\nu = 2.03$ pm/GHz; $\Delta\nu/\Delta\lambda = 493$ MHz/pm. Frequency width of

Fig. 9.5 Fine-structure of D_2 line of ^{87}Rb and ^{85}Rb. *Unit* MHz. Reproduced from [32] with permission

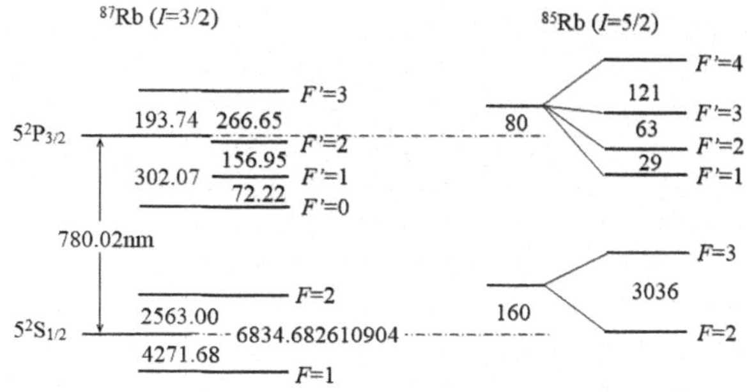

1 MHz corresponds to wavelength width of 2.03 fm (1 fm = 10^{-15} m). Doppler linewidth of D_2 line at room temperature is about 500 MHz.

(2) **Cesium (Cs)**

Atomic number $Z = 55$. Mass number 133 (abundance 100%); atomic weight 132.9054. Melting point: 28.4 °C. Boiling point: 669.3 °C. Specific gravity: 1.8785 g/cm^3 (solid state). Electron configuration: $1s^2$; $2s^2$, $2p^6$; $3s^2$, $3p^6$, $3d^{10}$; $4s^2$, $4p^6$, $4d^{10}$; $5s^2$, $5p^6$; $6s^1$. Valence = 1. The internal electron configuration is the same as the inertial gas xenon (Xe).

The transition related to laser cooling is the D_2 line from the ground state $6s_{1/2}$ to the first excited state $6p_{3/2}$ of the valence electron. Figure 9.6 shows the fine structure of D_2 line [12]. The quantum number of nucleus spin is $I = 7/2$. The quantum number F of total angular momentum is indicated in the figure. The working frequency of ^{133}Cs clock is 9,192,631,770 Hz exactly; the working wavelength of pump laser is 852.11 nm [19].

Relations between frequency and wavelength: Wavelength 852 nm corresponds to 352 THz. $\Delta\lambda/\Delta\lambda$ = 2.42 pm/GHz and $\Delta\nu/\Delta\lambda$ = 413 MHz/pm. Frequency width of 1 MHz corresponds to wavelength width of 2.42 fm.

(3) **Strontium (Sr)**

Atomic number $Z = 38$. Mass number 88 (abundance 82.74%), 87 (abundance 6.96%), 86 (abundance 9.75%), 84 (abundance 0.55%). Melting point: 769 °C. Boiling point: 1384 °C. Specific gravity: 2.54 g/cm^3. Electron configuration: $1s^2$;

Fig. 9.6 Fine-structure of D_2 line of ^{133}Cs. *Unit* MHz

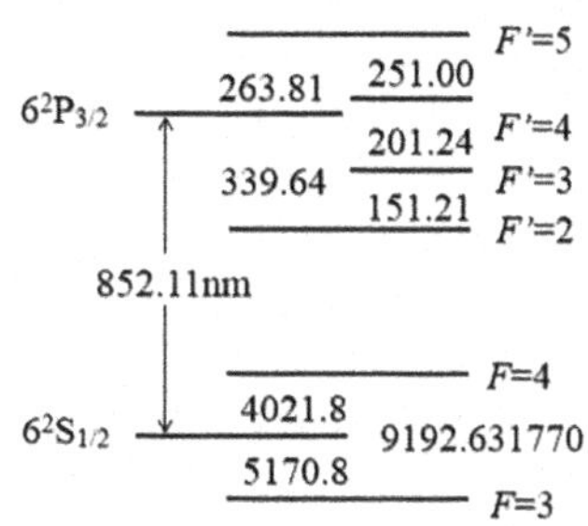

Fig. 9.7 Structure of ^{87}Sr energy levels

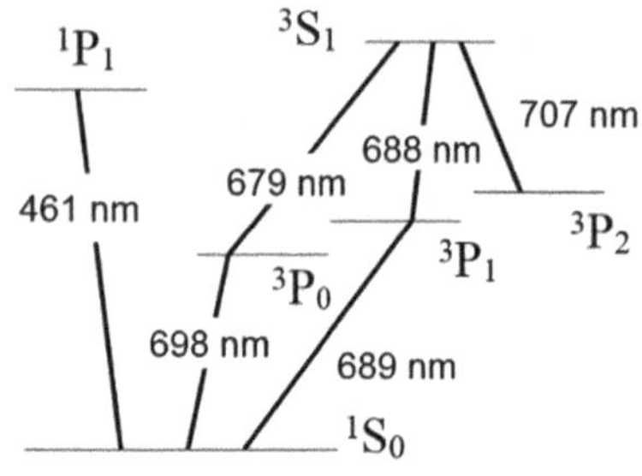

2s^2, 2p^6; 3s^2, 3p^6, 3d^{10}; 4s^2, 4p^6, 5s^2. Valence = 2. The internal electron configuration is the same as the inertial gas krypton (Kr).

Figure 9.7 shows energy level structure of ^{87}Sr, where the transitions with 461, 679, 689, and 707 nm are used for cooling; the transition from state $(5s^2)^1S_0$ to state $(5s5p)^3P_0$ is used for the optical clock. The working frequency is 429,228,004,229,873.4 Hz corresponding to λ = 698.445709612754 nm [17, 19].

(4) **Sodium (Na)**

Atomic number $Z = 11$. Mass number 23 (abundance 100%); atomic weight: 22.98977. Quantum number of nucleus spin is $I = 3/2$. The wavelengths corresponding to transitions from $3^2S_{1/2}$ to $3^2P_{3/2}$ and to $3^2P_{1/2}$ are 588.995 and 589.592 nm, respectively, the well-known sodium double yellow lines. The gap between the fine levels of the ground state is 1772.6261288 Hz, as shown in Fig. 9.8 [12].

(5) **Hydrogen (H)**

Atomic number $Z = 1$. Mass number 1 (abundance 99.985%); deuterium 2 (abundance 0.015%); tritium 3 (instable). Quantum number of nucleus spin of H is $I = 1/2$. The transition related to the microwave amplifier of stimulated emission (H-maser) is between the two sub-levels of the ground state, $F = 0$ and $F = 1$ with gap of 1,420,405,751.7667 Hz, which is just the working frequency of H-maser [12, 19].

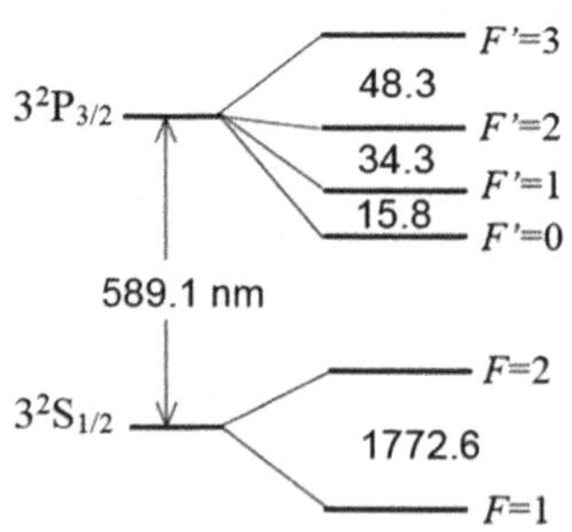

Fig. 9.8 Fine-structure of D_2 line of Na. *Unit* MHz

(6) Absorption lines of methane (CH_4) in near infrared band

Methane is the main component of natural gas, one of the most important energy sources, and also one of the origins of mine disasters. Its detection and sensing attract great attentions of researches and industries. The absorption spectrum of molecules covers usually a wide range. The absorption of methane is mainly in the band of 3 μm. Since the devices and components in the near infrared band are most mature and cheap due to the development of fiber communications, data in the band are given here. Figure 9.9 is the absorption spectrum at band $2\nu_3$ [34–36]. The data given by different sources may differ a little from each other. Table 9.1 shows the positions of several lines in band $2\nu_3$, given by Ref. [34].

(7) Absorption lines of acetylene (C_2H_2) in near infrared band

Acetylene is a flammable gas with high energy. Its absorption lines in the near infrared band are often used as reference frequencies for optical communication lasers. Figure 9.10 shows the absorption spectrum of *R* and *P* branches in the band near 1550 nm [37].

(8) Absorption lines of carbon dioxide (CO_2) in near infrared band

Carbon dioxide is regarded as the greenhouse gas of the earth. Its detection and measurement is an important mission for environment protection. Figure 9.11 shows its absorption spectra in the near infrared band [36].

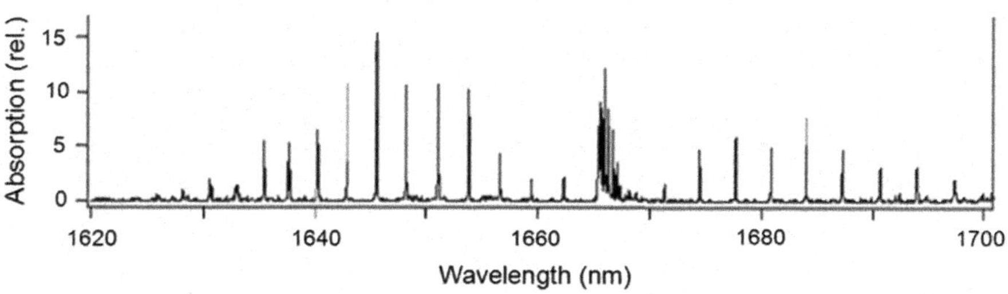

Fig. 9.9 Absorption spectrum of methane at band $2\nu_3$. Reprinted from Ref. [36]

Table 9.1 Wavelengths of absorption lines of methane at band $2\nu_3$ [34]

	λ (nm)		λ (nm)		λ (nm)		λ (nm)		λ (nm)
R0	1662.327	R3	1653.728	R4	1650.961	R5	1648.239	R6	1645.568
R1	1659.413		1653.726		1650.959		1648.234		1645.532
R2	1656.547		1753.722		1650.955		1648.221		
	1656.546				1650.948		1648.218		

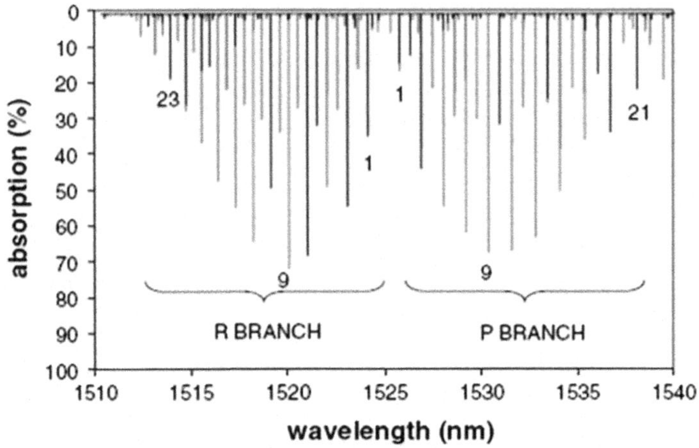

Fig. 9.10 Absorption spectrum of acetylene in the near infrared band. Reprinted from [37] with permission

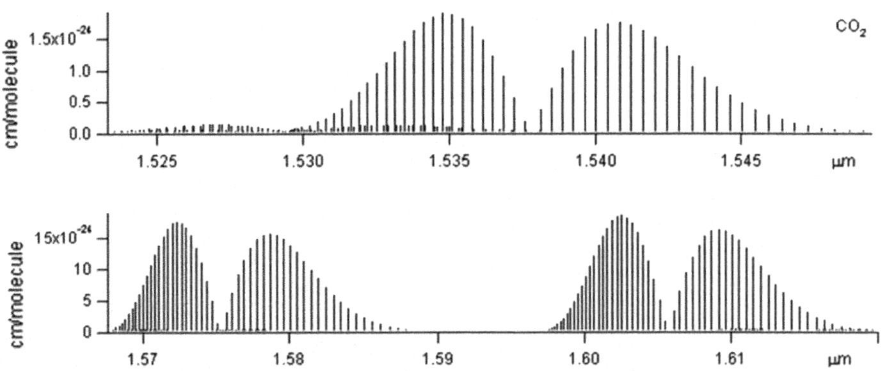

Fig. 9.11 Absorption spectra of carbon dioxide in near infrared band. Reprinted from [36]

9.2 Applications in Optical Communications

9.2.1 A Brief Review of Optical Fiber Communications

The invention of laser stimulated the birth of optical communication. Especially, since Sir Charles Kuen Kao proposed and expounded the usage of optical fiber in communications in 1960s [38], the fiber communications have achieved great progresses, and become one of the foundation stones of the information era. For the contribution, Kao was awarded the Noble Prize in physics 2009. The fiber communications basically uses the system of intensity modulation and direct detection (IM-DD) now. The information is carried by optical pulses encoded with standards of pulse code modulation (PCM), transported via optical fibers, and received by photodetector at the terminal. The waveform of detected signal is directly proportional to the optical intensity. The data transportation with code rate higher than 40 Gbit/s is now commercially available. The multiplexing of channels makes the capacity of communication system increased greatly. The most effective method is the dense wavelength division multiplex (DWDM). The spacing between channels used widely is 50 and 100 GHz; the wavelength covers a broad range from 1310, 1550 nm (C-band), through the ultra-long wavelength of 1670 nm. As the space technology developed, optical communications in free space between satellites and between satellite and ground station are playing more and more important roles.

The semiconductor laser is the key component in the optical communications, due to its advantages of high speed response, low cost, and the compatibility with the optical fiber. The limitation to the speed and capacity of communication is mainly from the fiber loss and its dispersion. The dispersion causes optical pulses broadened, and increases code errors. The main measure of avoiding the influence of dispersion is to use single longitudinal mode LD source, which should maintain narrow linewidth under high-frequency modulations. The DFB laser introduced in Chap. 4 is now used almost in all practical communication networks.

The lasers used in DWDM transmitters are required not only with a stable single mode, but also with the peak wavelengths stabilized at the standard formulated by the International Telecommunication Union (ITU), which is $193.4 \pm n * 0.1$ THz for 100 GHz DWDM system in the 1550 nm band. The number of LD for different channels is demanded up to hundreds or more in the system. The reference frequency for frequency stabilization is usually provided by thin film filters (TFF), and traced to the absorption line of acetylene (C_2H_2). Another important device used in DWDM is the erbium-doped fiber amplifier (EDFA), pumped by high power diode lasers. The pump wavelengths have also to be locked at the absorption lines of EDF at 980 or 1480 nm. The semiconductor lasers are now mass produced; their performances are ensured to meet with the requirement of the conventional DWDM optical fiber communication systems.

9.2.2 *Lasers for Coherent Optical Communications*

The demand of human society on the information has been increasing continuously, especially due to the development of Internet technology. As the applications of conventional system based on IM-DD and DWDM are widely adopted all over the world, the coherent optical communication (COC) becomes the target being pursued recent years, in which the optical frequency, phase, and polarization are used as data carriers, together with the pulse amplitude. Therefore, the density of multiplexing and the total capacity of communication are greatly increased [39]. The requirement of COC on laser's performances should be raised correspondingly.

Different from the IM-DD, the modulation formats in COC include the amplitude shift keying (ASK), frequency shift keying (FSK), phase shift keying (PSK), and polarization shift keying (PolSK) To demodulate the data carried by COC formats, heterodyne detection must be used instead of direct detection used in IM-DD. ASK is also called on-off keying (OOK) with waveforms similar to the intensity modulation (IM), but with the heterodyne detection used in receiver ends. The heterodyne signal is the result of correlation of the received optical wave and the local oscillation (LO). The correlation detection brings about remarkable advantages. First, the signal power will be amplified by LO. Second, the correlation detection with narrow linewidth LO suppresses greatly the extra-band noise. For the purpose, the transmitter and LO must use high coherent and low-noise lasers. The OPLL technology must be used in the correlation detection to demodulate the data carried by optical phase and/or frequency, as introduced in Chap. 8. The code impairment in the long distance propagation can also be corrected satisfactorily by OPLL.

More formats are developed based on the above-mentioned formats. PSK with phase shift of π is called the Binary PSK (BPSK); PSK with phase shift of $\pi/2$ is called the Quadrature PSK (QPSK); similarly, 8PSK, 16PSK, and more, can also be adopted. If the phase shift between the adjacent codes is marked as a symbol, such a scheme is called the differential PSK (DPSK). Figure 9.12a shows the waveforms of ASK, FSK, and PSK formats. The basic codes are divided into two forms, i.e., the return-to-zero (RZ) code and the non-return-to-zero (NRZ) code. For simplicity, Fig. 9.12a shows only NRZ code. Figure 9.10b shows the constellations of QPSK and 8PSK, where I and Q stand for phase shifts of In-phase and Quadrature.

Combination of ASK and PSK results in a new format termed the quadrature amplitude modulation (QAM), in which the carrier wave is expressed as

$$E_{i,j} = A_i \cos(\omega t + \phi_j), \quad (A_i = 0, 1; \phi_j = 0, \pi/2, \ldots). \tag{9.5}$$

It is easier to demodulate the signals of amplitude and phase, separately; the code speed of PSK is thus doubled. Accordingly, combination of ASK and BPSK is 4QAM, and combination of ASK and QPSK is 8QAM.

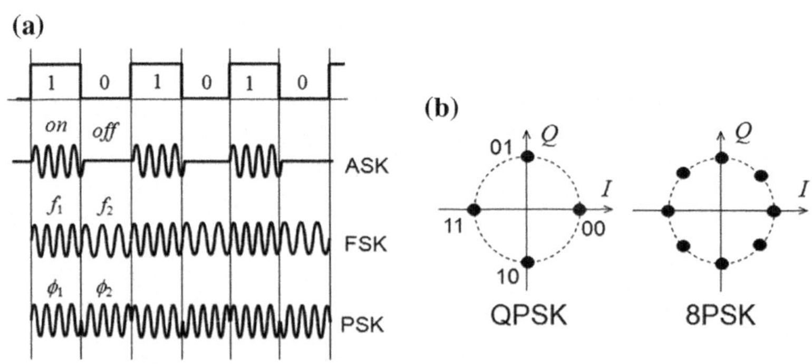

Fig. 9.12 **a** Waveforms of ASK, FSK, and PSK; **b** Constellations of QPSK and 8PSK

The de-multiplexing used in DWDM system is implemented by using DEMUX devices, such as the thin film filter (TFF), the arrayed waveguide grating (AWG), and the fiber Bragg grating (FBG). In coherence optical communications, the multiplexing density is much higher than that in DWDM. The technology of orthogonal frequency division multiplexing (OFDM) is used widely in wireless communications; with the idea referred, the coherent optical OFDM is developed successfully. The correlation detection allows the channel spacing reduced to the limit of mutual orthogonality between the adjacent channels. If the optical field of pulsed code is a square wave, the cross-correlation of adjacent channels is given by

$$R = \int_0^T e^{j2\pi(\nu_1-\nu_2)t}dt = \frac{\sin[\pi(\nu_1-\nu_2)T]}{\pi(\nu_1-\nu_2)}e^{j\pi(\nu_1-\nu_2)T}. \tag{9.6}$$

When the code length T meets condition of $(\nu_1 - \nu_2)\,T = 1$, the cross-correlation reaches zero, and the cross-talk between the adjacent channels can then be removed. The multiplexing density and the spectral efficiency are thus be raised greatly. Figure 9.13 shows the comparison between WDM and OFDM schematically. The OFDM technology has more advantages, as expounded in Refs. [39]. It can be seen from (9.6) that the laser linewidth and instability should be reduced to much less than the channel spacing, or otherwise the orthogonality would be impaired, and the cross-talk and code error would be increased.

Some more code formats and multiplexing methods are also developed, such as the optical time division multiplexing (OTDM), the polarization multiplexing (PM), and so on. In the different schemes of COC, the narrow linewidth, highly stable single-frequency semiconductor laser is the key device, both in the transmitter and in the receiver as local oscillator; and its quantity demand is tremendous for the huge networks. Compared with the applications in atomic physics and precision metrology, the requirement on coherence may be loosened in COC, but higher requirements on the dynamic characteristics and better reproducibility have to be satisfied. References [40, 41] give detailed analyses.

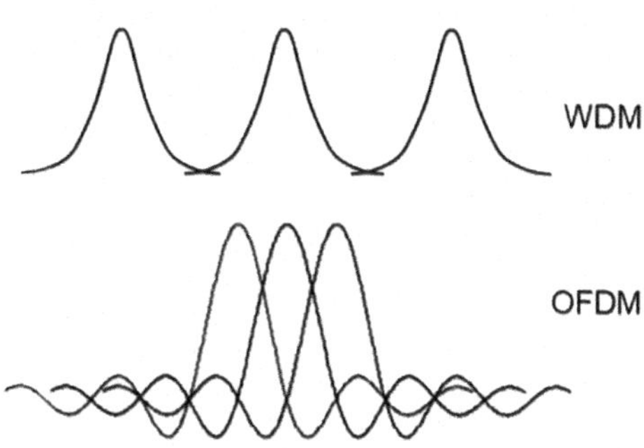

Fig. 9.13 Comparison of channel densities of WDM and OFDM

9.2.3 Microwave Photonics and ROF

In the ordinary microwave systems, the microwave signals, typically at X-band ($\sim$ 10 GHz) and millimeter wave band, are transported usually by waveguide tubes made of metal, or by coaxial cable. The components and devices for processing of microwave signals are also bulky. Since the waveband of optical communications has reached the centimeter and millimeter range, including the generation of high-frequency signals and their processing and transportation, it is possible to transport microwave signals by the optical fiber, instead of the cumbersome and expensive waveguide tube, and possible to process the signals in optical frequency domain. Especially, the mobile communication system is developed so much that almost everybody has his cellphone in hand. The huge information is actually transported through optical fibers. Combination of optical communication and wireless communication is obviously an optimized scheme; and the microwave photonics emerges as a new branch of information technology [42–45]. It is also termed radio over fiber (ROF) [46, 47], especially in local area networks (LAN), and access networks (AN) of communication systems, as shown in Fig. 9.14 schematically, where OADM is the optical add-drop multiplexer.

Researches of microwave photonics are concentrated on the following topics. First, the generation of microwave signals carried by optical waves, as introduced in Sect. 8.2.2, where OPLL plays important roles. Second, the processing of microwave signals. The microwave signals carried by optical wave will be distorted in propagation in fiber due to its dispersion; its signal-to-noise ratio is thus lowed, as analyzed in Refs. [48, 49]. The signal processing includes dispersion compensation, analog-to-digital conversion, frequency down-conversion and up-conversion, noise reduction, and other functions in optical domain [45].

The phased array radar is one of the important applications of microwave technology. The microwave photonics can greatly reduce its volume and weight to make it flexible and more effective. Figure 9.15 gives an example of phased array radar schemes. The combined beam of n antennas can be steered by adjusting the phase difference between the adjacent antennas, and the controlling signals of microwave phases are transported from the remote end to the antennas by optical fibers, instead of bulky metal waveguide or coaxial cable.

For a wide steering angles of the phased array, the phase difference between the antennas has to be scanned not only in a large range, but also constantly for

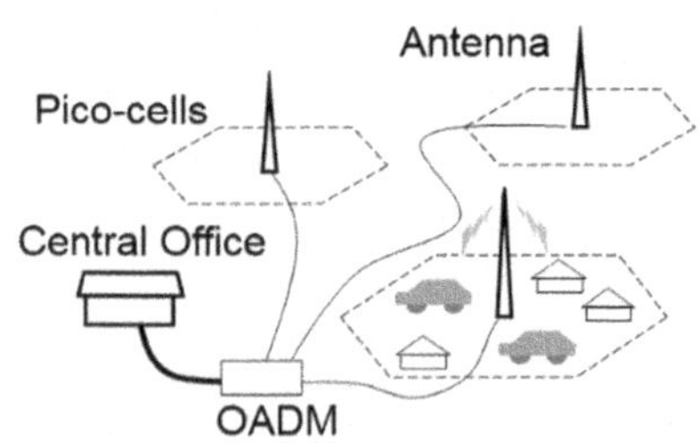

Fig. 9.14 Schematic ROF system connecting mobile communication with fiber networks

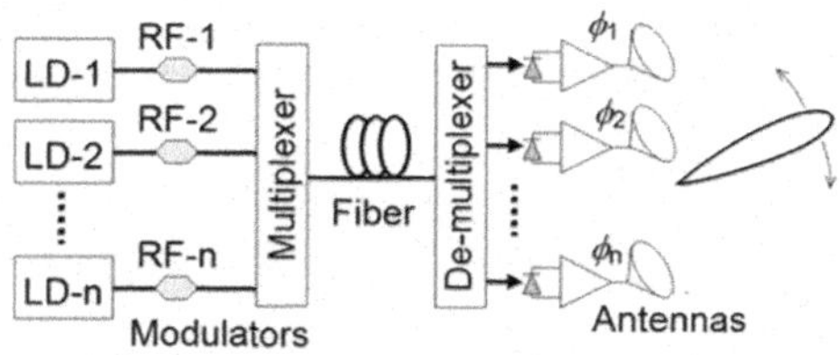

Fig. 9.15 Microwave antenna array controlled by signals transported via fiber

different central frequency. In optical domain the delay is generated usually by using a dispersive component based on relation of $\tau = \partial\phi/\partial\omega$. However, so generated delay is often not a constant in spectrum; in other words, it is not a true time delay. Several schemes have been proposed to generate true time delay; Ref. [50] reported a structure made of stack integrated micro-optical components. Experimental results demonstrate that based on the optical beamforming network the phased array antenna can accomplish RF-independent broadband beam steering without beam squint effect, and can achieve continuous angle steering. A fully photonics-based coherent radar system has demonstrated in field-trials [51]. Microwave photonics is a developing technology, readers can read more in literature.

9.2.4 *Inter-satellite Optical Communications*

Since the earlier stage of laser's invention and development, people have been attending to exploit the laser for spatial communications, just in the way similar to wireless and radar, which is now called the free-space optical communication (FSO). However, the laser beam in the air will be disturbed seriously by the air turbulence, causing great increasing of bit error rate (BER), so that the FSO in the atmosphere is used basically as an auxiliary tool in some situations. The exploitation of outer space has been one of the great events since the last century. The communication between the satellite and the earth and between satellites is one of the necessary parts of space technology; and the microwave communication has been playing the main role. However, its speed and capacity cannot meet the continuously growing demand; the optical communication is naturally regarded as a new candidate. Figure 9.16 gives a schematic diagram of outer space communication system, where GEO is the geostationary earth orbit satellite, LEO is the low earth orbit satellite, GS is the ground station.

Since no turbulence exists in the outer space, FSO communication can display its advantages. Since no dispersion exists there, the coherent optical communications will demonstrate its best performances [52]. In addition, no medium loss exists; the communication distance can be extended greatly so long as the divergence of laser beam is reduced as low as possible.

Compared with the fiber communication, features of outer space FSO are remarkable. First, the spans are much longer. GEO is 35,786 km away from the

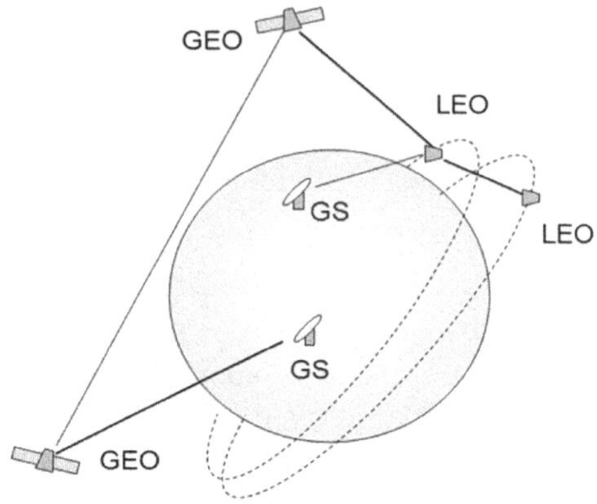

Fig. 9.16 Schematic diagram of outer space optical communication system

earth; the height of LEO is typically in the order of 1000 km; the inter-satellite distance may reach several ten million meters. The propagation loss comes mainly from the optical beam divergence. For such long distances, the divergence must be suppressed to the order of micro radiance (μrad) or less. According to the diffraction theory the divergence angle is $\theta = 1.22\lambda/D$, so the diameter D of laser transmitter telescope should be designed up to 1 m, even bigger; and the laser's spatial coherence, i.e., its beam quality, must be improved extremely. For the communication between satellites and the ground stations, the atmospheric turbulence and loss of the air and clouds still exist. A lot of new technologies have developed, such as the adaptive optics, techniques of store-and-forward, and so on [53, 54].

Second, since the satellites are moving very fast in space, whereas the laser beam is very narrow, the technology of acquisition-tracking-pointing (ATP) is extremely important. ATP involves a series of technologies on optics, electronics, and mechanics, which is beyond the scope of this book.

The third, Doppler frequency shift exists between the objects moving with high relative velocities: $\Delta\omega_{\mathrm{D}} = \boldsymbol{k} \cdot \boldsymbol{v} = \omega v \cos\theta/c$, where θ is the angle between the velocity and the optical wavevector. It is estimated that the Doppler shift may reach 1/380,000 of the optical frequency for communications between GEO and GS. If an 800 nm laser is used, the frequency shift is in the order of GHz. When the satellite moves around the earth, its velocity respect to GS varies continuously. If the transmitter is a single-frequency laser, the received optical signal is a chirping wave. A chirping-matched local oscillator has to be used in correlation detection, or otherwise the data will be lost. Therefore, narrow linewidth lasers with high speed controllable tunability are the must in the outer space FSO systems. Reference [55] demonstrated a laser system for coherent optical satellite links with Doppler shift compensated, in which the laser's frequency was swept in range of 24 GHz and at rate of 100 MHz/s.

9.3 Applications in Metrology and Sensors

9.3.1 High Precision Interferometry

The optical interferometer has made very important contributions to the progresses of science and technology. One of the bases of Einstein's theory of relativity is the results of measurement of ether by Michelson interferometer. Recent years, the gravitational wave, predicted by Einstein, was measured by the laser interferometer gravitational wave observatory (LIGO). The high precision interferometer is one of the fundamental tools in metrology. The fruits of interferometry are countless. As examples, the hydrophone and geophone are introduced briefly here.

Sensors to sound are used widely for both of civil use and military use. A famous example is the sonar invented in World War II. Since 1970s, after the invention of optical fiber, people have made great effort to develop the fiber hydrophone and geophone [56, 57]. Some of them have already been commercialized and used in many different fields, including sea fishery, marine biology, detection of seismic wave, well drilling, ultrasonic medicine, coast guarding, navigation of ship and submarine, and so on.

The fiber hydrophone is mainly composed of a sensor head and an interferometer. Figure 9.15a shows a typical structure of sensor head, that is a hollow mandrel with a fiber coil wrapped and fixed on it. Its lateral surface will be deformed under the sound pressure, and the fiber length changes accordingly. The sound wave is then transferred to a vibration of the optical path, expressed as [51]:

$$\Delta L = \left(1 - n^2 p_{12}/2\right) e_z n L = s\Delta p, \tag{9.7}$$

where L is the total length of fiber coil, n and p_{12} are the index and photo-elastic coefficient of silica fiber, respectively; $e_z L$ is the axial deformation of the fiber under the axial strain, which is proportional to the sound pressure, $e_z \propto \Delta p$. The sensitivity to sound pressure, s, is dependent on the design of sensor head. It is shown that a longer fiber is beneficial for higher sensitivity. Another important parameter is its frequency response bandwidth. Careful design and precise fabrication technology are necessary for higher performances. The structure of Fig. 9.17a is just an example; different structures have been developed for different situations. The structure of geophone will be different from that of Fig. 9.17a.

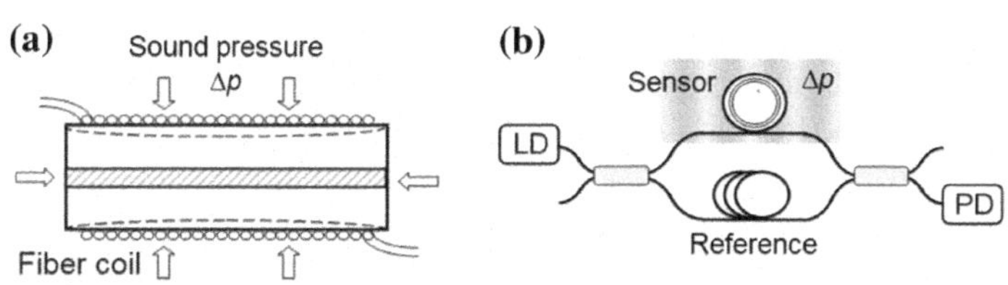

Fig. 9.17 Main parts of hydrophone: **a** Sensor head. **b** Interferometer

The interferometer is the best tool to detect the small phase change. Figure 9.15b shows a Mach–Zehnder interferometer as an example, where the sensor is inserted in one of its arms and a section of fiber is inserted in the other arm as the reference. The PD-detected signal is

$$I \propto \cos[k(nL + \Delta L - nL_{\mathrm{ref}})] = \cos(\phi_0 + ks\Delta p), \tag{9.8}$$

where $\phi_0 = nk(L - L_{\mathrm{ref}})$ is the biased phase. When it is adjusted to $\phi_0 = (m + 1/2)\pi$, the interferometer works at the quadrature point with the maximum sensitivity and a larger dynamic linear range: $I \propto \sin(ks\Delta p) \approx ks\Delta p$. The reference length should be designed and controllable to be optimized and stabilized.

Formula (9.8) holds for an ideal single-frequency light source. The finite linewidth δv of practical laser will lead to the interference phase noise:

$$\delta\phi = 2\pi n(L - L_{\mathrm{ref}})\delta v/c. \tag{9.9}$$

The laser's frequency noise is generally a white noise, i.e., its probability obeys Gaussian distribution; the interference phase noise is reduced to a visibility of the signal:

$$V = \exp(-\delta v/v_{\mathrm{int}}) = \exp(-\tau_{\mathrm{d}}/\tau_{\mathrm{c}}), \tag{9.10}$$

where $\tau_{\mathrm{d}} = (2\pi n|L - L_{\mathrm{ref}}|)/c = 1/v_{\mathrm{int}}$ is the delay corresponding to optical path difference of MZI; τ_{c} is the laser's coherent time. It is estimated that if a MZI with path difference of 1 m is used, the laser's linewidth of less than 30 Hz is required for a phase resolution of ≤ 1 μrad.

Laser's intensity noise causes also noise of MZI signal. By denoting δI as the intensity fluctuation, $\delta\phi$ as the interference phase induced by the intensity noise, they are related with each other as $\delta I/I \propto V \sin\delta\phi \approx V\delta\phi$. The interference phase noise induced by laser's intensity noise is then expressed as $\delta\phi_{\mathrm{RIN}} = \sqrt{\mathrm{RIN}}/V$, where $\mathrm{RIN} = \delta P^2/\bar{P}^2$ is the relative intensity noise of the laser source. Typically the averaged power detected by PD is about 100 μW. RIN of below −120 dB is then required for a phase resolution of ≤ 1 μrad.

The hydrophone and geophone systems are usually composed of a great amount of sensor heads and interferometers. The WDM and the optical time division multiplexing (OTDM) technologies used successfully in communications are naturally transplanted to the sensor systems. However, the frequency stabilizations have to be implemented with precisions higher than that for the ordinary communication. In addition the electronics and data processing must be optimized. With about 30-year development, great progresses on the fiber hydrophone have been achieved. As reported in journals, its sensitivity to pressure has reached 0.42 rad/Pa, equivalent to −127.5 dB respect to the unit of 1 rad/μPa. The performance is superior to that of commercial electrical sonar; its resolution reaches the limitation of noise at ocean bottom. Great scale hydrogen array systems have already been installed in the seabed [56–60].

9.3.2 Distributed Optical Fiber Sensors

The distributed fiber sensor is a big family of optical sensors; it is also the important user of single-frequency semiconductor lasers. The distributed fiber sensor is based on the optical scattering in fiber, i.e. Rayleigh scattering, and Brillouin scattering. Rayleigh scattering is caused by the particles in the medium, including the micro inhomogeneous distribution of molecules due to thermal fluctuations. The scattering is an elastic collision of photons and scatters, without changes of photon energy; the scattering coefficient is a function of photon frequency and the scattered angle. Lord Rayleigh explained why the sky is blue by the scattering after his name. Raman and Brillouin scatterings are inelastic collisions. Raman scattering causes change of internal states of molecules and related energy levels. Brillouin scattering corresponds to collision of photons and phonons. The latter is the description of sound wave in quantum mechanics. Due to Doppler Effect, the moving molecules will change the frequency of scattered photons.

(1) **Optical Time Domain Reflectometer Based on Rayleigh Scattering**

Characteristics of the three scatterings are affected by the internal properties of medium and their changes induced by environmental conditions. Therefore the scatterings can be utilized to make sensors; and the optical time domain reflectometer (OTDR) and the optical frequency domain reflectometer (OFDR) have been developed [61, 62]. The OTDR based on Rayleigh scattering is used to detect the loss and defects of fiber. Figure 9.18a shows schematically a OTDR system, where laser pulses are launched into the fiber under test, and the backscattered returned wave is received by the PD; ADC in the figure stands for the analog-to-digital converter. The returned signal shows an exponentially decayed waveform due to the fiber loss, as shown in Fig. 9.18b. The disconnected points are caused by the lumped loss or the reflections at fiber connections and fiber end. Their locations can be obtained by the propagation time.

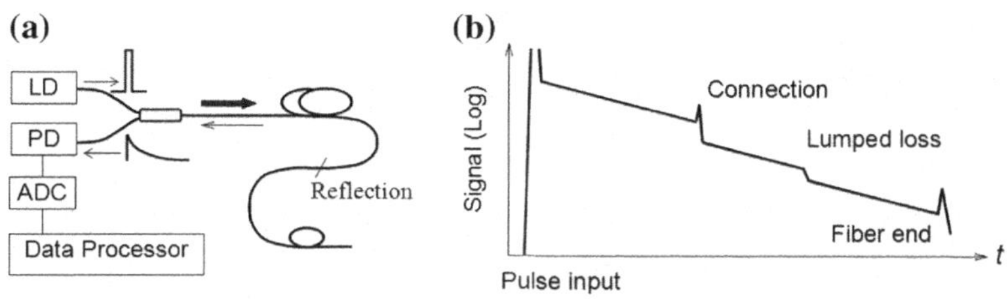

Fig. 9.18 **a** Schema of OTDR system; **b** Waveform of returned wave signal

The Rayleigh scattering is actually very weak; the usable signal is obtained by averaging of a great number of returned waves, and the working distance is limited. To overcome such a difficulty, the correlation detection is used in OTDR, in which a narrow linewidth laser is used to generate probe pulses; the returned wave is mixed with the local oscillation source (LO). The correlation detection brings about remarkable advantages. First, the beat signal is proportional to the product of returned field E_S and the LO field E_L. Its ratio to the signal of direct detection of scattering intensity is proportional to E_L/E_S, so that higher sensitivity can be obtained by higher LO power. Second, the noise band is limited within the frequency band determined by spectra of probe and LO. The noise out of the band is greatly suppressed. Such a sensor is termed the coherent optical time domain reflectometer (COTDR),

Furthermore, by using narrow linewidth sources and correlation detection, not only the amplitude of backscattering but also the changes of its optical phase can be acquired. More information is thus obtained. Such a system is termed the phase-sensitive OTDR (ϕ-OTDR) [62]. Figure 9.19 shows a schematic diagram of COTDR and ϕ-OTDR. In the structure a narrow linewidth DFB-LD working in continuous wave is taken as the source; its output is divided into two parts by the fiber coupler: the main part is chopped into pulses by the acousto-optic modulator (AOM) with the frequency shifted at the same time, used as the probe, and launched into the fiber under test (FUT) through an optical circulator. The other part is used as the local oscillation. The returned wave is mixed with LO, and detected by the double balanced photodetector (DB-PD). The beat frequency will be shifted to the modulation frequency of AOM, with low-frequency band noise and DC drift suppressed.

The returned wave by backscattering is written as

$$E_S = rE_0 e^{-\alpha z} \exp[j(2\beta z - \omega_S t)], \tag{9.11}$$

where r stands for Rayleigh scattering coefficient. The beat signals is obtained at the output ports of 3 dB fiber coupler, expressed as

$$E_{1,2} = \left(E_{L,S} + jE_{S,L}\right)/\sqrt{2}, \tag{9.12}$$

$$I_{1,2} = E_L^2 + r^2 E_0^2 e^{-2\alpha z} \pm 2rVE_L E_0 e^{-\alpha z} \cos\theta \sin(\Delta\omega t + \phi), \tag{9.13}$$

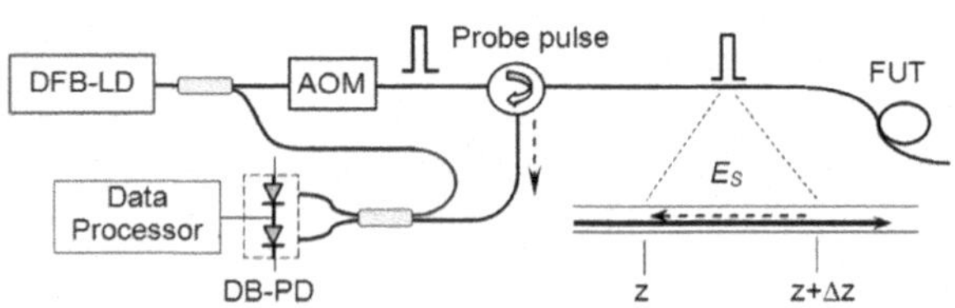

Fig. 9.19 Schematic structures of COTDR and ϕ-OTDR

where $V = \exp(-2nz/L_c)$ is the visibility of interference fringe, determined by laser's linewidth, characterized by its coherent length L_c. $\Delta\omega = \omega_S - \omega_L$ is the frequency shift of AOM. By using a differential amplifier or a double balanced photodiode (DB-PD), the DC components in (9.13) are canceled, giving sensed signal of $\Delta I = I_1 - I_2$ as a function of delay time $T = z/v$. The fiber loss $\alpha(z)$ can then be acquired from ΔI, so can the index variation $n(z)$, which is contained in the phase factor of $\phi = 2k\int_0^z F(\Delta z)n(z)\mathrm{d}z$, where $F(\Delta z)$ is the function of pulse waveform with spatial pulse width Δz. The phase factor is sensitive to external disturbances. In the data processing, ΔI is mixed with $\cos(\Delta\omega t)$ and $\sin(\Delta\omega t)$, giving the in-phase and the quadrature (I/Q) signals to demodulate the phase factor of ϕ.

Compared with the ordinary OTDR, COTDR and ϕ-OTDR show much longer working distance and higher sensitivity. The advantage relies mainly on the heterodyne detection, in which a single frequency laser with coherent length matching with the length of working fiber should be used to ensure the visibility high enough. However, the high coherence causes interference of backscattered waves within the spatial pulse width. Due to the index distribution along fiber, the interference displays random characteristics, called the interference fading, or the coherent Rayleigh noise (CRN). Detailed analyses and methods of mitigating CRN can be read in Refs. [63–65].

Due to the high sensitivity, ϕ-OTDR is suitable for dynamic sensing, since the data of every scan have SNR high enough used to rebuild vibrations; contrarily, in ordinary OTDR averaging of multiple scans is necessary to suppress noise. However, the response frequency f_{resp} of ϕ-OTDR is inversely proportional to the working distance, since a lower repetition rate is required for a longer sensing fiber, $f_{resp} = 1/(2T_{rt}) = c/(4nL)$, where T_{rt} is the round trip time. To overcome such a limitation, Ref. [66] proposed a system using a temporally sequenced multi-frequency pulses as $E_p = \sum_n \mathrm{rect}_\tau(t - n\tau_1)E_n e^{-j\omega_n(t-n\tau_1)}$, where τ is the pulse width, τ_1 is the repetition period of multiple pulses. Experimentally up to 0.5 MHz bandwidth of vibration detection over 9.6 km working distance was demonstrated by using a source with 100 frequencies. An agile precise tunable laser plays a crucial role in such a system.

The spatial resolution of OTDR is determined by the pulse width of probe: $\Delta z = c\tau/2n$. It is also dependent on the working distance, because the longer the distance, the less the energy of returned narrow pulse is received, and the lower the SNR is. Reference [67] proposed a system with frequency swept probe pulses and with matching filtering in data processing of receiver, which is transplanted from the scheme of pulse compression by linear frequency modulation (LFM) in microwave radar [68], as shown in Fig. 9.20 schematically. The spatial resolution of 30 cm was measured experimentally for sensing distance of around 20 km with a pulse width of 2 μs and sweeping range of 420 MHz. It is shown that although the LFM scheme works for lumped reflectors in radar, it is effective also for distributed dynamic sensors. Obviously, a frequency swept laser is indispensable in the scheme.

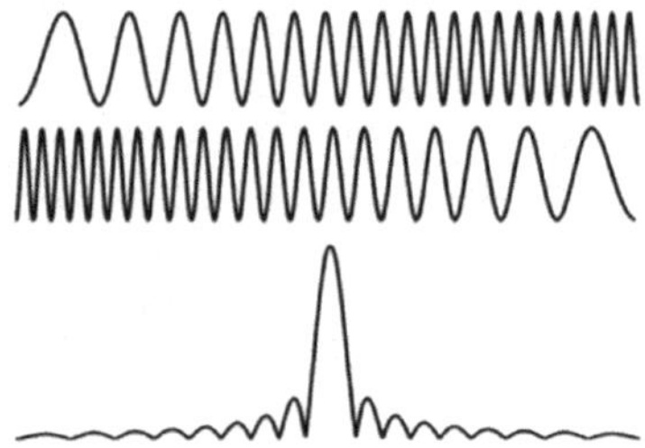

Fig. 9.20 Concept of pulse compression by linear frequency modulation

The polarization state of optical wave will vary along with the propagation randomly due to bending, twisting, lateral pressure, and other deformation of the fiber, which will lead to so-called polarization fading, described by the polarization angle θ of returned wave to that of LO in Exp. (9.13). It is an effect to be overcome; on the other hand, if the polarization state variation of the returned wave is measured and analyzed quantitatively, the system becomes another kind of OTDR, called POTDR.

(2) **Brillouin Optical Time Domain Reflectometer**

Brillouin optical time domain reflectometer (BOTDR) is developed based on the effect of Brillouin scattering. The collision of photons and phonons must obey the law of energy conservation and the law of momentum conservation:

$$\hbar\omega_S = \hbar\omega \pm \hbar\omega_a, \tag{9.14}$$

$$\hbar\boldsymbol{k}_S = \hbar\boldsymbol{k} \pm \hbar\boldsymbol{k}_a, \tag{9.15}$$

where the subscript a stands for the quantities of phonon, i.e., the acoustic quantities, and S stands for those of scattered photon; the quantities without subscript are for the incident photon. The function of BOTDR is to measure the Brillouin frequency shift along the fiber. The phonon's energy is much smaller than photon's energy; the Brillouin frequency shift in the silica fiber is about 11 GHz at room temperature. The frequency of scattered photon may shift to red side or blue side in spectrum, corresponding to creating a phonon or absorbing a phonon; the spectral lines are termed Stokes line and anti-Stokes line, respectively. Since the energy of phonons has a probability distribution, the frequency shift displays a Lorentzian line shape, whose width corresponds to phonon's lifetime. The experiments give a typical frequency width of 40 MHz in the silica fiber [69].

Due to the momentum conservation, only exists the backward Brillouin scattering in fiber, leading to $|\boldsymbol{k}_a| = 2|\boldsymbol{k}|$ and $\omega_a = V_a|\boldsymbol{k}_a| = 4\pi n V_a/\lambda$, where $V_a = \omega_a/k_a$ is the velocity of sound wave, which is determined mainly by the Young's Modulus and density of the medium, which are functions of temperature and strains of the material. Therefore, temperature changes and strains of the fiber can be reflected from measurement of Brillouin frequency shift, which is expressed as a function in the linear range [70]:

$$\nu_B(e, T) = \nu_B(0, T_r)[1 + C_e e + C_T(T - T_r)]. \tag{9.16}$$

The coefficients above can be obtained experimentally; their typical data are $C_T\nu_B$ = 1.1 MHz/K and $C_e\nu_B$ = 4.8 kHz/(μ), where $\mu\varepsilon = 10^{-6}$ is the micro-strain.

Different from the ordinary OTDR, the measurand of BOTDR is the frequency shift, rather than the intensity of scattered light. Obviously a narrow linewidth single frequency laser is necessarily used as the probe source. Figure 9.21 shows two basic structures of BOTDR [70–72]. In structure (a), the LO is a Brillouin fiber laser pumped by the same DFB-LD, whose frequency is $\nu_0 - \nu_B$, where ν_B is the Brillouin frequency shift of the laser fiber. The frequency of Brillouin backscattered light from the sensing fiber is $\nu_0 - \nu_S$. The beat frequency is then shifted to $|\nu_B - \nu_S|$ in a low-frequency band, easier to be detected and processed.

In structure (b), the returned wave from sensing fiber is mixed with the probe laser itself, giving beat signals in the band around 11 GHz; and then the beat signal is mixed with a microwave in the same band. This scheme needs a high-speed photodetector at the microwave band. Its advantage is that the noise and frequency drift of the Brillouin fiber laser can be avoided. In both of the two schemes the Brillouin frequency shift is read out by using the algorithm of fast Fourier transform (FFT), and usually by averaging of multiply detected data.

The spatial resolution of OTDR is one of the most concerned performances. It is determined by probe pulse width; basically, a narrow pulse width gives a higher resolution of localization. The precision of measured Brillouin frequency shift is related to the spectral width of detected signal. Specifically, the line shape of measured signal is the convolution of Brillouin line and the probe laser line. For laser pulses, the linewidth is larger than that in cw operation, determined by Fourier transform; the line shape of pulsed probes with square waveform is with sinc type. It is deduced theoretically [51] that the frequency width of convolution is $\Delta\nu \approx \Delta\nu_B$ in case of $\pi\tau\Delta\nu_B \gg 1$, where τ is the pulse width, $\Delta\nu_B$ is the Brillouin linewidth. For narrow pulses with $\pi\tau\Delta\nu_B \ll 1$, the frequency width of measured signal will be inversely proportional to the pulse width, $\Delta\nu \approx \sqrt{2}/\pi\tau$. It means if the pulse width is less than $(\pi\Delta\nu_B)^{-1}$, the uncertainty of frequency measurement will increase. Therefore, a tradeoff has to be taken between the spatial resolution and the precision of Brillouin frequency shift. The optimized pulse width is nearly equal to the

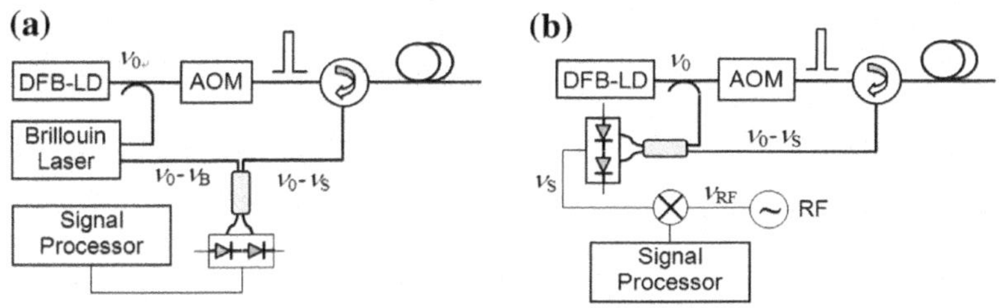

Fig. 9.21 Schema of BOTDR structure. **a** Based on a frequency shifted LO; **b** Based on heterodyne in microwave band

phonon's lifetime, ≈10 ns at the room temperature [71]. Obviously, the laser linewidth itself is a basic factor in the measurement of Brillouin line peak.

The intensity noise of laser will surely decrease the signal-to-noise ratio (SNR), and thus degrade the precision of frequency measurement. The spectral signal may be expressed as a parabolic function $u = u_0\left[1 - (v - v_B)^2/\Delta v^2\right]$ near its top, where u is the signal voltage. The peak is given by equation of $\delta u/u_0 = 2(v - v_B)\delta v/\Delta v^2 \approx 2\delta v^2/\Delta v^2 \to 0$. Since the SNR is defined as the power ratio of electrical signal, i.e., $\mathrm{SNR} = (\delta u/u_0)^{-2}$, the resolution of Brillouin frequency measurement is thus obtained with relation of [72]:

$$\delta v = \Delta v/\left[\sqrt{2}(\mathrm{SNR})^{1/4}\right]. \tag{9.17}$$

It is indicated that the laser with narrow linewidth and low intensity noise is one of the basic conditions of a high performance BOTDR.

(3) **Distributed Temperature Sensor Based on Raman Scattering in Fiber**

Raman scattering is related to internal movements of atoms and molecules with electron energy levels involved, including levels of vibration and rotation of molecules. The Raman spectroscopy is a powerful tool for analyses of compositions and impurities, because the Raman spectrum is a good characterization of internal structures and properties of materials. In the process of Raman scattering, the molecule absorbs a photon and transits to a virtual state and transits back to a lower level quickly, and emits a scattered photon at the same time. If the lower level is higher than the original ground level, red shift occurs in the scattering. If the molecule at an exited state absorbs a photon and then transits back from the virtual state to the ground level with a phonon absorbed, blue shift occurs. The red and blue shifts are called Stokes and anti-Stokes processes, respectively, as shown in Fig. 9.22a.

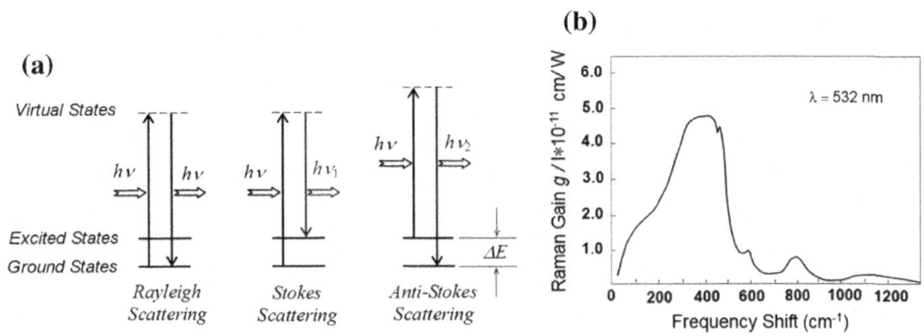

Fig. 9.22 **a** Schematic illustration of Stokes and anti-Stokes Raman scattering. **b** Raman gain spectrum of silica. Reproduced from [73] with the permission of AIP Publishing

The phonons involved are in optical frequency band of phonon's spectrum. Raman frequency shift is much larger than Brillouin frequency shift; it is about 13 THz for silica fiber, i.e. about 100 nm in 1550 nm wavelength band, and also with much wider bandwidth, as shown in Fig. 9.22b [73], which was generated by a 532 nm laser beam.

The intensity of Raman scattering depends on the number of molecules at related levels. In the medium under thermal equilibrium, the numbers of molecules at different levels obey Boltzmann distribution. The ratio of blue shift and red shift Raman scattering intensities is described by

$$I_2/I_1 \propto \exp(-\Delta E/k_B T), \tag{9.18}$$

where k_B is Boltzmann constant. By using probe pulses and measuring intensities of Raman scattering lines from the optical fiber, the distributed temperature sensor (DTS) is developed with the same principle of OTDR. The frequency shift and the spectral width of Raman scattering in fiber are quite large; the laser linewidth for DTS is not necessary be very narrow. But the intensity of Raman scattering is very weak, the probe must have a very high side mode suppression ratio (SMSR) with very low spontaneous emission background at the position of Raman line; or otherwise the Raman scattering will be drowned out by the Rayleigh scattering of probe. A proper notch-filter may help to mitigate the problem. The Raman scattering is so weak that averaging of a great number of acquired data is necessary to increase output SNR.

Apart from OTDR, the optical frequency domain reflectometers (OFDR) are also developed [74]. In the OFDR system, the frequency of high coherence laser is linearly swept in a serrated waveform as the probe; at the receiver end the backscattered wave mixed with the source wave, giving beat frequency proportional to the delay time. Its principle comes from the technology of frequency-modulated continuous wave (FMCW) used in radar. More discussion will be given in Sect. 9.4.

The fiber interferometers and fiber gratings have also been utilized to build quasi-distributed sensor systems [57, 75], including Mach–Zehnder interferometers, Michelson interferometers, and Sagnac interferometers. They put forward different technical requirements on laser diodes for different applications.

9.3.3 *Laser Spectroscopy*

(1) **Tunable Diode Laser Absorption Spectroscopy**

Narrow linewidth single-frequency semiconductor lasers with agile tunability are widely used in researches of laser spectroscopy and related spectroscopic apparatuses, mostly for the absorption spectroscopy and Raman spectroscopy. In the former area, the technology of tunable diode laser absorption spectroscopy

(TDLAS) is well developed. One of its applications is for methane detection. Methane is a dangerous gas in mines, and one of the gases giving harmful influence to the atmosphere environment. It is also a main composition of the natural gas, an important energy source. The methane detection by TDLAS is introduced briefly as an example here.

The absorption of CH_4 in the near infrared band is shown in Fig. 9.9. The InGaAsP-DFB laser in this band is commercially available, e.g., DFB-LD working at 1653.72 nm for CH_4 absorption line of band $2\nu_3$, branch R3. In case the concentration of CH_4 in the air is relatively low, its absorption line is in Lorentzian shape. As stated in Sect. 6.2, the modulation spectroscopy is categorized into two technologies of WMS and FMS. In methane detection and monitoring, WMS is usually used with modulation frequency Ω lower than the absorption linewidth $\Delta\omega$. It is necessary to find the position of absorption line peak in the modulated wavelength for measuring the concentration of CH_4. One of the methods is to modulate the laser's working current with a waveform combining sinusoidal wave and serrated wave. The first-order and second-order harmonics of absorbed signals are demodulated in dada processing. Ideally the zero point of the first order harmonic indicates the position of absorption peak; the second order is proportional to CH_4 concentration. However, as discussed in Sect. 6.2.3, the residue amplitude modulation (RAM) caused by accompanied power modulation has to be corrected [76].

Another method is to stabilize the laser frequency on CH_4 absorption peak by using a reference gas cell, as shown in Fig. 9.23. A gas cell as the sensor head is placed at the circumstances to be monitored and the gas to be monitored is pumped into the cell; the probe optical wave and absorbed wave are transported by fibers. For uses of an open space, the probe laser beam is collimated by a telescope; the diffused light by the background is collected by a receiving telescope. Appearance of methane can be sensed by the decreasing of received power at the absorption line. The working distance of this method is usually short.

(2) **Differential Absorption Lidar**

Carbon dioxide is another mostly concerned gas, because it is regarded as one of the greenhouse gases of the earth atmosphere. For the purpose the differential absorption lidar (DIAL) is developed, where the lidar stands for the lightwave radar. DIAL uses two lasers with different wavelengths; one of them is stabilized on one of CO_2 absorption lines; the other is controlled to be located out of the lines. For

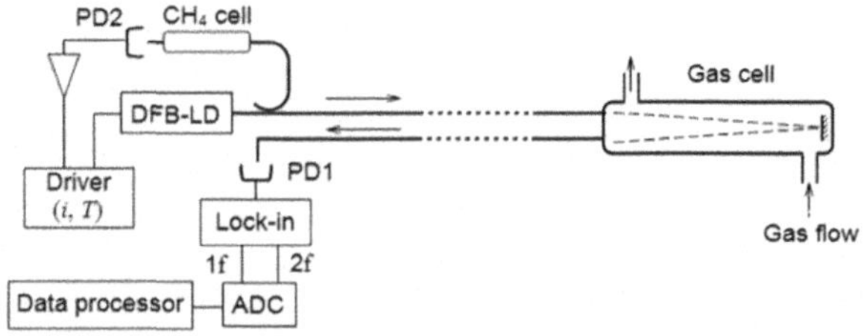

Fig. 9.23 Configuration of TDLAS for CH_4 detection

monitoring the pollution in atmosphere airborne and satellite-borne apparatuses are developed. When it is used to measure the spectra of returned wave from atmosphere, the difference between detected signals of on-line and out-of-line characterizes the integrated CO_2 concentration in the air column [77–79]. Quite a lot of CO_2 absorption lines exist in the band as shown in Fig. 9.11; the typical interval between lines is around 0.3 nm. The collision of molecules is the main factor of line broadening; the linewidth is thus related to the air pressure. Its FWHM linewidth is about 6 GHz under atmospheric pressure, and reduced to below 0.5 GHz under low pressure [80]. Compared with the linewidth of atom's absorption, the linewidth of CO_2 is not so narrow, but the requirement on frequency stabilization is quite high. Since the quantitative measurement of concentration is given by the absorption amplitude, the wavelength must be exactly locked at the line peak.

Figure 9.24 shows a typical structure of DIAL working in the near infrared band. The frequency of on-line laser in the DIAL is stabilized with a CO_2 cell for reference. Because the absorption of CO_2 is very weak, the path of gas cell should be long enough. Usually the path is extended by multiple roundtrip reflections in the cell. A gas cell with hollow core photonic crystal fiber is being developed [81], since the long fiber can be wound in a small volume. The power of laser must be very high for the airborne and satellite-borne DIAL. The master oscillator power amplifier (MOPA) technology is often used for the purpose.

The DIAL of Fig. 9.24 uses a Nd-doped YAG laser as the master laser; its power is amplified by the optical parameter oscillator (OPO), whose spectrum is stabilized by a CO_2-stabilized DFB-LD [78, 79]. The injection locked laser is also a good candidate with a frequency-stabilized laser as the master laser. The diffused light from the background and the atmosphere is collected by a telescope, and received by the photodetector; a small part of transmitted wave is divided as the local oscillation for correlation detection.

(3) **Raman Spectroscopy**

Raman spectroscopy is a powerful tool for identifying the composition and structure of materials, playing important roles in researches of chemistry and material science, and industries as well. The excited state depicted in Fig. 9.19a should be understood as a group of states, including rotation and vibration energy levels; and the ground state may also be composed of several sub-levels. The measured Raman spectrum will contain a lot of lines; the composition and structure can then be revealed. Raman spectrum can be measured by pumps with different wavelengths.

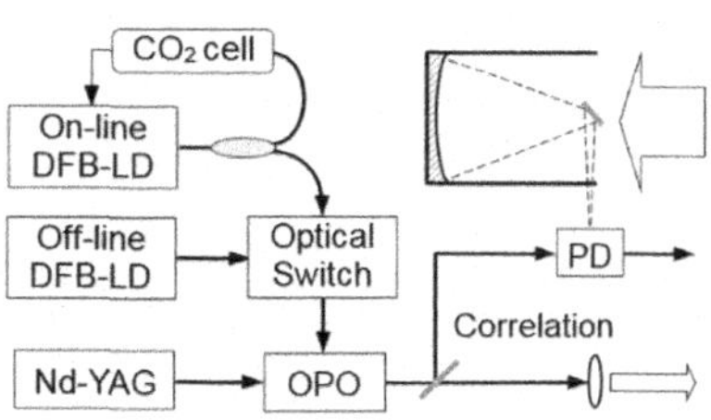

Fig. 9.24 Typical structure of DIAL for CO_2 monitoring

Solid-state lasers and gas lasers are often used as pump sources, because of their high power and high monochromaticity. However, the single frequency diode laser with high SMSR is attracting attentions, since compact apparatus of Raman spectroscopy are demanded. Reference [82] showed applications of narrow line LD in detecting compositions of gasoline by Raman spectroscopy. More advanced technologies are developed, especially the surface enhanced Raman spectroscopy (SERS) [83], and uses of micro probe made of optical fiber. Raman spectroscopy involves quite a lot of mechanisms and technologies; a lot of papers and monographs are published on the topic.

9.4 Applications in Lidar

9.4.1 Basic Concept of Lidar

The functions of radars include range finding, orientating, and imagination of target objects. Compared with the microwave radar, the lightwave radar (lidar) based on laser has unique features. Due to its much higher coherence and brightness, the precisions of range finding and orientating are enhanced greatly. In addition, the weight and volume of lidar are much lighter and smaller than those of microwave radar, an important advantage for airborne and satellite-borne applications. However, attenuation of lightwave in the air is more serious; it is harder for lidar to work in all-weather conditions. Combination of radar and lidar is favorable.

Two schemes of laser range finding are mostly used: one is based on the time of flight (TOF) of pulses; the other is based on the phase detection. In the former, the time t_{TOF} of round trip flight from the lidar to the target is counted and the distance between them is obtained as $L = 2ct_{\mathrm{TOF}}$, in the air with index of unity. Its spatial resolution is dependent on the pulse width and the response of detector. In the latter scheme, the laser intensity is modulated by a sinusoidal wave; the returned wave is in the same waveform with a phase delay. TOF can be obtained by the phase shift of received wave with respect to the transmitted wave. Its resolution is dependent on the precision of phase discrimination, usually much higher than that with laser pulses. However, the range of phase discrimination has to be limited in the range of 0–2π; the working distance is thus limited. Lasers used in the two schemes should meet certain requirements on their characteristics, whereas the requirement on the spectral property is not critical, it is not necessary to used single longitudinal mode lasers.

Another scheme is based on the technology of frequency modulation continuous wave (FMCW) developed for range finding in microwave radar [84], and utilized in lidar and in OFDR [68, 74], one of the distributed fiber sensors, as mentioned in Sect. 9.3.2. Its concept is illustrated in Fig. 9.25. The frequency of laser probe beam is modulated in a serrated waveform, with sweeping period of T, sweeping range of Δf, and sweeping rate of $s = \mathrm{d}f/\mathrm{d}t = \Delta f/T$. The reflected beam from the target is the same frequency-modulated wave, but delayed by the TOF $\tau = 2z/c$.

The frequency shift of $\mathrm{d}f = s\tau = 2sz/c$ is measured by correlation detection with the local oscillation which is synchronized with the transmitted probe wave. The two waves are written as

$$E_{\mathrm{LO}}(t) = A\ \exp j\left[\omega_0 t + \pi s t^2 + \phi(t)\right], \tag{9.19}$$

$$E_R(t,\tau) = B\ \exp j\left[\omega_0(t+\tau) + \pi s(t+\tau)^2 + \phi(t+\tau)\right]. \tag{9.20}$$

The correlation detection gives their beat signal:

$$I(t,\tau) = R_{\mathrm{PD}}\left[A^2 + B^2 + 2V\rho^2 AB\ \cos(2\pi s\tau t + \omega_0\tau + \pi s\tau^2)\right], \tag{9.21}$$

where R_{PD} is the responsivity of photodetector, $V = \exp(-\tau/\tau_{\mathrm{c}})$ is the visibility related to the laser's coherent time τ_{c}; ρ is a factor corresponding to the repetition period of multiple scanning, expressed as [74]:

$$\rho = \frac{\sin[\pi(f - \delta f)(T - \tau)]}{\pi(f - \delta f)(T - \tau)}, \tag{9.22}$$

where T is the period of scanning, $\mathrm{d}f = s\tau$ is just the beat frequency. The resolution of range finding by FMCW is deduced to be $\delta\tau = c/(2\Delta f)$. The signal-to-noise ratio caused by laser's linewidth is SNR = $V/(1 - V)$, besides the other noise sources. If the detectable level is characterized by SNR = 1, the maximum detectable distance is deduced to be $z_{\max} = (c\tau_{\mathrm{c}}/2)\ln 2$, proportional to laser's coherent time τ_{c}.

Apart from the serrated wave, FMCW lidar may use other frequency modulation waveforms, such as sinusoidal wave. The target distance can be calculated based on a complete Fourier frequency analysis of the correlation detected signal. Anyway, high coherence is the basic requirement on lasers in the different FMCW schemes.

The detection of target moving velocity is also one of the basic functions of radar and lidar. It can be simply calculated from the variation with time of the measured distance. For high speed moving targets, Doppler Effect is utilized for the purpose both in radar and lidar. With a single-frequency probe beam, the frequency of returned wave reflected from the target will be shifted in proportional to the relative velocity. The moving direction must be acquired with other assistant methods.

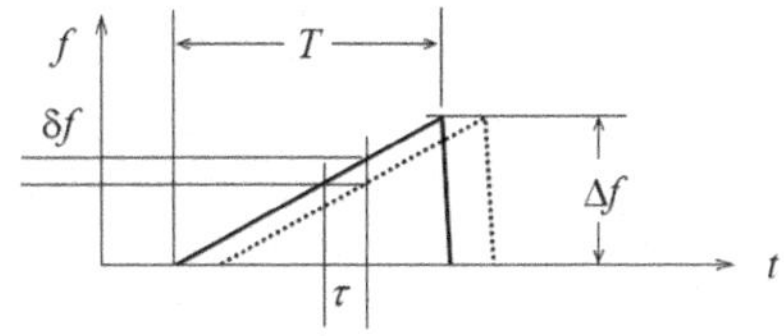

Fig. 9.25 Illustration of FMCW range finding

The lidar system is composed basically of a transmitter telescope, a receiver telescope, their rotating tables, and acquisition-pointing-tracking (APT) servos. These parts are necessary also for orientating targets. The angular velocity of mechanical rotator is not high enough for many applications. To exceed the limit the phased array (PA) antenna technology is developed, which can be regarded as a dynamic diffractive grating. The antenna of PA is divided into several elements; the phase shift between adjacent elements is modulated with controllable phase shift. The combination of radiations from all elements forms a beam in directions determined by the phase shift, based on the interference of multiple beams; the beam direction can be scanned by varying the phase shift, as shown in Fig. 9.13b schematically. The same idea can be used in optical domain, and the optical phased array (OPA) is then developed [85], usually with monolithically integrated optics components. Obviously the laser used for OPA should have linewidth narrow enough, or otherwise the combined beam will be broadened.

9.4.2 Brief Introduction to Synthetic Aperture Lidar

The function of imagination of radar relies on its spatially scanning, i.e., the capability of acquisition of the returned wave intensity distribution in azimuth and in pitching. The clearness of image is dependent on the diameter of receiving antenna, similar to that of a telescope. The diameter of single radar is limited, especially from viewpoints of cost, weight, and mechanical stability. To expand the size of receiving aperture, the antenna arrays composed of multiple antennas are built on the ground in big areas for applications of radio astronomy. The received radio wave signals are synthesized to rebuild images of celestial bodies. The synthetic aperture radar (SAR) is then developed with the same concept, which is installed on vehicles for airborne applications, as shown schematically in Fig. 9.26. The flying airplane emits radio wave and receives the reflected wave; the image of target is synthesized according to the time sequence of received signals. The data processing gives a similar effect to that of an expanded receiving antenna.

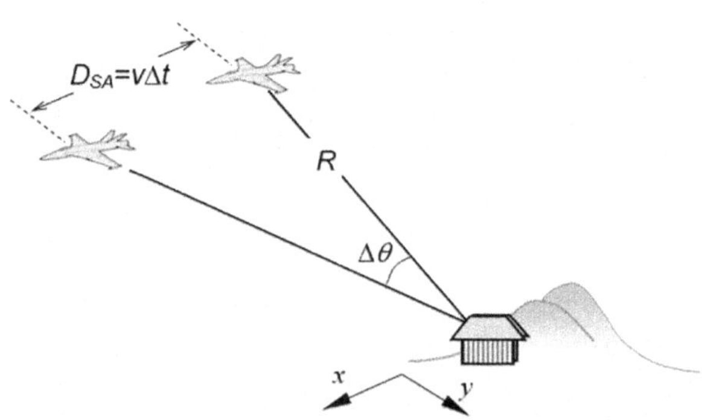

Fig. 9.26 Concept of an airborne SAR

The concept of SAR is also implanted in the lidar as an optical SAR (OSAR). The laser frequency of OSAR is swept linearly in a serrated form $f = f_0 + t\Delta f/T$; the resolution of range finding is $\delta R = c/(2\Delta f)$, according to analysis of FMCW lidar. If the plane flies in x-direction with velocity of v, the effective size of OSAR in x is $D_{SA} = v\Delta t$. The resolution of target image is determined by the diffraction limit, $\delta x \sim \lambda/(2\Delta\theta) = \lambda R/(2D_{SA})$. If two lidars are installed on the two wings of the plane, both resolutions of x and y directions will be improved. Detailed analyses of OSAR can be read in Refs. [86–88].

The critical component in OSAR is a high coherence laser with frequency swept in a high speed and in a broad range. In the experiment of OSAR reported by Ref. [87], a tunable external cavity semiconductor laser was used with frequency sweeping rate of 10 THz/s and swept range of 3.7 THz (1539–1568 nm) in period of 0.36 s. The laser linewidth is maintained 100 kHz, corresponding to coherent distance of $\sim$1 km.

The requirement of OSAR on laser's stability is quite high. The temporal variation of laser frequency and phases must be perfectly duplicated for different scanning periods; any deviations between the scans will result in noises in synthesizing the received data of different scan periods. OSAR requires a laser with high output power. The intensity fluctuation will also lead to noises. Fluctuations of laser frequency, phases, and intensity are inevitable in practice. The related hardwares and softwares should be optimized. Data processing of OSAR is required to produce results in multiple dimensions, the three-dimensional positioning, their variations in temporal and spectral domains, and the intensity of reflected waves. Research of the algorithm for OSAR is a big subject with a series mathematical methods and theories involved [87, 88].

References

1. Essen L, Parry JVL (1955) An atomic standard of frequency and time interval: a cæsium resonator. Nature 176:280–282
2. Beehler RE (1967) A historical review of atomic frequency standards. Proc IEEE 55:792–805
3. Sullivan DB, Bergquist JC, Bollinger JJ et al (2001) Primary atomic frequency standards at NIST. J Res Natl Stand Technol 106:47–63
4. Ramsey NF (1990) Experiments with separated oscillatory fields and hydrogen masers. Rev Mod Phys 62(3):541–552
5. Hänsch TW, Schawlow AL (1975) Cooling of gases by laser radiation. Opt Commun 13(1): 68–69
6. Clairon A, Laurent P, Santarelli G et al (1995) A cesium fountain frequency standard–preliminary results. IEEE Trans Instrum Meas 44(2):128–131
7. Chu S (1998) The manipulation of neutral particles. Rev Mod Phys 70(3):685–703
8. Cohen-Tannoudji CN (1998) Manipulating atoms with photons. Rev Mod Phys 70(3): 707–719
9. Phillips WD (1998) Laser cooling and trapping of neutral atoms. Rev Mod Phys 70(3): 721–741
10. Parker TE (2010) Long-term comparison of caesium fountain primary frequency standards. Metrologia 47(1):1–10

11. Laurent P, Abgrall M, Jentsch C et al (2006) Design of the cold atom PHARAO space clock and initial test results. Appl Phys B 84:683–690
12. Wang Y (2007). Laser cooling and trapping of atoms. Peking University Press (in Chinese)
13. Cohen-Tannoudji CN, Dupont-Roc J, Grynberg G (1992) Atom-photon interaction: basic processes and applications. Wiley, New York
14. Maitland A, Dunn MH (1970). Laser physics. Elsevier, Amsterdam
15. Liu P, Cheng H, Meng Y et al (2016) Improvement in medium long-term frequency stability of the integrating sphere cold atom clock. J Optical Society America B 33(7):1439–1443
16. Katori H, Takamoto M, Pal'chikov VG et al (2003). Ultrastable optical clock with neutral atoms in an engineered light shift trap. Phys Rev Lett 91(17):173005(1–4)
17. Targat RL, Baillard X, Fouche M et al (2006). Accurate optical lattice clock with ^{87}Sr atoms. Phys Rev Lett 97:130801(1–4)
18. Ludlow AD, Boyd MM, Ye J (2015) Optical atomic clocks. Rev Mod Phys 87:637–701
19. Wikipedia. https://en.wikipedia.org/wiki/Atomic_clock. Accessed 19 Mar 2017
20. NIST news (2014). NIST launches a new U.S. time standard: NIST-F2 atomic clock. Facebook Google plus Twitter: April 03
21. Dalton BJ, McDuff R, Knight PL (1985) Coherent population trapping. Optica Acta: Int J Opt 32(1):61–70
22. Boller KJ, Imamoglu A, Harris SE (1991) Observation of electromagnetically induced transparency. Phys Rev Lett 66(20):2593–2596
23. Harris SE (1997) Electromagnetically induced transparency. Phys Today 50(7):36–42
24. Fleischhauer M, Imamoglu A, Marangos JP (2005) Electromagnetically induced transparency: optics in coherent media. Rev Mod Phys 77(2):633–673
25. Wu H, Xiao M (2007) Cavity linewidth narrowing and broadening due to competing linear and nonlinear dispersions. Opt Lett 32(21):3122–3124
26. Lauprêtre T, Proux C, Ghosh R et al (2011) Photon lifetime in a cavity containing a slow-light medium. Opt Lett 36(9):1551–1553
27. Goorskey DJ, Wang H, Burkett WH et al (2012) Effects of a highly dispersive atomic medium inside an optical ring cavity. JMO 49(1/2):305–317
28. Ashkin A (1970) Acceleration and trapping of particles by radiation pressure. Phys Rev Lett 24(4):156–159
29. Neuman KC, Block SM (2004) Optical trapping. Rev Sci Instrum 75(9):2787–2809
30. Applegate RW, Squier J, Vestad T et al (2004) Optical trapping, manipulation, and sorting of cells and colloids in microfluidic systems with diode laser bars. Opt Express 12(19): 4390–4398
31. Simpson NB, Dholakia K, Allen L et al (1997) Mechanical equivalence of spin and orbital angular momentum of light: an optical spanner. Opt Lett 22(1):52–54
32. Iida S, Oono K, Kosaki H, Kumagaya H, Sawada S (1978) Tables of physical constants, 2nd edn. Asakura Publishing Co., Ltd., Tokyo (in Japanese)
33. Zhang J, Tao H, Wei D et al (2003) A laser diode system stabilized on the saturated absorption lines of rubidium atoms. Acta Optica Sinica 23(2):197–201 (In Chinese)
34. Margolis JS (1988) Measured line positions and strengths of methane between 5500 and 6180 cm^{-1}. Appl Opt 27(19):4038–4051
35. Rothman LS, Rinsland CP, Goldman A, et.al. (1998). The HITRAN molecular spectroscopic database and HAWKS (HITRAN atmospheric workstation): 1996 edition. J Quant Spectrosc Radiat Transfer 60(5):665–710
36. Molecular database. http://vpl.astro.washington.edu/spectra/allmoleculeslist.htm
37. Czajkowski A, Madej AA, Dube P (2004) Development and study of a 1.5 μm optical frequency standard referenced to the P(16) saturated absorption line in the (ν1 + ν3) overtone band of $13C_2H_2$. Opt Commun 234:259–268
38. Kao KC, Hockham GA (1966) Dielectric-fibre surface waveguides for optical frequencies. IEE Proc 113:1151
39. Ip E, Lau APT, Barros DJF et al (2008) Coherent detection in optical fiber systems. Opt Express 16(2):753–791

40. Shieh W, Bao H, Tang Y (2008) Coherent optical OFDM: theory and design. Opt Express 16(2):841–859
41. Ferrero V, Camatel S (2008). Optical phase locking techniques: an overview and a novel method based on single side sub-carrier modulation. ∑∏∯ ∫x∏∮∫ss 16(2):818–828
42. Seeds AJ (2002) Microwave photonics. IEEE Trans Microw Theory Tech 50(3):877–887
43. Capmany J, Pastor D, Ortega B (2005) Microwave signal processing using optics. In: Optical fiber communication conference, OThB1
44. Capmany J, Novak D (2007) Microwave photonics combines two worlds. Nat Photon 1:319–330
45. Yao J (2009) Microwave photonics. J Lightwave Technol 27(3):314–335
46. Cooper AJ (1990) Fiber/radio for the provision of cordless/mobile telephony services in the access network. Electron Lett 26(24):2054–2056
47. Smith GH, Novak D, Lim C (1998) A millimeter-wave full-duplex radio-over-fiber star-tree architecture incorporating WDM and SCM. IEEE Photon Tech Lett 10(11):1650–1652
48. Gliese U, Narskov S, Nielsen TN (1996) Chromatic dispersion in fiber-optic microwave and millimeter-wave links. IEEE Trans Microw Theory Tech 44(10):1716–1724
49. Hofstetter R et al (1995) Dispersion effects in optical mm-wave systems using self-heterodyne method for transport and generation. IEEE Trans Microw Theory Tech 43(9):2263–2269
50. Wang J, Hou P, Cai H et al (2015) Continuous angle steering of an optically controlled phased array antenna based on differential true time delay constituted by micro-optical components. Opt Express 23(7):9432–9439
51. Ghelfi P, Laghezza F, Scotti F et al (2014) A fully photonics-based coherent radar system. Nature 507:341–345
52. Chan VWS (1987) Space coherent optical communication systems–an introduction. J Lightwave Technol 5(4):633–637
53. Wilson K (2000) Optical communications for deep space missions. IEEE Commun Mag 38:134–139
54. Chan VWS (2003) Optical satellite networks. J Lightwave Technol 21(11):2811–2827
55. Chiodo N, Djerroud K, Acef O et al (2013) Lasers for coherent optical satellite links with large dynamics. Appl Opt 52(30):7342–7351
56. Nash P (1996) Review of interferometric optical fibre hydrophone technology. IEE Proc-Radar Sonar Navig 143:204–209
57. Fang Z, Chin K, Qu R et al (2012) Fundamentals of optical fiber sensors. Wiley, New York
58. Dandridge A, Cogdell GB (1991) Fiber optic sensors for navy applications. IEEE LCS:81–89
59. Beverini N, Falciai R, Maccioni E et al (2006) Developing fiber lasers with Bragg reflectors as deep sea hydrophones. Ann Geophys 49:1157–1165
60. Cranch GA, Nash PJ, Kirkendall CK (2003) Large-scale remotely interrogated arrays of fiber-optic interferometric sensors for underwater acoustic applications. IEEE Sens J 3(1): 19–30
61. Barnoski MK, Rourke MD, Jensen SM et al (1977) Optical time domain reflectometer. Appl Opt 16(9):2375–2379
62. Lu Y, Zhu T, Chen L et al (2010) Distributed vibration sensor based on coherent detection of phase-OTDR. J Lightwave Technol 28:3243–3249
63. Healey P (1984) Fading in heterodyne OTDR. Electron Lett 20(1):30–32
64. Pan Z, Liang K, Zhou J et al (2012) Interference fading-free phase-demodulated OTDR system. Proc SPIE 8421:842129(1–4)
65. Zhou J, Pan Z, Ye Q et al (2013) Characteristics and explanations of interference fading of ϕ-OTDR with a multi-frequency source. J Lightwave Technol 31(17):2947–2954
66. Wang Z, Pan Z, Fang Z et al (2015) Ultra-broadband phase-sensitive optical time-domain reflectometry with a temporally sequenced multi-frequency source. Opt Lett 40(22): 5192–5195
67. Lu B, Pan Z, Wang Z et al (2017) High spatial resolution phase sensitive optical time domain reflectometer with frequency-swept pulse. Opt Lett 42(3):391–394

68. Richards MA (2005) Fundamentals of radar signal processing. McGraw-Hill Education
69. Agrawal GP (2004). Nonlinear fiber optics. Elsevier Science, Amsterdam
70. Parker TR, Farhadiroushan M, Handerek VA et al (1997) Temperature and strain dependence of the power level and frequency of spontaneous Brillouin scattering in optical fibers. Opt Lett 22(11):787–789
71. Naruse H, Tateda M (1999) Trade-off between the spatial and the frequency resolutions in measuring the power spectrum of the Brillouin backscattered light in an optical fiber. Appl Opt 38(31):6516–6521
72. Horiguchi T, Shimizu K, Kurashima T et al (1995) Development of a distributed sensing technique using Brillouin scattering. J Lightwave Technol 13(7):1296–1302
73. Stolen RH, Tynes AR, Ippen EP (1972) Raman oscillation in glass optical waveguide. Appl Phys Lett 20:62–64
74. Uttam D, Culshaw B (1985) Precision time domain reflectometry in optical fiber systems using a frequency modulated continuous wave ranging technique. J Lightwave Technol 3(5):971–977
75. Juskaitis R, Mamedov AM, Potapov VT et al (1992) Distributed interferometric fiber sensor system. Opt Lett 17(22):1623–1625
76. Wang Y, Cai H, Geng J et al (2009) Logarithmic conversion of absorption detection in wavelength modulation spectroscopy with a current-modulated diode laser. Appl Opt 48 (21):4068–4076
77. Ehret G, Kiemle C, Wirth M et al (2008) Space-borne remote sensing of CO_2, CH_4 and N_2O by integrated path differential absorption lidar: a sensitivity analysis. Appl Phys B Laser Opt 90(3–4):593–608
78. Fix A, Budenbender C, Wirth M et al (2011) Optical parametric oscillators and amplifiers for airborne and spaceborne active remote sensing CO_2 and CH_4. Proceedings SPIE 8182:818206 (1–10)
79. Numata K, Chen JR, Wu ST et al (2011) Frequency stabilization of distributed-feedback laser diodes at 1572 nm for lidar measurements of atmospheric carbon dioxide. Appl Opt 50(7): 1047–1056
80. Barr JL, Humphries SD, Nehrir AR et al (2011) Laser-based carbon dioxide monitoring instrument testing during a 30-day controlled underground carbon release field experiment. Int J Greenhouse Gas Control 5:138–145
81. Huang C, Chen D, Cai H et al (2014) Transmission characteristics of photonic crystal fiber gas cell used in frequency stabilized laser. Chin Opt Lett 12(8):080602(1–5)
82. Ye Q, Xu Q, Yu Y et al (2009) Rapid and quantitative detection of ethanol proportion in ethanol–gasoline mixtures by Raman spectroscopy. Opt Commun 282:3785–3788
83. Haynes CL, McFarland AD, Van Duyne RP (2005) Surface-enhanced Raman spectroscopy. Anal Chem 77:338A–346A
84. Hymans AJ, Lait J (1960). Analysis of a frequency modulated continuous wave ranging system. Proceedings IEE 107-B:365–372
85. McManamon P (2005) An overview of optical phased array technology. Proceedings SPIE 5947:594701(1–10)
86. Rosen PA, Hensley S, Joughin IR et al (2000) Synthetic aperture radar interferometry. Proc IEEE 88(3):333–382
87. Bashkansky M, Lucke RL, Funk E et al (2002) Two-dimensional synthetic aperture imaging in the optical domain. Opt Lett 27:1983–1985
88. Beck SM, Buck JR, Buell WF et al (2005) Synthetic-aperture imaging laser radar: laboratory demonstration and signal processing. Appl Opt 44(35):7621–7629

Erratum to: Single Frequency Semiconductor Lasers

Erratum to:
Z. Fang et al., *Single Frequency Semiconductor Lasers*, Optical and Fiber Communications Reports 9, https://doi.org/10.1007/978-981-10-5257-6

In the original version of the book, the incorrect additional bibliography "Jointly published with Shanghai Jiao Tong University Press" and copyright text "Not for sale outside the Mainland of China (Not for sale in Hong Kong SAR, Macau SAR, and Taiwan, and all countries, except the Mainland of China)" have been now corrected as "Jointly published with Shanghai Jiao Tong University Press, Beijing, China" and "The print edition is not for sale in China Mainland. Customers from China Mainland please order the print book from: Shanghai Jiao Tong University Press", respectively, in Frontmatter.

The updated online version of this book can be found at
https://doi.org/10.1007/978-981-10-5257-6

Z. Fang et al., *Single Frequency Semiconductor Lasers*, Optical and Fiber Communications Reports 9, https://doi.org/10.1007/978-981-10-5257-6_10

Index

A

Acousto-Optic (AO) effect, 140
Acousto-Optic Modulator (AOM), 51, 178, 182
Agile laser, 205, 211
Allan deviation, 68, 70
Allan variance, 67–71
Amplitude Shift Keying (ASK), 280
Anti-stokes line, 290
Apodization of gratings, 99
Astigmatism, 20
Atomic clock, 267–269, 271
Autocorrelation, 42, 45, 51–53, 58, 64
Automatic Power Controller (APC), 65

B

Beer-Lambert law, 176
Bistability, 119
Blaze grating, 132, 133
Bragg diffraction, 83, 95
Bragg diffraction equation, 143, 151
Bragg grating, 117, 123, 142, 143, 149, 151, 162
Bragg wave vector, 82
Brightness, 2, 3
Brillouin OTDR, 290–292
Brillouin scattering, 287, 290
Brownian movement, 55

C

Carrier-envelope phase, 257
Characteristic temperature of threshold, 35
Chirped Bragg grating, 97
Cleaved facet, 16, 22, 27
Coherence, 2–5
Coherence length, 3, 4
Coherence time, 3
Coherent Optical Communication (COC), 2, 5, 239, 240
Coherent OTDR, 288, 289
Confinement factor, 23
Confocal Fabry-Perot (F-P) cavity, 156
Cooled atomic fountain clock, 267, 269
Cooled atom molasses, 269
Coupled mode theory, 82
Current modulation, 35, 36
Current tuning, 30

D

Degenerate four wave mixing, 254
Delayed self-heterodyne, 54, 59
Delayed self-homodyne, 59
Dense Wavelength Division Multiplex (DWDM), 89, 100
Differential Absorption Lidar (DIAL), 294
Differential gain coefficient, 24, 32
Diode laser, 16
Dispersion of refractive index, 169
Distributed Bragg Reflector (DBR) laser, 94
Distributed Feedback (DFB) laser, 1, 81
Distributed optical fiber sensor, 287
Distributer Temperature Sensor (DTS), 293
Doppler broadening, 170, 172, 174, 186
Doppler-cancellation technique, 244
Doppler cooling, 271
Doppler frequency shift, 271, 284
Doppler radar, 297
Double Balanced Photodiode (DB-PD), 236, 241
Double heterostructure laser, 1

E

Einstein coefficients, 10
Electric Absorption Modulator (EAM), 90, 100

Z. Fang et al., *Single Frequency Semiconductor Lasers*, Optical and Fiber Communications Reports 9, DOI 10.1007/978-981-10-5257-6

Electro-Optic (EO) effect, 140
Electro-optic ceramic, 140
Electro-Optic Modulator (EOM), 178, 182
Emission by carrier recombination between energy bands, 32
External (extended) cavity laser, 6
External (extended) Cavity Semiconductor Laser (ECDL), 142
Extinction coefficient, 169

F
Fabry-perot cavity laser, 6
Faraday Rotation Mirror (FRM), 54, 60
Fermi-dirac distribution, 9
Fermi level, 9, 13, 14, 36
Fiber Bragg Grating (FBG), 142
Finesse of F-P cavity, 118
Fourier convolution theorem, 57
Four Wave Mixing (FWM), 254
Free-Space Optical (FSO) communication, 283
Frequency chirping, 90
Frequency Discriminate (FD), 196
Frequency discriminator, 236, 251, 261
Frequency mixing, 254
Frequency Modulation Continuous Wave (FMCW), 206
Frequency Modulation Spectroscopy (FMS), 176, 196
Frequency noise, 6
Frequency pulling, 28
Frequency Shift Keying (FSK), 280
Frequency stability, 67, 70, 71
Frequency synthesizer, 268
Frequency swept laser, 212, 226
Fundamental mode, 18, 20

G
Gain coupled grating, 84, 93
Gain guiding, 19, 20
Gas cell, 168, 179–182, 191, 192
Gaussian–Hermit field distribution, 19
Gaussian line shape, 45, 56
Geophone, 285, 286
Geostationary Earth Orbit (GEO) satellite, 283
Grating equation, 131, 138

H
Harmonic oscillator model, 168
Heterodyne OPLL, 238, 247
Heterostructure, 2
Hilbert transform, 210
Hollow-Core Photonic Crystal Fiber (HCPCF), 181
Holographic photolithography, 93
Homodyne OPLL, 236, 238
Hydrogen maser, 268
Hydrophone, 285, 286

I
Index coupled grating, 83, 84
Index guiding, 19, 20, 33
Index of Energy-to-Data Ratio (EDR), 109
Index of Heat-to Bit Rate ratio (HBR), 109
Inherent linewidth, 4
Inherent noise, 41, 45
Injection locked laser, 259
Injection locking by side mode, 129, 130, 248
Injection locking of semiconductor laser, 127
In phase/quadrature (I/Q) demodulator, 241
Intensity Modulation-Direct Detection (IM-DD), 279
Internal quantum efficiency, 23, 25
I/Q bridge, 242

J
Junction temperature, 35, 37
Junction voltage, 14

K
Kerr effect, 214, 218
Kramers–Kronig (K-K) relations, 28, 168, 184

L
$\lambda/4$ phase shift, 87, 93
Lamb dip, 172
Langevin forces, 49, 72
Laser cooling, 3, 5, 6, 267, 269–271, 274, 275
Laser Diode (LD), 16
Laser frequency sweeping, 6
Laser line width, 4
Lateral mode, 19, 20
LD array, 21
Lead Lanthanum Zirconate Titanate (PLZT), 218
Light-current curve kink, 73, 121
Light Emitting Diode (LED), 13
Lightwave radar (Lidar), 296
Linewidth enhancement factor, 47, 48
Liquid Crystal (LC), 140, 214, 217, 222
Lithium Niobate (LiNbO3), 215
Lithium Tantalate (LiTaO3), 215
Littman-Metcalf structure ECDL(Littman structure), 134
Littrow structure ECDL, 134
Longitudinal electro-optic effect, 216
Longitudinal mode, 4, 5
Longitudinal mode spacing, 22
Longitudinal relaxation time, 169

Loop filter, 241
Lorentzian line shape, 24, 28, 44, 58
Loss coefficients in-cavity, 24
Low Earth Orbit (LEO) satellite, 283
Low Pass Filter (LPF), 258

M
Mach–Zehnder Interferometer, 51, 60, 63
Magneto-induced dichroism, 190
Magneto-optical Trap (MOT), 271
Master laser, 127–130
Master Oscillator-Power Amplifier (MOPA), 141
Maxwell–Boltzmann distribution, 171
Metal-Organic Vapor Phase Epitaxy (MOVPE) or Chemical Vapor Deposition (MOCVD), 91
Michelson interferometer, 51, 60
Micro Electric Mechanical System (MEMS), 142
Micro Electronic Mechanical System (MEMS), 107
Microwave photonics, 207
Mode hopping, 119, 121, 133, 136, 146
Mode locked laser, 253–255
Mode partition noise, 75, 84
Modulation Current Efficiency Factor (MCEF), 111
Molecular Beam Epitaxy (MBE), 91, 102
Monolithically integrated laser, 81
Multiple layer dielectric film, 102
Multiple Quantum Wells (MQW), 101

N
Non-equilibrium carrier, 9, 11, 14

O
1/*f* noise, 45, 56, 57, 211, 212
On-Off Keying (OOK), 280
Optical Coherent Tomography(OCT), 207
Optical frequency comb, 5, 244, 252, 255, 256, 258, 262
Optical Frequency Domain Imaging (OFDI), 207
Optical frequency domain reflectometer, 287
Optical Frequency Domain Reflectometry (OFDR), 213
Optical frequency transfer, 247
Optical interferometry, 4, 5
Optical negative feedback with rotated polarization, 184, 191, 192
Optical Path Deference (OPD), 4
Optical Phased Array (OPA), 298
Optical Phase Locked Loop (OPLL), 6, 235
Optical pressure, 274
Optical Time Domain Reflectometer (OTDR), 287
Optical tweezers, 274
Orthogonal frequency division multiplexing, 281

P
Passive waveguide grating, 95
Peltier cooling, 38
Phased array, 21
Phase Locked Loop (PLL), 235
Phase-sensitive OTDR, 288, 289
Phase shift DFB laser, 84, 86, 87, 90, 91
Phase Shift Keying (PSK), 280
Photoelastic effect, 224
Photon density, 10, 23, 25, 30
Photon lifetime, 5, 25, 32
Pining of carrier concentration, 26
Planar diffraction grating, 117, 130
P-N junction, 1, 2
P-N junction laser, 13, 17, 26
Pockels effect, 214, 215
Polarization Controller (PC), 54
Polarization fading, 54, 60
Polarization spectroscopy, 184, 186, 187
Polarization shift keying (PolSK), 280
Pound-Drever-Hall (PDH) frequency stabilization, 193
Power Spectral Density (PSD), 42, 44, 55, 61, 69
Proton or ion implantation, 105
Pump and re-pump lasers, 271

Q
Quadrature Amplitude Modulation (QAM), 280
Quantum noise, 46
Quantum Well (QW) laser, 2
Quartz clock, 267
Quasi-continuous tuning, 100
Quasi-fermi level, 9, 14

R
Rabi oscillation frequency, 170
Radio Over Fiber (ROF), 207
Raman scattering, 287, 292, 293
Ramsey fringes, 268, 271
Random drift, 6
Random flicker, 55
Random walking, 55, 57
Rayleigh scattering, 287, 288, 293

Reactive Ion Etching (RIE), 93
Relative Intensity Noise (RIN), 45
Relaxation oscillation, 33, 34
Residual Amplitude Modulation (RAM), 181
Ring cavity, 119, 156, 160–162

S
Sampled waveguide grating, 94, 96
Saturable absorption spectroscopy, 170, 172, 174, 179, 184, 186
Saturated absorption spectroscopy, 6
Saturation intensity, 170
Schawlow-townes formula of linewidth, 48
Self-mixing interference, 200, 201
Self-pulsation, 34
Self-referenced frequency stabilization, 200
Self-referencing, 257
Semiconductor Optical Amplifier (SOA), 100
Side Mode Suppression Ratio (SMSR), 88, 98, 129
Single Side Band (SSB) modulator, 229
Slave laser, 128–130
Slow-light effect, 163
Spatial hole burning of gain, 90, 98
Spectral line shape, 44
Spectrum of beat signal, 52
Spontaneous emission factor, 24, 26, 29
Spontaneous emission lifetime, 32
Standard deviation, 65
Stokes line, 290
Stop band, 83, 84, 97
Strained layer quantum well laser, 2
Strong feedback, 122, 127
Sub-Doppler cooling, 271
Super-Luminescence Diode (SLD), 27
Super-mode, 21
Super structure grating, 93
Synthetic Aperture Lidar (OSAR), 299
Synthetic Aperture Radar (SAR), 298

T
Temperature tuning, 30
Thermal relaxation time, 38
Thermal resist, 37
Threshold current density, 26
Time of Flight (TOF), 296
Total Internal Reflection (TIR), 187
Transparence carrier concentration, 24
Transparent electro-optic ceramic, 218
Transverse Electric (TE) polarization mode, 18
Transverse electro-optic effect, 216, 227
Transverse Magnetic (TM) polarization mode, 18
Transverse mode, 4, 5, 16–18
Transverse relaxation time, 170
Tunability, 6
Tunable Diode Laser Absorption Spectroscopy (TDLAS), 294

V
Variance, 43–45, 47, 48, 51, 56, 57, 60, 61, 67–69, 71
Vertical Cavity Surface Emitting Laser (VCSEL), 2, 101
Visibility of interference fringes, 4, 58
Voltage-Controlled Oscillator (VCO), 231
Volume Bragg Grating (VBG), 142

W
Wafer bonding (fusing) technology, 103, 105
Waveguide Bragg Grating (WBG), 142
Wavelength Modulation Spectroscopy (WMS), 176
Weak feedback, 120, 122, 127, 148
White noise, 44, 53, 55–57, 65

Z
Zeeman effect, 180, 184, 190

Printed by Printforce, the Netherlands